K. Herholz · P. Herscovitch · W.-D. Heiss

NeuroPET

Positron Emission Tomography
in Neuroscience and Clinical Neurology

Springer-Verlag Berlin Heidelberg GmbH

K. Herholz · P. Herscovitch · W.-D. Heiss

Neuro PET

Positron Emission Tomography in Neuroscience and Clinical Neurology

With 66 Figures and 11 Tables

Springer

Karl Herholz, Prof. Dr.
W.-D. Heiss, Prof. Dr.
Neurologische Universitätsklinik
Max-Planck-Institut
für Neurologische Forschung
Joseph-Stelzmann-Str. 9
D-50931 Köln
Germany

Peter Herscovitch, MD

Chief, PET Imaging Section
National Institutes
of Health Clinical Center
10 Center Drive MSC 1180
Bethesda, MD 20892-1180
USA

Additional material to this book can be downloaded from http://extras.springer.com.

ISBN 978-3-642-62283-0   ISBN 978-3-642-18766-7 (eBook)
DOI 10.1007/978-3-642-18766-7
Library of Congress Control Number: 2004105242

Dr. Herscovitch's work as author was performed outside the scope of his employment as a U.S. government employee. This work represents his personal and professional views and not necessarily those of the U.S. government.

springeronline.com

The use of general descriptive names, registered names, trademarks, etc. in this publications does not imply, even in the absence of a specific statement, that such names are exempt from the relevant protective laws and regulations and therefore free for general use.

Product liability: The publishers cannot guarantee the accuracy of any information about dosage and application contained in this book. In every individual case the user must check such information by consulting the relevant literature.

Editor: Ute Heilmann, Heidelberg
Desk editor: Dörthe Mennecke-Bühler, Heidelberg
Production editor: Bernd Wieland, Heidelberg
Cover design: F. Steinen, ᵉStudio Calamar, Spain
Typesetting: AM-productions GmbH, Wiesloch

21/3150/Wd – 5 4 3 2 1 0
Printed on acid-free paper

Acknowledgments

Positron emission tomography provides unprecedented insights into human brain function in both health and disease, but is a complex endeavor involving many scientific disciplines. It would therefore have been impossible to write this book without the support of many friends and colleagues. We are deeply indebted to them all.

Several colleagues made essential contributions to this book. Lutz Kracht and Johannes Klein at the Max Planck Institute in Cologne selected and prepared many of the images used in the figures. We are also grateful to the staff of the Department of Radiology at the University of Cologne, who performed most of the MRI scans that were used for co-registration. Many colleagues from around the world allowed us to reprint figures from their publications or contributed images from their current work, as indicated in the figure legends. We are especially grateful to Vjera Holthoff, who made essential contributions to the chapter on psychiatric diseases. Of course, our entire work depended upon the professional dedication and collaboration over many years of the physicians, physicists, radiochemists, engineers, technologists and nurses in our clinics and laboratories.

The production of the CD-ROM that accompanies this book would not have been possible without Stefan Vollmar. He developed the visualization software package VINCI that is included in the CD-ROM and that was used to prepare most of the figures and movie clips. He was also responsible for the digital formatting and layout, and gave his expert advice whenever there were problems with the computer software or hardware.

Last but not least, we wish to express our gratitude to Ute Heilmann, Dörthe Mennecke-Bühler, Bernd Wieland and Thomas Wurm of Springer-Verlag for their encouragement and professional support.

March 2004

Karl Herholz
Peter Herscovitch
Dieter Heiss

Contents

List of Abbreviations

5-HT	5-Hydroxytryptamine (serotonin)
AAAD	Aromatic amino acid decarboxylase
ACHC	^{11}C-labeled aminocyclohexane carboxylate
AChE	Acetylcholine esterase
AD	Alzheimer disease
AIDS	Acquired immune deficiency syndrome
ALS	Amyotrophic lateral sclerosis
AMT	^{11}C-Alpha methyl-tryptophan
ANOVA	Analysis of variance
ApoE4	Apolipoprotein E4
APP	Amyloid precursor protein
BA	Brodmann area
BBB	Blood–brain barrier
BGO	Bismuth germanate
BOLD	Brain oxygen level-dependent contrast
BP	Binding potential
CB	Cannabinoid (receptor)
CBD	Corticobasal degeneration
CBF	Cerebral blood flow
CBV	Cerebral blood volume
CCD	Crossed cerebellar diaschisis
CFT	^{18}F-2-β-Carbomethoxy-3-β-(4-fluorophenyl)tropane
CJD	Creutzfeldt-Jakob disease
CMRglc	Cerebral metabolic rate of glucose
CMRO$_2$	Cerebral metabolic rate of oxygen
CO	^{11}C-carbon monoxide, ^{15}O-carbon monoxide
CO$_2$	^{15}O-carbon dioxide
COMT	Catechol-O-methyltransferase
CSF	Cerebrospinal fluid
CT	Computed tomography
DASB	^{11}C-3-amino-4-(2-dimethylaminomethyl-phenylsulfanyl)-benzonitrile
DAT	Dopamine transporter
DBS	Deep brain stimulation
DLB	Dementia with Lewy bodies
DMO	^{11}C-5,5-dimethyl-2,4-oxazolidinedione

DPN	^{11}C-diprenorphine
DRD	DOPA-responsive dystonia
DTBZ	^{11}C-dihydrotetrabenazine
DV	Distribution volume
EEG	Electroencephalography
FDG	^{18}F-2-fluoro-2-deoxy-D-glucose
FDOPA	L-6-^{18}F-fluoro-3,4-dihydroxyphenylalanine
FESP	^{18}F-Fluoroethylspiperone
FET	O-(2-^{18}F-fluoroethyl)-L-tyrosine
FIAU	^{124}I-2'-fluoro-2'-deoxy-5-iodo-1-β-D-arabinofuranosyluracil
FLT	3'Deoxy-3'-^{18}F-fluorothymidine
fMRI	Functional magnetic resonance imaging
FMZ	^{11}C-Flumazenil
FORE	Fourier rebinning
FOV	Field of view
FTD	Frontotemporal dementia
FWHM	Full width at half maximum
GABA	γ-Aminobutyric acid
GDNF	Glial cell line-derived neurotrophic factor
GFP	Green fluorescent protein
GLUT	Glucose transporter
GPi	Internal part of globus pallidus
GSO	Gadolinium oxyorthosilicate
HD	Huntington disease
HSV	Herpes simplex virus
HSV-TK	Herpes virus thymidine kinase
ICD	International Classification of Diseases
LOR	Line of response
LSO	Lutetium oxyorthosilicate
mAB	Monoclonal antibody
mAChR	Muscarinergic acetylcholine receptor
MAO	Monoamine oxidase
MAP	Maximum a posteriori probability algorithm
MCA	Middle cerebral artery
MCI	Mild cognitive impairment
MDMA	3,4-Methylene-dioxymethamphetamine
MET	^{11}C-Methionine
MMSE	Mini-Mental Status Examination
MP4A	N-^{11}C-Methyl-4-piperidyl-acetate
MP4P	N-^{11}C-Methylpiperdin-4-yl propionate
MPTP	N-Methyl-4-phenyl-1,2,3,6 -tetrahydropyridine
MRI	Magnetic resonance imaging
MSA	Multiple system atrophy
NAA	N-Acetyl-aspartate

NMDA	N-Methyl-D-aspartate
NMPB	^{11}C-N-methyl-4-piperidylbenzilate
NMSP	^{11}C-N-methylspiperone
O$_2$	^{15}O-oxygen (gas)
OCD	Obsessive-compulsive disorder
OEF	Oxygen extraction fraction
OMFD	3-O-methyl-6-^{18}F-fluoro-L-dopa
OMG	^{11}C-O-methyl-glucose
OPCA	Olivo-ponto-cerebellar atrophy
OSEM	Ordered subsets expectation maximization
PD	Parkinson disease
PET	Positron emission tomography
PK	^{11}C-PK-11195
PS	Presenilin
PSP	Progressive supranuclear palsy
PSR	Protein synthesis rate
PTSD	Posttraumatic stress disorder
PVC	Partial volume correction
RAC	^{11}C-Raclopride
SAB	Subarachnoid hemorrhage
SCA	Spinocerebellar atrophy
SERT	Serotonin transporter
SND	Striatonigral degeneration
SPECT	Single photon emission computed tomography
SPM	Statistical parametric mapping
SSQ	Sum of squared residuals
STN	Subthalamic nucleus
THC	Tetrahydrocannabinoid
TIA	Transient ischemic attack
TK	Thymidine kinase
TLE	Temporal lobe epilepsy
VD	Vascular dementia
VEGF	Vascular endothelial growth factor
VMAT2	Vesicular monoamine transporter type 2
WAY	^{11}C-WAY-100635
WHO	World Health Organization

Introduction

Positron emission tomography (PET) provides unbiased in vivo measurement of local tracer activity at very high sensitivity. This is a unique property unmatched by other imaging modalities. When PET was introduced into medicine more than 25 years ago, the first organ of major interest was the brain. Since then, PET has flourished as an extremely powerful and versatile tool in scientific brain studies, whereas its use as a diagnostic tool in clinical neurology remains limited. This is in contrast to its use in other applications, particularly oncology, where its value in clinical diagnosis is more widely appreciated. We think this situation is unfortunate, because PET can contribute more to clinical neurology and clinical neuroscience than is generally perceived today. Realization of its potential will require very close cooperation between PET experts and clinicians and the integration of PET into clinical studies. Thus, in this book we review PET in neuroscience, with particular emphasis on findings that indicate its potential for improving diagnosis and treatment in neurology and psychiatry. We want to improve the transferability of the enormous scientific advances in brain PET into clinical care so as to produce tangible human benefit [1]. We wish to guide both nuclear medicine specialists and also neurologists and psychiatrists in the use of PET. We therefore focus on practical and potentially clinically relevant issues, identifying solid ground as well as open questions that require further research, and we see this targeted presentation as complementary to more general PET textbooks and reviews.

Brain PET is often classified as a method of functional brain imaging. In that area, it is in competition with other methods, such as single photon emission computed tomography (SPECT) and functional magnetic resonance imaging (fMRI), which are generally more accessible and less costly. Yet PET is not only functional, but also quantitative. Thus, it should be seen as a method of *measuring* (rather than merely *imaging*) physiological parameters locally in the brain at a spatial resolution that provides more than 1000 such independent local measurements in a single session. This aspect has been well recognized in the neuroscience field but is often neglected or even intentionally pushed aside in the clinical arena when, for the sake of convenient display, we present these measurements as color-coded images, rather than in the format of local quantitative measures with associated reference values and normal ranges. Probably because convenient methods of handling these vast amounts of data are not commonly available in clinical routine, the potential of quantitation is grossly undervalued and is often unrecognized.

Clinical examination is more powerful in neurology than in many other medical specialties with respect to its ability to localize a lesion precisely, probably because focal lesions in the highly structured nervous system lead to well-described combinations of symptoms, the classic syndromes that are often specific for lesion location. Residual uncertainty about lesion location can readily be removed by MRI [and often computed tomography (CT) will suffice]. Thus, there is very little need for additional techniques to locate macroscopically visible lesions. Identification of lesion etiology is less straightforward, and improved methods would be welcome for many diseases. Nonetheless, a major problem in clinical neurology is the irreversibility of many types of lesions, which poses severe limits on the effectiveness of treatment. Thus, identification of functional changes that precede structural lesions and could be reversible with effective treatment remains a major challenge for functional and quantitative imaging methods. Such changes may be subtle and therefore difficult to differentiate from normal variation in brain function, which also depends on the functional state ("resting," or performing mental or motor tasks) during a PET study. These factors have to be taken into account for proper interpretation of the results, which generally cannot be evaluated without detailed information on the patient's clinical status, the functional condition during the examination, and any structural brain alterations. Thus, very close cooperation between specialists in radiology, nuclear medicine, and neurology or psychiatry is crucial if diagnostic misinterpretation is to be avoided. This book aims at bridging the medical specialties to illuminate the huge potential for improvement that could be achieved in clinical science and practice by close cooperation and interaction.

The acceptance of PET in the clinical neurological arena also depends crucially on scientific evidence to substantiate its diagnostic powers. This evidence has been reviewed and published by several authors and committees in the U.S. [2, 3], in the UK [4], and in Germany [5]. Reid et al. [6] published criteria that can be applied to judge the quality of diagnostic procedures. In order to obtain reimbursemant for clinical studies, it is essential for not only the scientific community but also regulatory authorities to be convinced that procedures are cost effective [7]. Research on the economic implications of improved diagnostic techniques has still not been developed to a satisfactory level [8, 9]. In addition to technical performance and diagnostic accuracy, data are also required on diagnostic, therapeutic, health, and organizational impacts [4]. To achieve all this, brain PET studies would need to be more closely integrated into prospective clinical studies than is currently the case. It is hoped that this book will contribute to better integration by alerting nuclear medicine specialists to the clinical issues, and neurologists to the potential of PET.

On the basis of currently available data, it might be expected that the contribution of PET to diagnosis at the level of ICD clinical diagnosis will always be limited, simply because much cheaper standard procedures provide those diagnoses with high accuracy, leaving little room for substantial cost-effective improvement. Yet there is a great need for individual optimization of therapy, which cannot be achieved by standard criteria and procedures, and its impact needs to be studied in adequately designed studies. The uncertain diagnostic status of PET in many areas is in contrast to the situation with structural imaging (CT and MRI), where many indications are regarded as more or less self evident. This is because structural images are perceived as an in vivo anatomical investigation, and therefore as an essential and basic part of medical practice, and because the detection of structural lesions may potentially lead to surgical therapy. To demonstrate the impact of functional investigation, which often does not change the main diagnosis but may provide essential information for optimal therapy, these studies have to be closely integrated into the neurological work-up. In this book, we therefore also indicate possibilities for PET research to identify and clarify that potential.

Early diagnosis based on functional and thus reversible changes could open up new therapeutic possibilities to prevent irreversible damage. This may be particularly true for the *neurodegenerative diseases*, which are discussed in Chapter 2 (sections 2.1, 2.2). That this possibility has not yet become a widespread reality is probably because large studies that employ efficient tools for early diagnosis of neurodegenerative disease are lacking, with a consequent failure to develop efficient treatment or even to detect any neuroprotective potential of drugs that are already available. Since most clinical researchers have little access to PET and little awareness of its potential, there is still a lack of adequate prospective studies that would firmly establish its role in early diagnosis. We will therefore review those data that already provide ample evidence of its clinical potential for early and differential diagnosis and identify the open questions that need to be studied prospectively on a larger scale.

With the progress in CT and MRI, *brain tumors* are now much more easily diagnosed and localized than a few decades ago. Nonetheless, the most frequent primary brain tumors, the gliomas, still often pose especially difficult diagnostic questions and, mainly due to their invasive growth, there has been little improvement in their treatment. There are specific clinical situations in which metabolic and functional imaging has an obvious potential for providing important information for optimum individual treatment, and this should be confirmed by the integration of these techniques into future clinical trials (section 2.3).

Diagnosis of *cerebrovascular disease* usually relies on the high sensitivity of MRI, in particular with perfusion and diffusion imaging, and the widespread use of ultrasound techniques to identify arterial stenoses. In spite of this progress, treatment of acute cerebral infarction is still far from meeting the challenge to save as much brain as ideally possible. PET can make important contributions to improvements in therapy by elucidating the in vivo pathophysiology of acute ischemic stroke and chronic cerebrovascular disease (section 2.4).

Modern imaging techniques have revolutionized the clinical practice of diagnosing and treating *epilepsy*, because they are able to demonstrate in vivo the often small focal lesions, such as hippocampal sclerosis or migration disorders, that can cause epileptic seizures and can be accessible to surgical therapy if medication fails. PET had a leading role in this area in the 1980s, but localization and identification of these lesions is now mostly achieved with MRI. It is increasingly recognized that further progress might depend on a better understanding of the functional changes that are associated with these lesions, and some of these open questions can be addressed by the use of specific PET tracers (section 2.5).

Functional imaging also opens up new perspectives in *psychiatry*, because we can now study the local functional changes that are associated with such disorders as schizophrenia, depression, and drug addiction, where we usually do not find morphological lesions in CT and MRI (apart from possible changes in structure volumes). Measurements of receptor and transmitter changes by PET are beginning to contribute to improvements in drug therapy by way of more specific targeting and objective dose finding, although the implications of the new findings for diagnosis are not yet clear (section 2.7). It is mainly in the field of psychiatry that the pharmaceutical industry is becoming increasingly aware of the potential of PET to measure to what extent and at what doses new drugs actually bind to their target receptors. Thus, PET is becoming an essential tool for *drug development* in preclinical trials and in the early phases of clinical trials [10, 11].

The main strength of PET is its ability to measure *local physiological processes*. Many scientific studies have been devoted to normal regulation and to disease-related alterations of cerebral blood flow and energy metabolism, which have laid the ground for the clinical use of ^{18}F-2-fluoro-2-deoxy-D-glucose (FDG) and ^{15}O-water. Similarly ongoing developments in molecular imaging of tissue proliferation and neurotransmitter systems may open up new clinical perspectives. Therefore, Chapter 3 of this book presents information about radiopharmaceuticals that are particularly relevant for neurologists and psychiatrists who wish to explore new clinical and scientific applications.

Many neurological diseases probably have a genetic cause or there is at least a genetic predisposition to them, and the molecular techniques of genomics and proteomics are therefore increasingly used to disentangle their etiology and pathophysiology. Molecular techniques are not limited to the test tube but, to some extent, have also been introduced to PET under the name of *molecular imaging*. Many PET tracers bind very specifically to certain receptors or are metabolized and trapped by specific enzymes and are therefore are well suited to measurement of the local expression or functional activity of these receptors and enzymes. In gene therapy trials, measurement of the local expression of a transferred gene can provide one of the most critical parameters that determine potential treatment efficacy. Recent technical advances have led to improvements that allow the study of the brain of rats and mice at a spatial resolution in the order of 1 μl with microPET [12]. There is a direct research path of the endophenotype from genetic experimental animal models to human research. The huge potential of this approach and its special aspects and limitations in the brain, owing to the separation of the brain from the circulation by the blood–brain barrier, are detailed in section 3.9.

Accurate measurements can be obtained only if we master the technical aspects of PET. Measurements depend on the production of isotopes, mostly by bombardment of stable isotopes with protons or deuterons from a cyclotron, followed by radiochemical tracer synthesis. The relevant physics and chemistry exceed the scope of this book, however, which is focused on the medical and neurological aspects of PET. Expertise in engineering and physics is also essential for image acquisition and reconstruction, but there are also medical aspects involved in these steps, and this is the first topic that is addressed in Chapter 4. Coregistration of PET images with structural images (CT and MRI) is essential for accurate evaluation of functional changes associated with structural lesions. Matching of images to digital atlas templates that provide normative values from representative reference samples may enhance quantitative evaluation substantially. Chapter 4 also includes a description of the physiological models and mathematical procedures for extraction of physiological parameters from kinetic data (which can be skipped by readers who do not want to go into these technical details). Selection of appropriate statistical procedures to allow valid inferences is another important aspect of PET that is often not covered routinely by professional specialists in clinical laboratories. We therefore describe the essential concepts of all those physical measurement aspects and *data processing methods* that are necessary for adequate medical interpretation of PET findings.

Clinical Studies

2.1   Dementia and Memory Disorders

2.1.1   Clinical and Research Issues

Dementia is a clinical syndrome characterized by impaired short- and long-term memory and associated with deficits in abstract thinking, judgment, and other higher cortical functions, or personality change; the disturbance interferes significantly with the activities of daily life. Alzheimer's disease (AD), the most common dementing disorder, affects approximately 4 million people in the United States. Although younger individuals can be affected, the age of onset is typically over 65, and the prevalence increases with age. The aging of the population in the developed world, coupled with the devastating impact of the disease and the costs of long-term care, have made AD an increasing public health concern. Therefore, it is not surprising that AD has long been a subject of functional brain imaging studies.

The first reports of PET in the study of AD appeared in the early 1980s [13–17]. These studies followed earlier literature reports describing decreases in whole-brain blood flow and metabolism measured with the Kety-Schmidt technique in dementia. The initial observation of reduced cerebral metabolic rate of glucose (CMRglc) in temporal and parietal association cortices in AD was followed by a tremendous increase in the frequency of PET studies of dementia, with the aims of studying the pathophysiology of AD and other dementing disorders and, more recently, of defining a potential role for ^{18}F-2-fluoro-2-deoxy-D-glucose (FDG) imaging in diagnosis and management.

Pathophysiological studies initially focused on the use of FDG to characterize local abnormalities in CMRglc in relation to clinical symptoms, disease progression, and postmortem findings. More recently, other tracers have been used to study receptor systems, particularly the cholinergic system, and patterns of brain activation. PET has the potential to provide information on the interaction of neurotransmitter systems with the function of limbic, frontal, temporal, and parietal association cortices to extend our understanding of the cognitive and behavioral manifestations of AD. Radiotracers for use in imaging amyloid, which is specifically associated with disease pathology, are actively being developed. The possibility of using PET in the diagnosis of dementia was a logical extension of the initial finding of temporoparietal hypometabolism in AD. The fact that the diagnosis of AD and other dementing disorders is based on clinical assessment and is essentially one of exclusion provided a strong motivation for the proposal of a clinical role for PET. Although the promise of FDG-PET in clinical diagnosis has been recognized by the Quality Standards Subcommittee of the American Academy of Neurology [18] and other professional organizations [19, 20], its routine use is not recommended. Despite many years of study, this issue remains controversial.

As the field of clinical dementia research has progressed, the potential role of PET has evolved beyond the simple question of distinguishing AD patients from control subjects. Advances in the diagnostic criteria for AD have improved clinical diagnosis and raised the performance threshold required for any diagnostic test, including PET. In addition, there has been growing recognition of other dementing illnesses, for which diagnostic criteria have been developed. Other types of dementia, such as frontotemporal dementia (FTD), vascular dementia (VD), and dementia with Lewy bodies (DLB), need to be distinguished from AD. This is usually done by clinical examination and neuropsychological tests, and in the case of VD also by structural imaging. Although some of the non-AD dementia types exhibit distinct findings with PET (see Table 2.1, 2.2), it is not clear whether PET can provide better diagnostic separation than current clinical approaches. Research in this area must also take into account the uncertainty in the neuropathological "gold standards" with respect to the borders between the various types of neurodegenerative disease, especially as they often have overlapping or partially nonspecific neuropathological characteristics [21].

A very promising application of PET is in the preclinical and early stages of AD. PET studies in cognitively normal subjects at genetic risk for AD, e.g., those homozygous for the *e4* allele for apolipoprotein E (apoE), show decreases in CMR-glc in the same brain regions as in patients with probable AD [22]. Preclinical studies are also of great interest in the relatively fewer subjects at risk for monogenetically inherited forms of AD, owing to mutations in presenilin *(PS)1*, *PS2*, or the amyloid precursor protein *(APP)* genes. Recently, attention has focused on what appears to be a transitional clinical stage preceding early AD: mild cognitive impairment (MCI) [23]. MCI is a characterized by impaired memory function that is not accompanied by the criteria for dementia. FDG-PET may be able to predict in which individuals MCI will progress to clinically manifest AD [24]. These studies suggest a role for PET in the selection of subjects for clinical trials of treatments to prevent AD. In addition, PET may be a useful surrogate marker in treatment trials designed to slow AD progression in established cases.

2.1.2   Alzheimer's Disease and Mild Cognitive Impairment

2.1.2.1   Cerebral Glucose Metabolism

2.1.2.1.1   Main Findings

A consistent finding that has been noted since the earliest PET studies in AD is hypometabolism affecting the temporal and parietal association cortex [13–16], with the angular gyrus usually being the center of the metabolic impairment (Fig. 2.1). The frontolateral association cortex is also frequently involved, to a variable degree [25–28], while there is relative sparing of the primary motor, somatosensory, and visual cortical areas. In general, this pattern corresponds to the clinical symptoms, with impairment of memory and associative thinking, including higher order sensory processing and planning of action, but with relative preservation of primary motor and sensory function. These changes differ from those of normal aging, which leads to predominantly mesial frontal metabolic decline and may cause some apparent dorsal parietal and fronto-temporal (perisylvian) metabolic reduction resulting from partial volume effects caused by atrophy [29–33] (Fig. 2.2).

More recent studies that used voxel-based comparisons against normal reference data clearly show that the posterior cingulate gyrus and the precuneus are also impaired early in the course [34]. This was not directly obvious from inspection of FDG-PET scans, because metabolism in that area is typically above the average level normal in the cortex [35]. When impairment develops the metabolism decreases to the level of surrounding cortex but there is no visually apparent hypometabolic lesion. Thus, this potential diagnostic sign is easily missed by standard visual interpretation of FDG scans (Fig. 2.3).

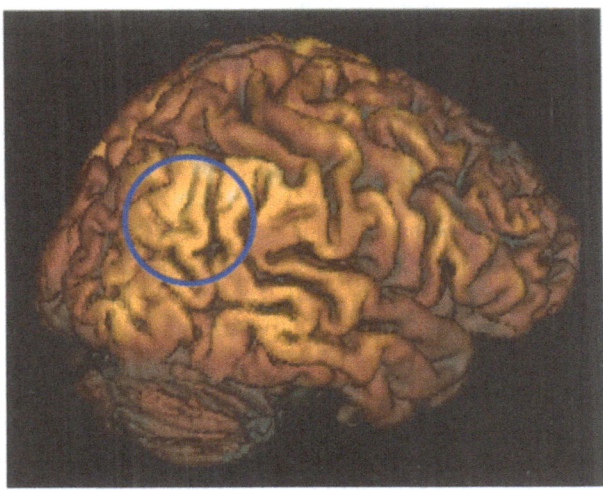

Fig. 2.1. Surface rendering of cortical metabolic abnormalities in Alzheimer disease (AD). *Bright areas* have impaired metabolism, which is typically most severe in the angular gyrus region (marked by *blue circle*). A video clip showing more of the cortical surface and its metabolic abnormalities is provided on the attached CD-ROM

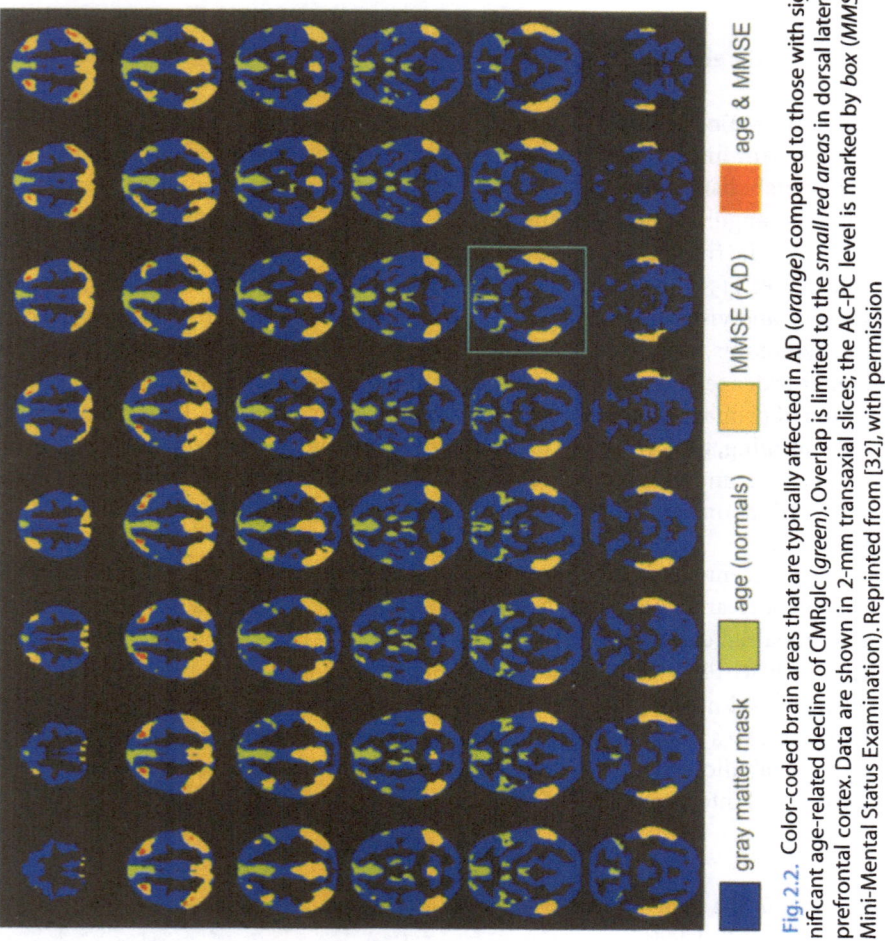

Fig. 2.2. Color-coded brain areas that are typically affected in AD (*orange*) compared to those with significant age-related decline of CMRglc (*green*). Overlap is limited to the *small red areas* in dorsal lateral prefrontal cortex. Data are shown in 2-mm transaxial slices; the AC-PC level is marked by box (*MMSE* Mini-Mental Status Examination). Reprinted from [32], with permission

Of considerable interest is the metabolic status of the mesial temporal cortex, including the entorhinal cortex. Tangles appear there very early in the course of AD, and substantial atrophy subsequently develops, as demonstrated in many MRI studies [36]. Memory impairment is, of course, an early and cardinal manifestation of AD. It has been reported that mesial temporal (entorhinal) metabolism is reduced very early in APOE4-positive individuals [37], but the more common observation is that hippocampal hypometabolism is associated with more severe dementia. In our experience, FDG uptake in AD may be reduced in the mesial temporal lobe, but not to the same degree or with the same frequency as in neocortical association areas. It seems that the mesial temporal lobe atrophy in AD detected on MRI is correlated with remote metabolic deficits in the temporoparietal and frontal association cortex and in the posterior cingulate cortex [38].

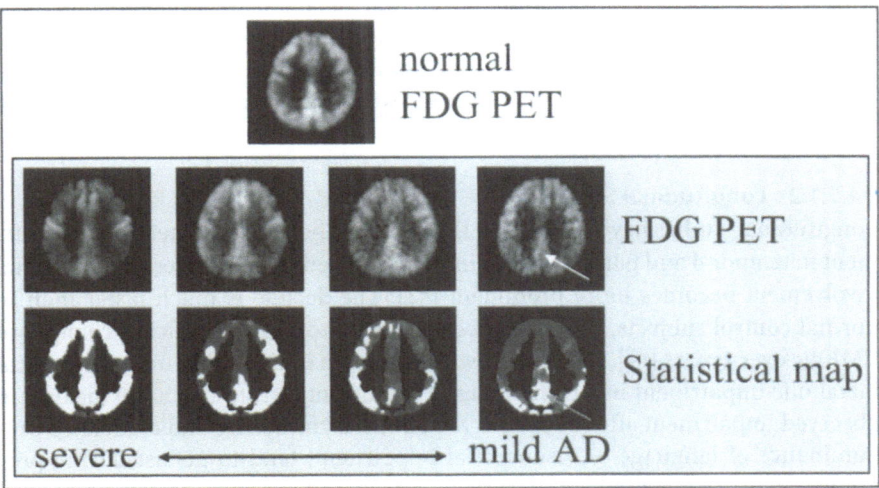

Fig. 2.3. Fluor-18-fluorodeoxy-D-glucose (FDG)-positron emission tomography (PET) (*top row* transaxial slices at level of semioval center) and corresponding t-maps of with metabolically abnormal regions identified by *white* color (*bottom row*) in four patients with AD of different grades of severity. Note the difficulty of recognizing the metabolic impairment of the posterior cingulate in the mild cases (*arrow*) on original scans

In early AD there is often distinct hemispheric asymmetry, which usually corresponds to the predominant cognitive deficits (i.e., language impairment in the dominant and visuospatial disorientation in the nondominant hemisphere) [39]. Frontal and parietal asymmetry is often associated with mild mirror asymmetry of cerebellar metabolism. This indicates crossed-cerebellar diaschisis rather than primary cerebellar involvement in the disease process [40]. An unexplained observation is that the pattern of temporoparietal hypometabolism is more prominent in younger than in older AD patients [41, 42]. This could be because younger patients have a greater cognitive reserve and therefore require a more advanced pathologic process to become symptomatic. Patients with Down syndrome, who are at increased risk for premature AD, develop the metabolic changes of AD in association with their cognitive decline [43, 44].

Dynamic PET studies have been performed to determine whether changes in FDG transport and phosphorylation are associated with cortical hypometabolism. Both glucose transport and phosphorylation have been found to be reduced [45–47]. The findings are consistent with those recorded in an in vitro study that showed lowered levels of glucose transporters in AD brain [48]. Studies of glucose transport are of interest because of findings that resistance to insulin and insulin-like growth factor may occur in AD [49].

Several nootropic drugs intended to enhance cognitive function alter the cerebral glucose metabolism (CMRglc). Increased CMRglc has been demonstrated with piracetam (a putative enhancer of cerebral metabolism) [50], propentofylline (an adenosine uptake blocker) [51], and metrifonate (a long-lasting cholinesterase inhibitor) [52]; the last two studies showed an associated improvement in certain cognitive measures. Reductions in CMRglc were found with physostigmine [53], however, in spite of improved attention.

2.1.2.1.2 Longitudinal Studies

Longitudinal studies have shown that the severity and extent of metabolic impairment in temporal and parietal cortex increases as dementia progresses and frontal involvement becomes more prominent [54]. The decline is much faster than in normal control subjects, who have been reported to show significant changes of CMRglc over 3 years [55]. Our own longitudinal data indicate that the total cortical metabolic impairment in AD increases at a constant rate, to reach the maximum observed impairment after 8–9 years. Metabolic asymmetries and associated predominance of language or visuospatial impairment tend to persist during progression [56, 57]. Metabolic rates in basal ganglia and thalamus remain stable and are unrelated to progression [55]. Thus, in late dementia there is typically a pattern of severe hypometabolism in temporoparietal and frontal association cortices, with relative sparing of primary cortical areas.

2.1.2.1.3 Relation to Neuropsychological Tests
and Behavioral Symptoms

Owing to the widespread metabolic deficits and pervasive cognitive impairment in AD, it can be difficult to relate metabolic changes in specific brain regions to specific types of cognitive disturbance. Memory impairment, which occurs early in the disease, correlates with a decline in CMRglc in virtually all limbic and association areas affected, and also with cerebral atrophy [58]. The strongest associations of memory abnormality are usually seen with temporal cortex [59]. In early AD, episodic memory impairment correlates most closely with reduced parahippocampal and retrosplenial CMRglc [60]. It is noteworthy, however, that the initial metabolic impairment in AD is not limited to hippocampal structures, more prominent neocortical metabolic abnormalities tending to occur early and sometimes preceding nonmemory cognitive impairments [61, 62].

The metabolic asymmetry that is often seen, especially in early AD, correlates to some extent with the neuropsychological profile. Patients with more severe impairment of the dominant (usually left) hemisphere have more language problems [39, 63], whereas patients more severe impairment of the nondominant hemisphere have more visuospatial deficits. These neuropsychological and metabolic asymmetries tend to persist during progression of the disease [56].

When they have controlled for age, sex, education, and severity of illness, researchers have extracted more regionally specific relations between CMRglc and cognitive function. For example, impaired semantic processing of language has been found to be related to glucose hypometabolism in the inferior temporal gyrus and inferior parietal lobule [63]. It has been difficult to design neuropsychological tests to assess frontal lobe function in AD [64], and so far the results of such tests have not been correlated with frontal metabolism.

Behavioral symptoms, such as apathy, disinhibition or paranoia, are not typically prominent in early AD, but may significantly contribute to the clinical syndrome with disease progression and cause a heavy burden for caregivers. As in frontotemporal dementia, where these symptoms appear early, there is evidence that the behavioral deficits are related to frontal metabolism. For example, a relationship between severity of delusional thought and CMRglc was found in the right superior dorsolateral frontal cortex (Brodmann's area 8), the right inferior frontal pole (Brodmann's area 10), and the right lateral orbitofrontal region (Brodmann's area 47) [65]. Lack of insight into disease severity appears to be related to fronto-orbital hypometabolism [66], while agitation, disinhibition, and irritability have been associated with decreased anterior cingulate and orbitofrontal CMRglc [67, 68].

2.1.2.1.4 Relation to Autopsy Data
Autopsy studies are potentially valuable for our understanding of the mechanisms of decreased CMRglc in AD, but are limited by the difficulty of obtaining a PET scan shortly before death. Also, autopsy patients typically have advanced disease, so that a spectrum of the severity of neuropathological change may not be available for correlation with PET. Many studies have used autopsy findings to confirm a PET diagnosis recorded before death. In an autopsy series of 22 patients, the sensitivity, specificity, and diagnostic accuracy of bilateral temporoparietal hypometabolism associated with AD were 93 %, 63 %, and 82 %, respectively [69]. There have been several small series with typical PET findings and pathologically confirmed AD [25, 70]. Findings in 138 patients examined in multiple centers have been combined and indicate a sensitivity of 94 % and a specificity of 73 % for FDG-PET to allow correct prediction of a pathological diagnosis of AD [71].

Among the histopathological hallmarks of AD, intraneuronal neurofibrillary tangles correlate significantly with severity of dementia, whereas there is only a loose correlation or none at all for amyloid plaques [72]. Thus, it might be expected that neurofibrillary tangles should also correlate with hypometabolism, because this is also closely related to dementia severity. However, studies comparing the spatial relation of premortal FDG uptake with postmortem histopathology have not generally found consistent relationships, suggesting that alternative mechanisms underlie the metabolic defect [70, 73]. For example, in the report of a case in which PET was performed 8 hours before death in a subject with advanced AD there was an inverse correlation between FDG uptake and amyloid protein in parieto-occipital and prefrontal tissue, but not in temporal lobe tissue [74]. In another study in which PET was performed a mean of 2.5 years before death in six patients, there was a significant negative correlation between local metabolism and neurofibrillary tangles in only two cases [75]. Neurofibrillary tangles in neocortical association areas occur relatively late in the progression of AD [76], but the metabolic changes are present prior to clinical onset, so that it is unlikely that hypometabolism is directly related to the tangles. Thus, the microstructural or molecular changes that cause the functional deficits seen with FDG remain to be determined. Potential candidates are dendritic or synaptic changes [77] and tissue deposition of beta amyloid [78]. Another possibility is that hypometabolism in association cortex reflects decreased activity from distant afferent inputs, resulting in decreased local synaptic activity that is unrelated to local pathology.

2.1.2.1.5 Preclinical and Early AD

"Mild cognitive impairment" (MCI) is the term used for an isolated memory deficit in the absence of other criteria for dementia in individuals who are otherwise functioning well [23, 79]. This condition may represent a transitional phase before clinically recognizable AD. Approximately 15% of patients with MCI progress to dementia per year, but many remain stable for several years or even prove to have had a reversible memory deficit. The clinical diagnosis of MCI is often uncertain, because results may depend on the particular memory test used [80]. The development of a practice parameter for identification and management of MCI should stimulate research in this important area. PET could help elucidate the pathophysiological relationship between MCI and AD, and could potentially identify those patients who are at high risk of developing AD. Early diagnosis could become a major clinical issue if drugs that slow or prevent disease progression become available. Candidate drugs have already been identified, and PET could have an important role in clinical trials.

Metabolic reductions in MCI are focused in limbic structures, and particularly in parahippocampal gyrus and posterior cingulate [81]. Data are accumulating which show that presence of the AD metabolic pattern predicts conversion from MCI to clinical dementia of Alzheimer type, and therefore indicates "incipient AD" (Fig. 2.4). In a longitudinal study of APOE ε4-positive, nondemented subjects with memory complaints, FDG-PET findings predicted subsequent cognitive decline 2 years later [82]. In Cologne, we studied patients who had mild cognitive deficits mostly limited to impaired memory and who scored higher than 23 on the Mini-Mental Status Examination (MMSE) but did not yet fulfill the criteria for probable AD [83]. Most of these patients would have fulfilled the criteria for MCI. We found that in 60–70% of patients with moderate or severe metabolic impairment of association cortices the MMSE scores declined by 3 or more points within 2 years (mostly leading to clinical dementia), whereas only 10–20% of patients without such metabolic impairment had a significant cognitive decline [84] (Fig. 2.5). Several other recent reports have demonstrated the predictive value of temporoparietal metabolic impairment for conversion of MCI to AD [85–88].

Abnormal cortical glucose metabolism has also been observed in asymptomatic subjects at high risk of AD [22, 37, 89, 90]. For example, cognitively normal subjects homozygous for the ε4 allele for APOE had a pattern of cortical hypometabolism similar to that of AD. Asymptomatic ε4 heterozygotes had declines in CMRglc in several brain regions affected in AD over a 2-year period. These were group analyses that did not study the diagnostic use of FDG-PET to predict

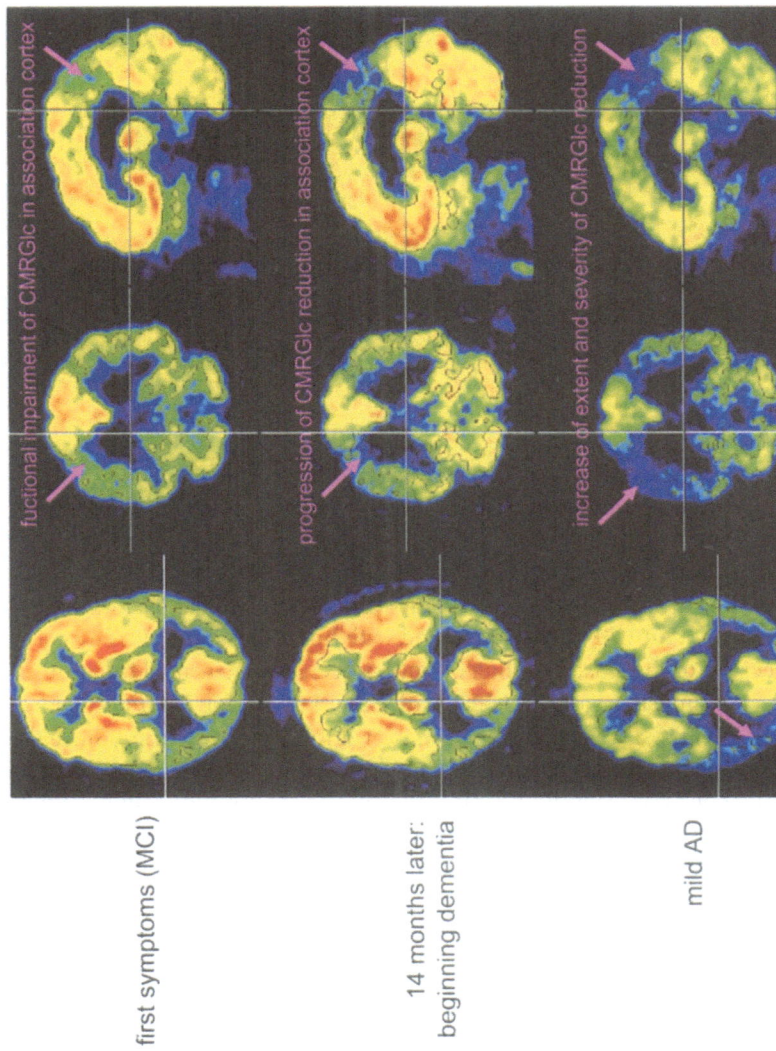

first symptoms (MCI)

14 months later: beginning dementia

mild AD

Fig. 2.4. Decline of cerebral metabolic rate (glucose) (*CMRGlc*) in association areas with progression of AD from the clinical stage of mild cognitive impairment (MCI) to mild dementia (three follow-up FDG-PET scans, each showing same orthogonal slices at position marked by *crosshairs*)

dementia in individual subjects, and it is unlikely, for obvious practical reasons, that PET could be used as a general screening procedure for AD in asymptomatic subjects [22]. These studies, however, do provide support for using PET to select subjects for clinical trials of treatments to prevent AD. From longitudinal data in cognitively normal ε heterozygotes, it was estimated that between 50 and 155 subjects would be needed in each placebo and treatment group to detect a 25% attenuation of CMRglc decline over 2 years [22]. In patients with established AD, it was estimated that 61 patients per treatment group would be needed to detect a similar drug effect over 1 year [91].

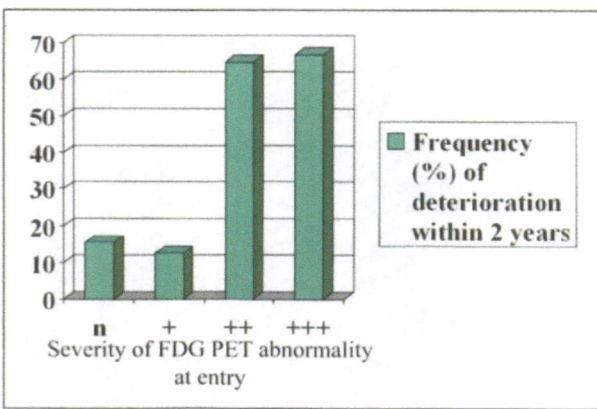

Fig. 2.5. Frequency of dementia progression in patients with MCI or very mild AD (MMSE 24)

2.1.2.1.6 Comparison with Other CBF and Metabolic Studies

An early study indicated that oxygen metabolism ($CMRO_2$) was altered in AD, with a similar pattern to that of CMRglc [17]. In a direct comparison of FDG and ^{15}O tracers [92], $CMRO_2$ appeared to be reduced by less than FDG and CBF in association areas. The typical pattern of abnormality in association areas seen with FDG has been described for CBF images obtained with both PET and SPECT [93, 94], probably because of the close coupling of both processes to neuronal function. In the few direct comparisons of FDG-PET with SPECT that have been performed, however [95–97], FDG-PET has always been shown to be more accurate. Since the greatest benefit of functional imaging in dementia diagnosis will probably not be in cases with typical AD symptoms but rather in those with more subtle symptoms, e.g., MCI [86, 98–100], the higher accuracy is certainly needed. Whether PET could be more cost effective than SPECT in this situation in spite of its higher acquisition and operating costs has not yet been studied. Usually, ^{15}O studies provide less favorable image statistics than FDG, which may be one reason why a clinical study of ^{15}O-water PET [94] led to the conclusion that it is not useful for clinical diagnosis of AD.

Alterations in CMRglc are also correlated to some extent with changes in local blood volume that influence magnetic susceptibility functional MR imaging [101, 102]. Only a few studies have been done so far with this technique, which is sensitive to susceptibility artifacts that limit its clinical potential.

2.1.2.1.7 FDG-PET for Clinical Diagnosis of AD

The definitive diagnosis of AD requires histopathological verification, which is usually not possible until after death. Pathological studies have demonstrated that the histopathological features of AD disease can be found even in cognitively normal individuals many years before the manifestation of dementia [103]. The diagnosis of AD and other dementias is based on clinical assessment, however, and is a diagnosis of exclusion. A major advance was the development of explicit clinical criteria, such as the NINCDS-ADRDA criteria [104], which, when carefully applied, accurately predict the presence of AD [105]. Using these criteria does require care and time, however. The identification of a metabolic pattern in AD, and the motivation for basing diagnoses on tests rather than clinical observation, especially for a disease as devastating as AD, have led to the proposal that PET could be useful to assess patients with the symptoms of dementia. Professional organizations have not yet recommended this use of PET as a standard clinical tool, however [18–20]. The Quality Standards Subcommittee of the American Academy of Neurology (AAN) recommends using standardized clinical diagnostic criteria that can be checked with structural neuroimaging (either CT or MRI) to rule out relatively rare treatable conditions such as frontal brain tumor, chronic subdural hematoma, and normal pressure hydrocephalus, plus screening for depression, hypothyroidism, and vitamin B12 deficiency [18]. PET has been noted as a promising technique, but it is felt that further prospective studies are still needed to determine whether it adds anything to the clinical diagnosis.

As noted above, several studies have compared premortem diagnoses with autopsy findings to determine the accuracy of PET [25, 70]. In one series of 22 patients with memory loss or dementia that was difficult to characterize by means of clinical criteria, the sensitivity and specificity for bilateral temporoparietal hypometabolism associated with AD were 93 % and 63 %, respectively; the clinical diagnosis was more highly specific (88 %), but its sensitivity was lower, at 64 % [69]. A large study combining data from eight centers compared PET findings (visual assessment of the FDG uptake pattern), clinical course, and autopsy results [71]. In the 146 subjects in whom longitudinal clinical follow-up was possible, PET predicted clinical progression of dementia with a sensitivity of 91 %, but had a lower specificity for discrimination from other neurodegenerative disorders of 75 %. Results were similar in the 138 patients with regard to autopsy diagnoses (sensitivity 94 %, specificity 73 %).

Whether these results support the clinical use of PET in the work-up of patients presenting with possible dementia has yet to be established. Questions have been raised about the design of these studies, their applicability in hospital practice, and their cost effectiveness [106–108]. Reports comparing premortem and postmortem diagnoses have generally not fulfilled the methodological standards that are desirable for diagnostic test research [6, 109]. The critical issue of whether PET has an incremental benefit when added to a careful clinical assessment has so far not been assessed.

Cost–benefit analyses have also not been definitive. One study [98] estimated that use of PET could decrease rates of both false-negative (from 8.3 % to 3.1 %) and false-positive (from 23.0 % to 11.9 %) diagnoses, with a subsequent reduction in costs of additional care for patients with advanced dementia who could have benefited from earlier diagnosis and for unnecessary drug therapy of patients in whom it was not indicated. Other studies have calculated that general use of PET in patients with a clinical diagnosis of dementia is not cost effective [108, 110]. It has been concluded that current treatment with acetylcholinesterase inhibitors,

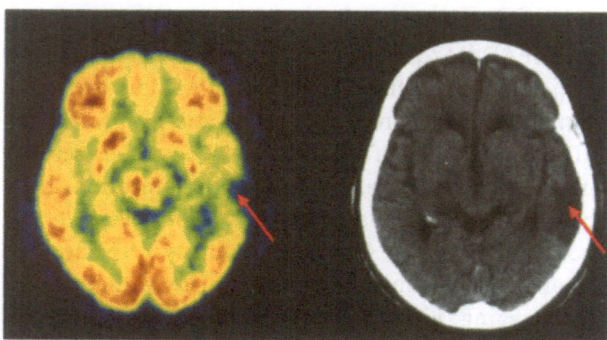

Fig. 2.6. Temporal metabolic impairment in a patient with memory deficit and aphasia due to a left temporal ischemic infarct (*arrow*)

which have few side effects, should be offered to patients identified as having AD according to standard clinical criteria. PET would be warranted when treatments become available that are effective in slowing disease progression but have the potential for moderate to severe complications.

Although the pattern of FDG uptake in AD is quite a sensitive indicator of the diagnosis, it can also be seen in other conditions. One limitation of the uncritical use of FDG-PET is that patients with Parkinson disease (PD) can show a very similar metabolic pattern [111, 112], even in the absence of major cognitive deficits [113]. These changes may even be reversible with electrical stimulation of the subthalamic nucleus [114]. Yet, with appropriate clinical information, the "pseudo-AD" metabolic pattern should not be a major problem, because it is seen only in patients with long-standing PD who have the typical motor symptoms. The question of whether the occurrence of that pattern in PD indicates incipient dementia (of Alzheimer or Lewy body type) requires further study.

In theory, any functional impairment of cortical association areas, regardless of the underlying etiology, could mimic AD in FDG-PET and in the clinical symptoms. To exclude nondegenerative lesions as causes of focal FDG reductions that could mimic AD, it is always wise to compare the PET images with structural images, which are part of any comprehensive diagnostic work-up (Fig. 2.6). Conversely, sparing of structures that are typically not affected may not be perfect, and the resulting atypical FDG distribution patterns may lead to erroneous exclusion of AD.

Patients with late-onset AD may show less difference between typically affected and nonaffected brain regions, which could potentially lead to reduced diagnostic accuracy of FDG-PET [41, 115]. This could reflect the fact that at higher age multifactorial damage to the brain is likely to be cumulative and, actually, in neuropathological studies too the proportion of diagnoses of unclassifiable dementia is increasing among the oldest old. Thus, according to general clinical wisdom, FDG-PET probably has less potential for diagnostic use in very old patients.

2.1.2.1.8 Technical Issues

During a scan, subjects are often asked to keep their eyes closed for at least 30 minutes during FDG uptake, to avoid visual stimulation which could induce variability of occipital uptake. Some patients may forget this or need repeated reminders, which interfere with pure resting conditions. Alternatively, studies can be performed with the eyes open. This potential confounder mainly affects occipital glucose metabolism and is more of a consideration for scans obtained for research, rather than clinical, purposes. PET is generally well tolerated in demented patients. Sedation (if needed) with benzodiazepines leads to a global reduction of CMRglc but does not change the metabolic pattern typical for AD [116] and may permit the study of patients who are otherwise unable to cooperate.

Early PET studies of AD were typically performed with arterial blood sampling to obtain plasma levels of FDG and glucose for use in the operational equation to calculate CMRglc (see section 4.9.2.2 for methodology). Because the level of FDG uptake is approximately proportional to CMRglc, images of tissue counts are often used to avoid blood sampling. In clinical applications especially, visual analysis of the pattern of FDG uptake is usually sufficient. It has long been recognized that even in normal subjects there is variability in both global and regional PET measurements, which reflects both physiological variation and methodological inaccuracies. Therefore, techniques are often applied to facilitate detection of regional changes in spite of global variability by "normalizing" regional data. Regional values are divided by the global value or by the value in a structure presumed to be minimally involved in the disease process, to adjust for the effect of global variation, e.g., motor cortex, thalamus, cerebellum, or brain stem in AD [117–119]. A combination of all typically preserved regions (cerebellum, brain stem, putamen, motor and visual cortex) provides the most robust reference to scale brain FDG uptake in AD [120]. Thus, blood sampling to obtain absolute CMRglc is not necessary for the diagnosis of AD, and evaluation of the FDG uptake pattern with normalization to typically preserved reference regions is sufficient and the most sensitive method.

Voxel-based comparisons of scans in early AD patients against data obtained in normal controls show decreased FDG uptake in the posterior cingulate gyrus that is not apparent on visual inspection of images. This shows the value of automated methods of image analysis [34] in both research and clinical applications in AD. It is becoming standard to base the interpretation of patient studies not only on visual interpretation of the tracer distribution, but also on quantitative mapping with reference to an appropriate normal sample [32, 121–124] (Fig. 2.3). This technical advance also overcomes the ambiguities of qualitative image interpretation and reduces the variability introduced by differing physician experience and expertise.

Activity in normal cerebral cortex, with a thickness of 3–5 mm, cannot be measured accurately with current standard PET equipment with a spatial resolution of 4–8 mm, because of incomplete recovery of counts and partial volume averaging [125]. Although techniques for atrophy correction are available, [126, 127], their use is limited by the requirement of high-quality coregistered MRI and the sensitivity of the procedure to noise and bias. Most probably the issue will be settled by the next generation of PET scanners with higher spatial resolution of approx. 2.5 mm (see section 4.2).

The impact of cerebral atrophy depends upon the context in which PET is being used. For clinical studies, cortical atrophy and partial volume averaging with metabolically inactive, enlarged cerebrospinal fluid spaces will further decrease the apparent FDG uptake, actually increasing the diagnostic sensitivity for AD. On the other hand, other atrophic processes will also appear to have reduced FDG uptake, thus reducing specificity. Many degenerative brain diseases (and even normal aging) lead to global brain atrophy, with associated partial volume effects that are typically most pronounced in dorsal parietal and perisylvian cortex. Therefore, reduced FDG uptake in the angular gyrus and posterior cingulate provides better specificity for identification of the typical AD pattern than does the reduction in dorsal parietal cortex. Whether CMRglc in AD is actually reduced per unit volume of remaining brain tissue is an important research question. Studies that have applied a correction for partial volume averaging have demonstrated that the metabolic declines persist, although they are not as great [126, 128]. This indicates that the declines in CMRglc in AD are not due solely to cerebral atrophy.

2.1.2.2 Cholinergic Degeneration

Degeneration of cholinergic neurons has been observed in several neurodegenerative diseases, most notably in AD and PD [129, 130], whereas it may be mostly intact in vascular dementia (VD) [131] and in frontotemporal dementia (FTD) [132]. These neurons express acetylcholine esterase (AChE) as an enzyme for degradation of acetylcholine. In recent years, the piperidine analogues ^{11}C-labeled N-methyl-4-piperidyl acetate (MP4A) [133] and N-methyl-4-piperidyl propionate (MP4P) [134] have been developed for in vivo imaging of cerebral AChE with PET. Reduced AChE activity in AD has been observed in several studies with these tracers [119, 135, 136]. It is reduced in all cortical areas, most severely in occipital and temporal cortex, but is still preserved in mild to moderate AD in the nucleus basalis of Meynert [137], suggesting that cholinergic impairment begins in cortical regions and affects the basal cell nuclei slowly as the result of a "dying back" phenomenon (Fig. 2.7). Compared with PD, in which AChE may be reduced without dementia, the reduction in AD is more severe in the parieto-temporo-occipital association cortex [138].

The additional AChE inhibition resulting from the action of choline esterase inhibitors, which are used therapeutically in AD to enhance synaptic acetylcholine levels by inhibition of hydrolysis, can also be measured. Studies suggest that donepezil at standard doses inhibits cerebral cholinesterases by only 27 % [139], or more specifically, inhibits AChE by 39 % [140].

Labeled ChE inhibitors have also been used to measure cerebral AChE expression. ^{11}C-Physostigmine and ^{11}C-CP-126,998, an N-benzylpiperidine benzisoxa-

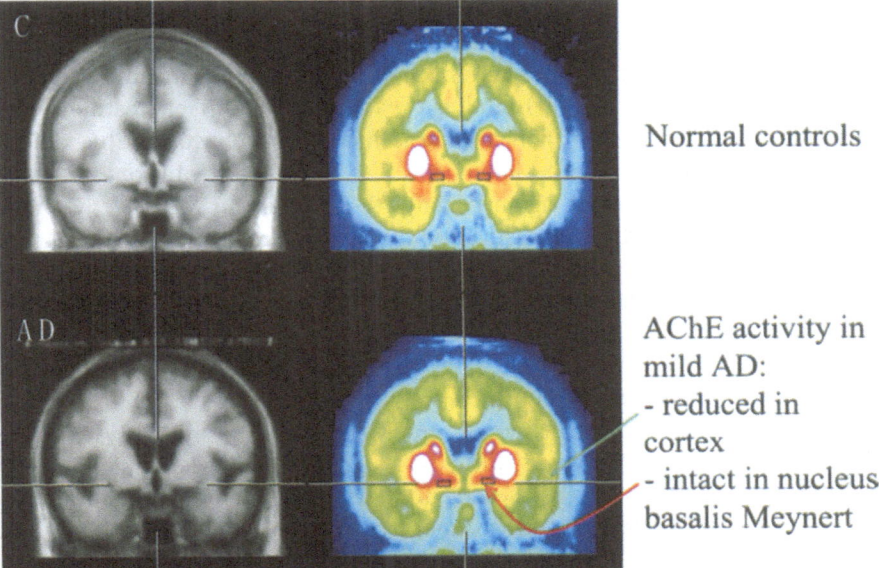

Normal controls

AChE activity in
mild AD:
- reduced in
cortex
- intact in nucleus
basalis Meynert

Fig. 2.7. MP4A-PET (coronal slices 2 mm posterior to AC) demonstrates reduced cortical acetyl-choline (AChE) activity in mild AD, but high activity in basal forebrain is still preserved at that stage

zole, showed a good correlation with the known AChE concentrations measured in human brain [141, 142]. ^{11}C-Methyl-tetrahydroaminoacridine easily crosses the blood–brain barrier and is highly concentrated in the brain, but the regional brain distribution of the tracer does not parallel that of in vivo AChE concentrations [143].

Unfortunately, there is no tracer for the main acetylcholine synthetic enzyme, choline acetyl transferase. The vesamicol derivative ^{18}F-fluoroethoxy-ben-zovesamicol is a tracer for the vesicular acetylcholine transporter in cholinergic synapses [144], but has not been used in humans. Its SPECT analogue demonstrates reduced binding in AD [145], but its use may be limited by toxicity that requires very high specific activities.

Autopsy studies reveal that nicotinic receptors are altered and reduced in AD, and nicotinic agonists may exert beneficial cognitive effects. Therefore, in vivo studies are of potential diagnostic and therapeutic interest [146]. Epibatidine is a highly specific ligand, but is also very toxic [147, 148] and therefore not likely to be of clinical diagnostic use. Much research is being done in attempts to develop less toxic derivatives and related tracers [149–152]. One agent that appears promising is A-85380 [153], which has been used to demonstrate the competing effects of nicotine and metabolites in the monkey [154]. It seems promising for human PET studies if labeled by ^{18}F in the 6-position [155, 156]. ^{11}C-Nicotine has also been used to image nicotinic receptors in the brain [157], but practical use and quantitation is limited by relatively high levels of unspecific binding.

So far, no consistent changes have been observed with regard to muscarinic acetylcholine receptor (mAChR)-binding capacity in AD [158]. Yet, treatment with AChE inhibitors leads to an agonist-induced increase of in vivo mAChR radioligand binding [159]. Increased binding was also induced by infusion of a muscarinic cholinergic receptor agonist [160]. ^{18}F-FP-TZTP, an M2 selective muscarinic agonist, showed higher distribution volumes in cognitively normal, older subjects with an APOE ε4 allele. A possible explanation for this finding is a lower concentration of acetylcholine in the synapse of APOE ε4-positive subjects [161].

2.1.2.3   Other Receptors and Transmitters

In addition to the cholinergic system, multiple other transmitter systems are impaired in AD. In a study with the 5-HT$_{2A}$ ligand ^{18}F-altanserin, significantly reduced binding was observed in several brain regions, including the anterior cingulate, prefrontal cortex, and sensorimotor cortex [162]. A related finding of potential clinical relevance is an association between 5-HT$_{2A}$ receptor polymorphism and psychotic symptoms in AD [163].

Dopaminergic innervation of the striatum, as measured by FDOPA, is largely intact in AD (in contrast to dementia with Lewy bodies, see section 2.1.3), even in the presence of mild "parkinsonoid" rigidity [164]. In the striatum, there is a modest reduction of dopamine D1 receptors but not of D2 receptors [165], whereas D2 receptors are reduced in hippocampus [166]. In contrast to CMRglc, benzodiazepine receptors are largely preserved in AD, suggesting that synaptic dysfunction rather than mere neuronal loss underlies CMRglc reduction [167].

Alzheimer plaques are associated with inflammatory responses, and patients taking certain nonsteroidal antiphlogistics seem to have a lower incidence of AD. Imaging of inflammation can be achieved with ^{11}C-PK-11195, a ligand for the peripheral benzodiazepine receptor that is expressed on leukocytes and activated microglia [168]. Increased binding has been demonstrated in AD, including mild and early forms, suggesting that microglial activation is an early event in the pathogenesis of the disease [169].

2.1.2.4   Amyloid Imaging

Since most dementia researchers regard amyloid deposition as the most important and specific pathophysiological event in AD, newly developed tracers that label amyloid plaques and neurofibrillary tangles are likely to play an increasingly important part [170–173]. 2-(1-(6-(2-^{18}F-fluoroethyl)(methyl)amino)-2-naphthyl)ethylidene)malono nitrile (FDDNP) is sufficiently lipophilic to enter the brain and bind to amyloid beta(1–40) fibrils [174]. FDDNP competes with the chemically related anti-inflammatory drugs naproxen and ibuprofen for binding to amyloid beta(1–40) fibrils [175]. In a preliminary report, observation of greater accumulation and slower clearance of FDDNP was described in subjects with AD [172]. The tracer also binds to prion plaques in vitro [176].

Another approach is based on histological dyes that are known to bind to amyloid, such as thioflavin and congo red [177–180]. Very promising data have been presented for the thioflavin-based compounds [181], especially for [N-methyl-^{11}C]2-(4'-methylaminophenyl)-6-hydroxybenzothiazole (BTA-1), which has also been named Pittsburgh compound B (PIB) [182]. The real-time biodistribution kinetics have been studied in transgenic mouse models of AD using multiphoton microscopy. PIB entered the brain quickly and labeled amyloid deposits within minutes. The nonspecific binding was cleared rapidly, whereas specific labeling was prolonged [183]. First human data have also been reported, demonstrating increased accumulation in AD in temporal, parietal and occipital cortex as well as in striatum [184]. First human data have also been reported, demonstrating increased accumulation in AD in frontal, temporal, parietal and occipital cortex as well as in striatum [184]. Recent data in transgenic mice suggest that a relatively simple stilbene derivative, N-^{11}C-methylamino-4'-hydroxystilbene, may also be useful as a PET imaging agent for mapping amyloid beta plaques [185].

2.1.2.5   Activation Studies

Activation studies have been performed in AD to correlate the level of performance of specific cognitive tasks with the loci and degree of neuronal activity as reflected in measurements of CBF and CMRglc and to determine how the patterns of activation differ from those in normal subjects. In principle, impairment of activated CMRglc could be a more sensitive index of the functional/metabolic failure of neuronal systems than metabolism at rest [186], but high interindividual variability limits any potential clinical use of this approach [187]. Metabolism and blood flow in impaired association cortex can be activated by memory tasks [187]. Both higher and lower activation than in normal controls have been described. It seems that with simple tasks that can still be well-performed by AD subjects, increased CBF activation may be observed [188], whereas activation is decreased if performance is impaired in more difficult tasks [189, 190]. Impairment of olfaction is a frequent finding in AD, and therefore an olfactory memory task may be of particular interest. Mesial temporal lobe activation during such a task was reduced in AD patients [191]. In normals, less CMRglc activation is seen in memory tasks with successful priming, whereas in AD patients priming and the associated attenuation of the metabolic response may be reduced [192].

Patterns of activation across the brain can be altered in AD. For example, during word retrieval, failure of temporoparietal activation in AD may be associated with increased frontal and cerebellar activation, perhaps reflecting compensatory mechanisms [193]. In a detailed rCBF study of visual memory, AD patients showed altered patterns of activation, suggesting that there is a functional disconnection between hippocampus and prefrontal cortex [194].

2.1.3 Dementia with Lewy Bodies

Lewy bodies, intracellular deposits consisting of alpha synuclein and ubiquitin in the midbrain, are the pathological hallmark of PD. In dementia with Lewy bodies (DLB) they also occur in cerebral cortex. DLB is now recognized as the second most common neurodegenerative disease causing dementia after AD. DLB patients typically have fluctuating levels of cognition and consciousness, visual hallucinations early in their course, falls, and parkinsonism [195, 196]. There are clinical and pathological overlaps between DLB and PD, and between DLB and AD. Up to 30% of patients with PD develop dementia and often hallucinations, especially when treated with dopaminergic drugs. It is likely that PD and DLB are manifestations of the spectrum of Lewy body disorder, with possible intermediate manifestations and progression of primary midbrain lesions to involve cortical areas and vice versa. The dementia in PD also seems to have Lewy bodies as its pathological basis [197]. The nosological classification of DLB is further complicated by the fact that cortical Lewy bodies may also occur to a variable extent in AD [198]. FDG-PET studies in DLB have shown a pattern of temporal and parietal hypometabolism similar to that in AD. The distinguishing feature is reduced metabolism in primary visual cortex, which is usually spared in AD [199–201]. Hypometabolism in visual cortex may be the correlate of impaired visual processing and visual hallucinations. The diagnostic reliability of this finding is not yet clear, however.

A more specific PET finding in DLB is probably reduced ^{18}F-DOPA uptake in the putamen [202, 203], an abnormality that is absent in AD [164] (Fig. 2.8). In contrast

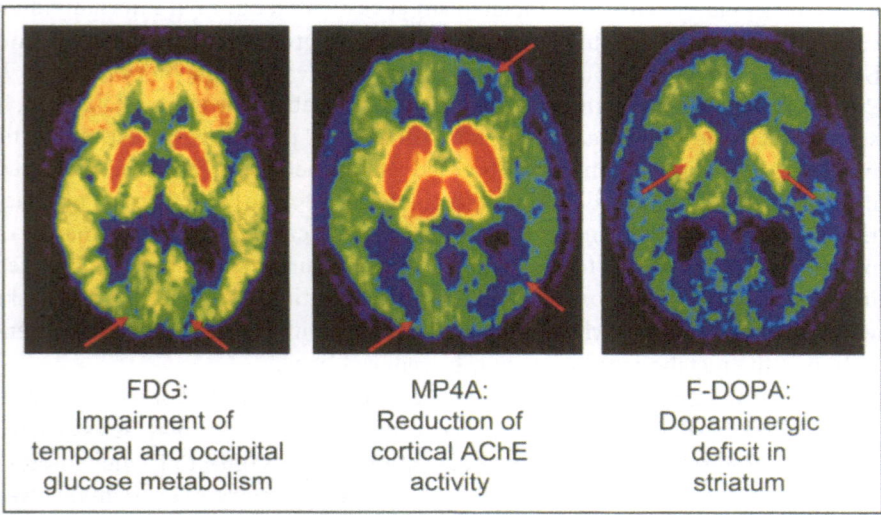

FDG:	MP4A:	F-DOPA:
Impairment of temporal and occipital glucose metabolism	Reduction of cortical AChE activity	Dopaminergic deficit in striatum

Fig. 2.8. CMRglc (FDG), AChE activity (MP4A), and dopamine synthesis (FDOPA) in DLB, with main abnormalities marked by *red arrows*

to PD, where dopaminergic dysfunction is most severe in the posterior putamen, in DLB the caudate and the anterior and posterior parts of the putamen seem to be equally affected.

2.1.4   Frontotemporal Dementia and Related Disorders

Frontotemporal dementia (FTD) is not a specific clinical or pathological diagnosis, but rather refers to a heterogeneous group of disorders clinically characterized by early changes in personality and behavior, or by aphasia, with memory impairment absent or less prominent [204]. Abnormalities include social disinhibition, inappropriate behavior, and impulsiveness. The spectrum of FTD includes classic Pick's disease, but also primary progressive aphasia and semantic dementia [205]. Clinical differentiation from AD is usually not very difficult, but FTD is usually not diagnosed at an early stage because mild symptoms are difficult to verify objectively. FTD is often associated with impairment of language production, and in the variants of semantic dementia and primary progressive aphasia the language disorder is the initial symptom and dementia may initially be absent [206]. FTD is characterized by neuronal degeneration primarily affecting the frontal and anterior temporal lobes, without Lewy bodies, plaques or tangles; Pick bodies are present in some cases. Familial FTD may be associated with parkinsonism and with mutations in the *tau* gene on chromosome 17 [207].

FTD is identified on FDG-PET scans by a distinct frontal or frontotemporal metabolic abnormality [208, 209]. Apparently, mesial frontal hypometabolism is the most common and can be found in nearly every case [66] (Fig. 2.9). Very frequently, there is also prominent focal atrophy of the frontal and temporal lobe in one hemisphere, corresponding to a metabolic deficit that is also strongly asymmetric, is centered in the anterior pole of the temporal lobe, and extends to other association areas (Fig. 2.10). It seems that FTD can also be differentiated from corticobasal degeneration with predominant parietal metabolic reduction [210], although histopathological features may overlap [211]. Frontal metabolic impairment occurs in many other conditions, including progressive supranuclear palsy (in combination with midbrain impairment) [212], spinocerebellar atrophy [213], and cocaine abuse [214]. There are also cases with combinations of frontal and temporoparietal hypometabolism that could represent either AD or FTD. Furthermore, there are a few cases with the clinical syndrome of AD but isolated frontal hypometabolism [25].

Semantic dementia, a clinical syndrome that is probably related to FTD (and also called temporal variant of FTD), is characterized by fluent dysphasia with severe anomia, reduced vocabulary, and impaired single-word comprehension, progressing to a stage of virtually complete dissolution of the semantic components of language [215]. The main metabolic impairment is located in temporal cortex of the dominant hemisphere, where there is also focal atrophy [215]. With a semantic decision task, patients activate some areas similar to those activated in

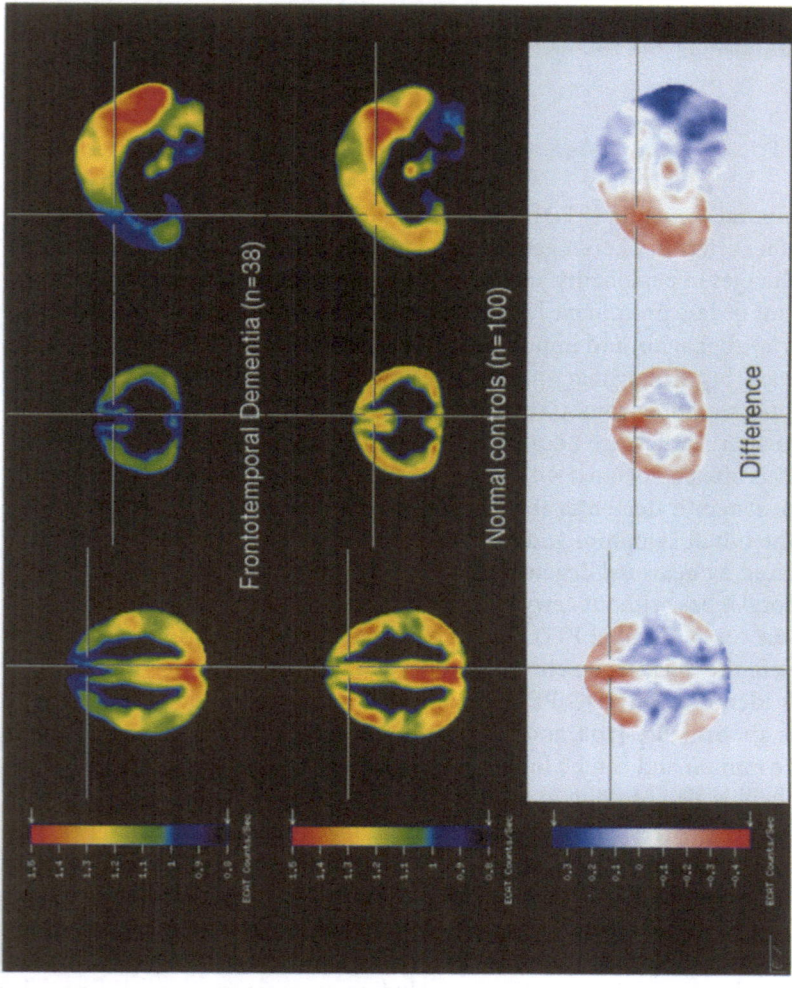

Fig. 2.9. CMRglc in FTD (*top row*) compared with controls (*middle row*) and difference between FTD and controls (*bottom row*), average data (spatially normalized brains, three orthogonal slices shown) from the NEST-DD study (www.nest-dd.org)

normal subjects, including regions of significant atrophy, but show substantially reduced activity particularly in the left posterior inferior temporal gyrus [216].

Primary progressive (nonfluent) aphasia is another related syndrome associated with left frontal and temporal hypometabolism [217–221] that can also affect additional brain areas to a lesser degree, suggesting that it is not a strictly focal impairment. A similar condition also seems to exist for the right hemisphere, clinically consisting of progressive prosopagnosia [222] (Fig. 2.11). Most cases seem to be histopathologically related to FTD, and they may progress to FTD or dementia of Alzheimer type after several years.

Frontal dysfunction with impairment of executive function and verbal fluency may also be present in amyotrophic lateral sclerosis (see also section 2.2.3.5.). In

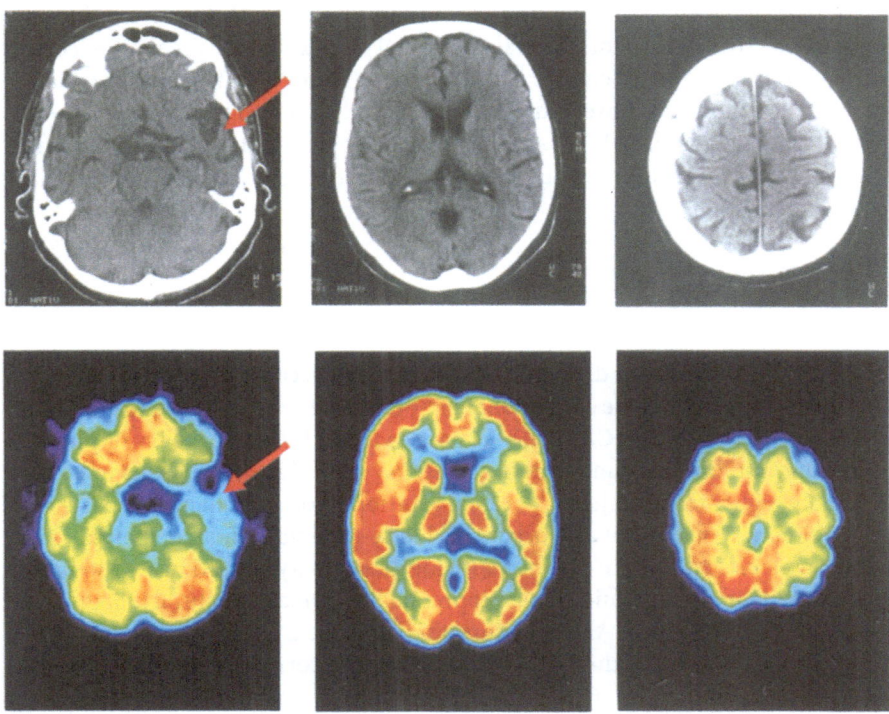

Fig. 2.10. Left focal anterior temporal atrophy (*arrow*) and extended impairment of cortical CMRglc with pronounced hemispheric asymmetry in frontotemporal dementia (FTD)

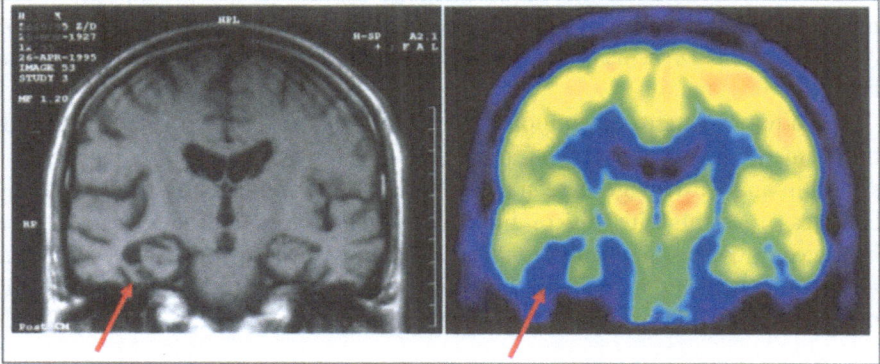

Fig. 2.11. Progressive prosopagnosia with atrophy and metabolic impairment of right lingual and fusiform gyrus (*arrow*). The patient had severe difficulties in recognizing familiar faces and had mild verbal and visual memory impairment, but was otherwise cognitively normal (MMSE 30)

such patients a frontal reduction of resting CMRglc [223] and significantly impaired activation in dorsolateral prefrontal cortex, lateral preemptor cortex, medial prefrontal and preemptor cortices, insular cortex and the anterior thalamic nuclear complex was observed [224, 225]. Moderate frontal dysfunction and hypometabolism is also observed in other neurological and psychiatric disorders (see Table 2.2 and section 2.7) and therefore cannot be regarded as a specific diagnostic feature.

2.1.5   Vascular Dementia

The diagnosis of vascular dementia (VD) is difficult because there is no consensus on clinical criteria, and the correspondence between existing criteria (e.g., ICD-10, DSM-IV, NINDS-AIREN, CAMDEX) is poor [226, 227]. The frequency of pure VD is low in most European and American autopsy series [228], but seems to be considerably higher in Japan [229]. Cerebral arteriosclerosis frequently coexists with AD in elderly subjects (but also occurs in elderly subjects without dementia). Also, cerebrovascular lesions are detected with high sensitivity on MRI (on T2-weighted images, with or without fluid signal suppression), whereas structural imaging provides no specific signs of AD. Therefore, there may be a tendency to diagnose VD clinically on the basis of the MRI findings when the correct diagnosis would otherwise have been mixed dementia, as seen from the finding of neuropathologies in 20–40 % of dementia cases, at least in Europe and the U.S [228]. There seem to be no distinctive features of VD in FDG-PET (except in those patients who have multiple cortical ischemic infarcts that are seen as corresponding lesions on MRI/CT and PET). Several studies suggest that a diffuse global reduction of cerebral glucose metabolism is a typical finding in VD (Fig. 2.12) and that the degree of that reduction in association cortex is similar to that seen in AD [230, 231]. Thus, the contrast between metabolic impairment in association areas and preserved metabolism in primary areas, basal ganglia, and cerebellum, which is typical for AD but not for VD, seems to provide some means of distinguishing between these two types of dementia [230].

VD often has a strong component of frontal dysfunction with corresponding impairment of working memory and frontal metabolism [59]. Severe leukoaraiosis (white matter damage as seen on MRI and CT) is an indicator of microvascular disease. Patients with this condition and a metabolic pattern similar to AD also have a cognitive profile similar to that of AD (more severely impaired recognition memory), whereas patients without the AD pattern seen on PET have predominant attention and working memory deficits [232].

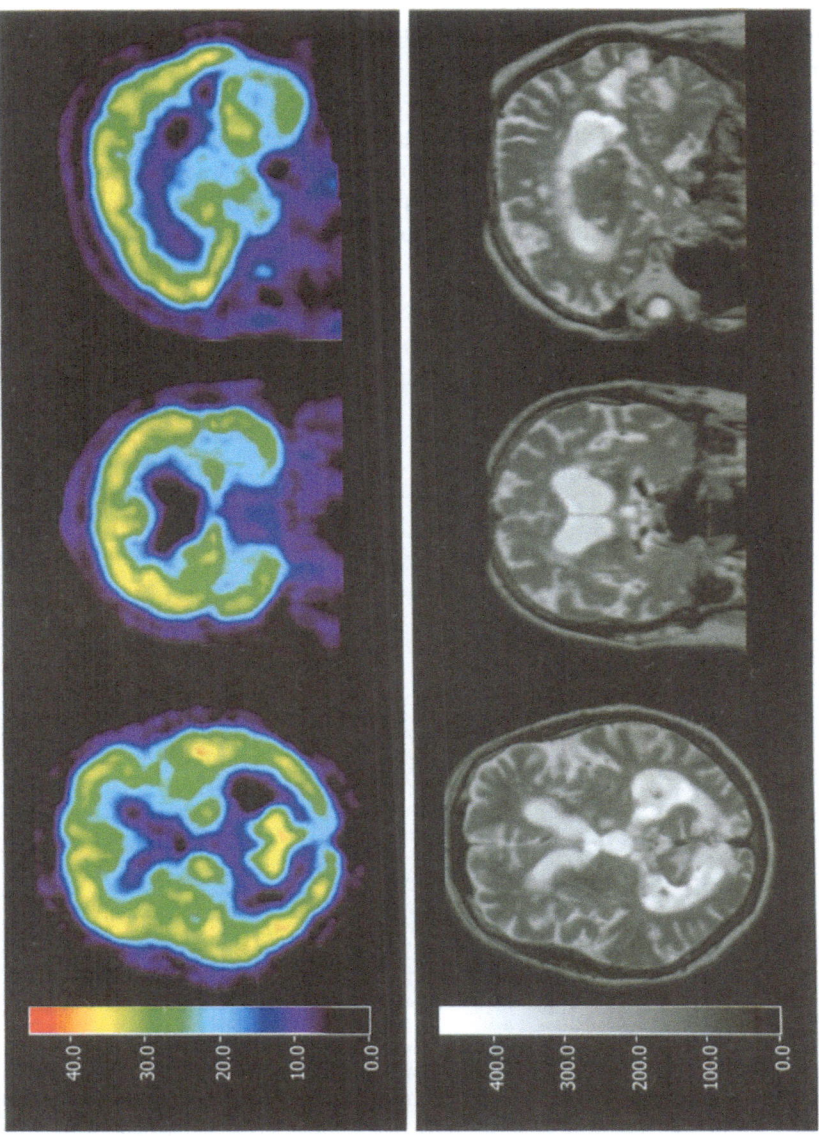

Fig. 2.12. Widespread reduction of CMRglc including striatum in vascular dementia (orthogonal slices through basal ganglia with coregistered T2-weighted MRI)

2.1.6   Creutzfeldt-Jakob Disease

Creutzfeldt-Jakob Disease is characterized by rapidly progressive dementia, often accompanied by insomnia, myoclonus, and other extrapyramidal disorders. In all cases reported so far, CMRglc was severely reduced in a multifocal fashion [233–237] (Fig. 2.13). It is not clear whether these changes, which apparently occur in variable locations, can be reliably differentiated from AD and other dementing diseases. Another clinical manifestation of prion disease, progressive fatal insomnia, is accompanied by thalamic hypometabolism [238, 239].

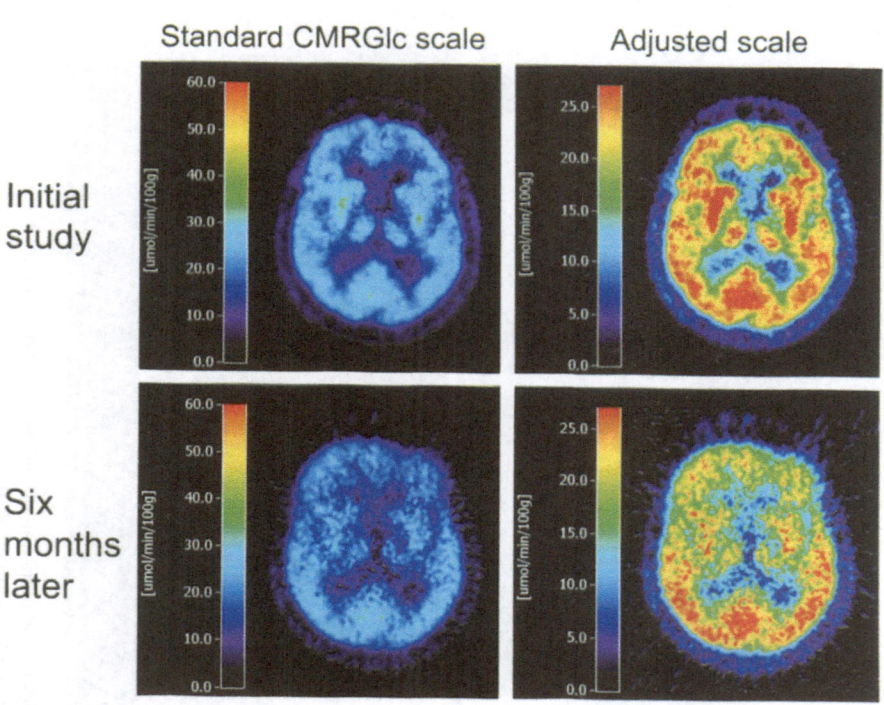

Fig. 2.13. FDG-PET in a patient with Creutzfeldt-Jakob disease(CJD) demonstrating severe impairment of global brain metabolism at initial presentation (*top left*) and at follow-up (*bottom left*). Note that after adjustment of the color scale (which shows the images as they would be presented without quantitation of CMRglc) little focal abnormality is seen (*top right*), with some frontal predominance of metabolic impairment at follow-up (*bottom right*). Owing to very poor patient cooperation a movement artifact appeared and distorted the frontal brain contour

2.1.7   Other Memory Disorders

Memory functions are distributed across many brain areas, and probably all cortical association areas also have some memory functions. Nonetheless, the structures that form Papez's circuit, in particular the hippocampus, thalamus, fornix, and the mamillary bodies in the basal forebrain, are essential for most memory functions and bilateral damage to them will cause severe amnesia. Not all parts of this circuit are large enough to be imaged with PET, but a reduction of CMRglc in thalamus, cingulate cortex, and basal frontal cortex has been seen in most patients with amnesia, regardless of the etiology [240, 241]. This pattern is different from that typically seen in dementia, i.e., hypometabolism in temporoparietal or frontolateral association neocortex. There are many disorders that may cause memory dysfunction, which are treated in this book in the appropriate sections. The most characteristic memory disorders are presented in section 2.1.7.1.

2.1.7.1   Hypoxic and Global Ischemic Brain Damage

One day after a hypoxic insult and resuscitation, brain metabolism is severely impaired, with a global reduction of CMRglc in gray matter by an average of approx. 50 %. Perfusion is variable and may be disturbed by a loss of vascular tone. No clear relation to prognosis has been established [242]. After several days, cerebral lesions may be demarcated on CT and MRI. Bilateral hippocampal lesions may lead to severe anterograde amnesia. Since the hippocampal CA1 sector is sensitive to hypoxia, amnesic syndromes may persist after circulatory arrest. In such cases, often widespread hypometabolism is found, which may be particularly severe in hippocampal regions, thalamus, posterior cingulate cortex, and mesial prefrontal cortex [240, 243–245]. Bilateral thalamic infarcts may be a consequence of global ischemia and may also lead to severe amnesia, associated with widespread cortical hypometabolism [246–248] (see also section 2.4.4). Impairment of memory and other cognitive functions may also occur after major cardiac surgery and may be associated with widespread reduction of CMRglc [249]. If patients do not regain consciousness after hypoxic brain damage, the outcome is often a vegetative state (see section 2.6.2).

2.1.7.2   Other Amnesic Syndromes

Korsakoff syndrome is a severe memory disorder characterized by profound anterograde amnesia associated with confabulation; it is usually a chronic defect state after severe abuse of alcohol and delirium. Reduction of CBF and energy metabolism has been observed in cingulate, precuneus, thalamus, frontal, temporal, and parietal cortex [250–252] (Fig. 2.14). Hypometabolism in most of these structures has also been observed in other amnesic syndromes, but in Korsakoff syndrome hippocampal metabolism is intact [253], whereas the metabolism in the upper brain stem is impaired [240]. In dementia associated with Marchiafava-Big-

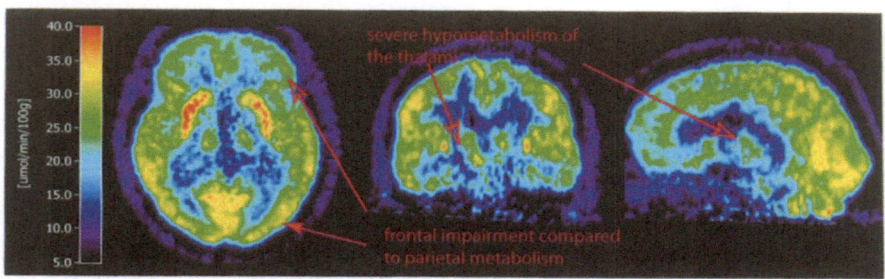

Fig. 2.14. Korsakoff syndrome with moderately reduced cortical CMRglc and impairment of thalamic CMRglc

nami disease, another alcohol-related disorder associated with damage to the corpus callosum, CMRglc is markedly reduced in the frontal and temporo-parieto-occipital association cortices, as in AD [254].

Intoxication with domoic acid, a contaminant of mussels that is structurally related to the excitatory neurotransmitter glutamate, can lead to chronic residual memory deficits and motor neuronopathy or axonopathy. It is associated with decreased glucose metabolism in the medial temporal lobes [255]. Excessive abuse of 3,4-methylene-dioxymethamphetamine (MDMA) and other substances can also lead to brain damage, including amnesic syndromes and associated reductions of local CMRglc (see section 2.7.5.3).

Few patients have been studied in the acute phase of transient global amnesia, which usually lasts for a few hours and then gradually resolves. In two patients PET findings disclosed mild but significant changes in the amygdala (right or left) and left posterior hippocampus, which could account for anterograde memory disturbance [256]. Most PET studies performed after recovery found normal CBF and energy metabolism [257]. In some instances, however, mild memory deficits may persist, with associated reductions of metabolism in hippocampal and association cortices [240].

Psychogenic amnesic syndromes without brain damage are of considerable interest with regard to both diagnosis and their neurobiological correlates [258]. In a case report on activation by an explicit memory task, increased activation of the right anterior medial temporal region including the amygdala was found, but hippocampal activation was decreased, suggesting that information-processing pathways were altered substantially [259].

2.2   Movement Disorders

Parkinsonian syndromes can be divided into three groups: (1) idiopathic PD, (2) multiple system atrophy (MSA) and similar disorders, and (3) Parkinsonian syndromes secondary to other diseases or exogenous agents. The most frequent

Table 2.1. Recommendation for a standard fluor-18-fluorodeoxy-D-glucose (FDG)-positron emission tomography (PET) procedure to assist with early diagnosis of dementia (based on recommendations in [32])

Typical clinical situation: Patient with mild cognitive impairment or progressive subjective memory impairment or personality changes that are suggestive of incipient dementia but do not allow a definitive clinical diagnosis, major structural lesion (apart from possible atrophy) excluded by CT or MRI

Patient preparation: Examination requires normal fasting plasma glucose, which is usually achieved by fasting for 6 hours before examination

PET study under standard conditions (e.g., resting with eyes closed, ears unplugged; see chapter 3.5.3 for alternatives) with i.v. injection of 200–370 MBq FDG and data acquisition 20–60 min after injection

Image reconstruction with appropriate corrections for scatter and attenuation

Image evaluation:

- Visual examination of images:
  – If changes in the uptake pattern typical for Alzheimer disease (AD), dementia with Lewy bodies (DLB), or frontotemporal dementia (FTD) are present and clinical symptoms are consistent with beginning disease, PET confirms the clinical impression
  – If such PET changes are present but not consistent with clinical symptoms, findings are inconclusive and additional tests may be required
  – If there are PET abnormalities (that cannot be explained by partial volume effects attributable to atrophy alone) but they are not typical for AD, DLB, or FTD, other diagnoses should be considered (additional tests may be required)
  – If FDG-PET is normal in presence of mild clinical symptoms, these are probably not due to neurodegenerative disease
- To increase accuracy, abnormalities should be checked quantitatively by comparison with a normal data base (as provided by some software packages or www.nest-dd.org, see also section 4.10 for details)
- If absolute quantitation of CMRglc was done, global level should be checked (could be reduced in vascular dementia (VD) without much change in the uptake pattern, but can also be reduced as a result of other conditions or sedating drugs; see section 3.5)

It is recommended that the term "dementia" not be used to describe PET findings, because dementia is defined as a clinical syndrome that cannot be seen on any images. The PET findings relate to the pathophysiological alterations of AD, DLB, or FTD that may lead to the respective clinical syndromes

disorder is PD, which is relatively easy to diagnose if it presents with the typical combination of clinical symptoms, slowly progressive course, and response to dopaminergic therapy. It is due to degeneration of dopaminergic neurons located in the substantia nigra. It may involve degeneration of other aminergic neurons, but generally leaves the postsynaptic striatal and cortical neurons intact. Clinical symptoms usually arise only when approximately 50 % of dopaminergic neurons are already gone. Symptomatic treatment is effective initially but cannot prevent disease progression, which usually leads to severe disability and death. The real-

Table 2.2. Characteristic FDG-PET findings in neurodegenerative diseases

Disease	Brain regions with reduced FDG uptake
Alzheimer disease (AD)	Temporoparietal association cortex, posterior cingulate cortex and precuneus, sometimes also frontolateral association cortex
Dementia with Lewy bodies (LBD)	As in AD, plus primary visual cortex (FDOPA is abnormal, in contrast to AD)
Frontotemporal dementia (FTD)	Predominantly frontomesial, also frontolateral and anterior part of temporal lobe
Parkinson disease (PD)	FDG-PET usually normal (apart from atrophy effects), but cortical impairment similar to that in LBD possible in late stages of disease
Olivo-ponto-cerebellar atrophy	Putamen, brain stem, cerebellum, often also cerebral cortex
Striatonigral degeneration	Putamen
Progressive supranuclear palsy	Frontal, basal ganglia and midbrain
Corticobasal degeneration	Mainly parietal, central and frontal cortex, striatum and thalamus, often very asymmetrical (contralateral to clinical symptoms)
Spinocerebellar degeneration	Variable, probably depending on subtype, may be similar to OPCA
Huntington disease	Caudate nuclei, putamen, with progression also thalamus and cortex

ization that the neurotoxin 1-methyl-4-phenyl-1,2,3,6-tetrahydropyridine (MPTP) specifically damages dopaminergic neurons allowed the development of primate models of PD [260]. PET allows the study of similarities in pathophysiology between this model and the human disease by using the same tracers in vivo in experimental animals, PD patients, and drug abusers who have been accidentally exposed to MPTP [261].

Classification of MSA has a complicated history, and definitive neuropathological diagnosis is currently based on the neuropathological core findings of glial cytoplasmatic inclusions [262]. There are three main clinical manifestations: olivo-ponto-cerebellar atrophy (OPCA), striatonigral degeneration (SND), and Shy-Drager syndrome. Clinical diagnosis is often difficult, especially if the patient is seen only once. For instance, the clinical symptoms with predominant ataxia in OPCA may overlap with those of hereditary spinocerebellar atrophy (SCA). SND is distinct from PD in the presence of autonomic failure and, in most cases, the absence of tremor, but descriptions of variants of all MSA types abound in the literature. Objective means of diagnosing these relatively rare diseases in vivo would be highly welcome as more certainty would be helpful in counseling patients and relatives, even though there is currently no effective treatment.

There are also other degenerative diseases that can present clinically as Parkinsonian syndromes, often in combination with dementia. Progressive supranuclear palsy (PSP), also known as Steele-Richardson-Olszewski syndrome, is combined with postural instability leading to severe falls and vertical gaze palsy. Corticobasal degeneration (CBD) is characterized clinically by the association of dystonia, apraxia, and the strange feeling of an "alien limb." Both diseases are related to pathological deposition of hyperphosphorylated tau protein in neurofibrils and are therefore related to frontotemporal dementia (see section 2.1.4), with some overlap of clinical symptoms in addition. To complicate matters, there may also be overlap of PD and dementia with Lewy bodies (DLB). Both these diseases are related to the same intracellular inclusion bodies containing α-synuclein that are called Lewy bodies, the only difference being that they are located mainly in the midbrain in PD and in the cortex in LBD.

Parkinsonian syndromes secondary to other diseases (e.g., cerebrovascular disease, Wilson's disease) and exogenous agents (e.g., neuroleptic drugs, carbon monoxide poisoning) may be a diagnostic challenge insofar as the physician needs to be aware of these conditions and to exclude them, which is usually possible by standard clinical means.

2.2.1   Idiopathic Parkinson Disease

2.2.1.1   Diagnostic Issues

In a similar way to "Alzheimer disease", "Parkinson disease" is a clinical diagnosis and standard diagnostic procedures including CT and MRI mainly serve to exclude other diseases that may lead to parkinsonism. Functional imaging with PET can demonstrate the disturbance of dopamine synthesis that is the hallmark of PD and thus allow confirmation of the diagnosis [263], which is of particular interest if the clinical features are somewhat atypical (e.g., an unusually early age of onset).

If symptoms are mild and clinically uncertain, positive PET findings may make it possible to exclude psychogenic movement disorders, which can otherwise be difficult to discriminate from early PD. It may also be difficult to differentiate clinically between monosymptomatic resting tremor, which is probably a subtype of PD, and essential tremor, which is a different disease with a better prognosis. As with all neurodegenerative disorders, the clinical importance of that distinction is likely to increase as soon as specific drugs become available that could prevent disease progression [264].

If symptoms are severe and accompanied by features that are atypical in PD, such as incontinence or cognitive impairment early in the progression, pyramidal signs, ataxia, or lack of response to L-DOPA, differentiation of PD from MSA and related disorders is required. The established clinical criteria of PD have been demonstrated to be rather restrictive, and a recent comparison with neuropathological diagnoses allowed the conclusion that PD can present with a broader clinical picture of disease than previously thought acceptable [265]. Degeneration of dopaminergic neurons with lowering of the levels of their respective PET markers has been described in all diseases that cause parkinsonism, but the relative involvement of the rostral and caudal parts of the striatum (i.e., caudate vs anterior and posterior putamen) provides for some distinction between different diseases. Reduction of dopamine receptor binding and CMRglc in putamen is not a feature of mild to moderate PD, but is seen in MSA and in diseases that damage the basal ganglia [266–271]. Thus, FDG-PET has a clear potential for improvement of the clinical distinction [272, 273] (see Table 2.3).

With further advances in the genomics of Parkinson disease and detection of a steadily increasing number of polymorphisms of questionable clinical significance [274], PET may also be used as a tool to detect subclinical changes of neurotransmitter and receptor levels in vivo and thus contribute to the identification of the clinical phenotype.

Table 2.3. Pre- and postsynaptic striatal PET findings in major movement disorders

Disease	Presynaptic markers (FDOPA)	Postsynaptic markers (RAC, FDG)
Parkinson disease (PD)	Reduced, least in head of caudate, most in caudal putamen ("anterior-posterior gradient")	FDG high normal, D2 up-regulated
Olivo-ponto-cerebellar atrophy (OPCA)	Reduced, usually involving caudate and putamen	Reduced FDG also reduced in cerebellum
Striatonigral degeneration (SND)	As in OPCA	RAC like OPCA, but FDG reduced only stratum
Progressive supranuclear palsy (PSP)	Mildly reduced	Striatum mostly normal, but FDG reduced in frontal cortex and midbrain
Corticobasal degeneration (CBD)	Mildly reduced	RAC may be reduced, FDG reduced in striatum and parietal cortex on affected side
Huntington disease	Normal (apart from atrophy)	Severely reduced (most in caudate)

RAC: ^{11}C-Raclopride imaging of D2 dopamine receptors

2.2.1.2   Dopamine Synthesis

The most widely used PET tracer in Parkinson disease is L-6-^{18}F-fluoro-3,4-dihydroxyphenylalanine (FDOPA), which was introduced into clinical use in 1983 [275]. It is an amino acid that is transported across the BBB by the carrier system for large neutral amino acids (see section 3.7). It is then decarboxylated to ^{18}F-dopamine by aromatic amino acid decarboxylase (AAAD) and stored in dopamine vesicles (see section 3.10.1). Accumulation in the putamen is decreased in experimental parkinsonism induced by MPTP [261] and declines with the number of vital dopaminergic neurons [276, 277]. It is also correlated with striatal dopamine levels soon after even a single-dose MPTP lesion [278]. It is not strictly specific for dopaminergic cells but also accumulates in regions with relatively high concentrations of norepinephrine and serotonin.

In normal subjects, FDOPA images taken 30–90 min after injection of ^{18}F-FDOPA with blockade of peripheral tracer decarboxylation by carbidopa show high tracer uptake in the striatum (caudate and putamen) and the midbrain, whereas uptake in cerebral cortex and cerebellum is much lower. In patients with PD, striatal tracer uptake and retention is reduced, most markedly on the side opposite to the major motor signs [279, 280]. The posterior parts of the putamen are most affected and the heads of the caudate nuclei least, which is consistent with preferential degeneration of dopaminergic neurons in the caudal and mediolateral part of the substantia nigra pars compacta [281]. This typical differential intrastriatal distribution of reduced uptake is often referred to as the "rostrocaudal gradient" [282]. The difference between caudate and posterior putamen may be even wider in reality, because underestimation resulting from partial volume effects is stronger in the smaller caudate nucleus [283]. A longitudinal FDOPA study over 5 years confirmed that the disease process in PD first affects the posterior putamen, followed by the anterior putamen and the caudate nucleus [284] (Fig. 2.15).

The FDOPA uptake deficit in putamen is related to indices of motor function in PD, such as clinical Hoehn and Yahr grades [285] and finger tapping [286]. The deficit in the caudate heads seems to be related more to speed of mental function [287] and to memory [288, 289]. This is consistent with the view that this part of the striatum, with its intense connections to frontal cortex, mainly supports mental rather than motor function.

FDOPA uptake in the putamen is also reduced in monosymptomatic resting tremor, which seems to be the purest form of tremor-dominant PD [290, 291]. In contrast, dopaminergic neurons and dopamine transporters are intact in essential tremor [292]. Essential tremor may be clinically similar to tremor-dominant PD, but does not respond to L-DOPA and usually does not progress to severe disability.

It is not possible to distinguish between PD and other disorders with parkinsonian symptoms on the basis of FDOPA studies alone, because many of these disorders are associated with some degree of decreased FDOPA uptake [272]. The distinction is much better if imaging for indicators of postsynaptic receptors and

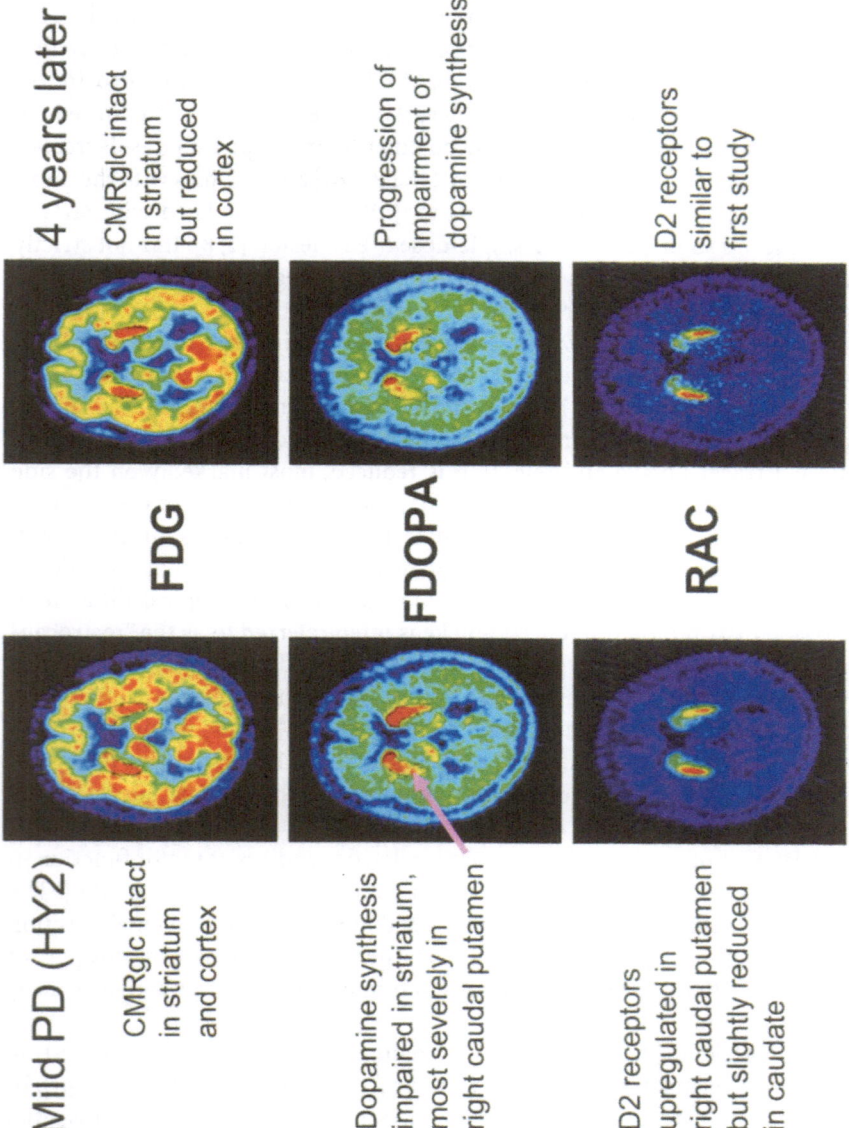

Fig. 2.15. Progression of FDG, FDOPA, and RAC findings in Parkinson disease (PD)

neuronal function, such as ^{11}C-raclopride or FDG, is performed, because these scans are normal or even elevated in PD but abnormal in most other disorders [266].

Most of the initial studies with ^{18}F-FDOPA employed kinetic analysis of uptake curves with arterial blood samples, including correction for plasma metabolites (see section 4.9.2.3). These more complex studies show that simplified models using occipital cortex as a reference tissue region do not provide accurate estimates of AAAD activity, but discriminant analyses indicate that simple estimates, such as the striatum-to-occipital ratio or the graphically derived unidirectional transport rate constant K_i with tissue reference [293, 294], separate normals from PD patients at least as accurately as estimates of striatal AAAD activity based on plasma input function [295, 296].

Although FDOPA uptake in pallidum is low and difficult to measure accurately, a compensatory increase in the internal segment of pallidum was observed in PD [297]. FDOPA is also taken up by the pineal gland, and this uptake is even increased in PD, which may indicate compensatory up-regulation of monoaminergic transmitter systems outside the basal ganglia [298]. FDOPA uptake in midbrain may be due not only to dopaminergic, but also to noradrenergic and serotonergic, cells [299, 300]. A study indicates that impairment of serotonin $5-HT_{1A}$ receptors in midbrain raphe nuclei may be associated with tremor in PD [301].

2.2.1.3   Dopamine Release and Turnover

There have been several studies indicating that the main mechanism of the dopaminergic deficit in PD is a failure in vesicular storage, resulting in a reduction of dopamine release. Displacement of ^{11}C-raclopride (RAC; see section 3.10.6) was used to show a decrease in dopamine release in the contralateral putamen during a unilateral foot extension/flexion movement [302] and during sequential finger movement [303]. Reduced putaminal dopamine release that is correlated with reduced FDOPA uptake is also seen in PD patients after metamphetamine challenge, whereas frontal dopamine release appears to be intact [304]. In contrast to dopamine release, dopamine turnover may in fact be increased, as suggested by increased displacement of RAC after acute levodopa administration [305]. Further evidence for increased dopamine turnover comes from kinetic analysis of FDOPA data in MPTP-lesioned monkeys [306] and in patients with PD [307]. Changes in dopamine turnover may be related to age at onset [308]. FDOPA uptake indices may not be a sensitive means of detecting very early damage to the dopaminergic system, because they may be maintained at near-normal levels as a result of increased dopamine turnover [309].

2.2.1.4   Dopamine Transporters

Imaging of the dopamine transporter (DAT) (see section 3.10.2) provides another way of studying the function and integrity of dopaminergic neurons. Early studies often used ^{11}C-nomifensine [310], which has relatively low affinity and selectivity for DAT but is still able to demonstrate reduced striatal binding in PD [311, 312] and in MSA [313]. With recently developed high-affinity tracers based on tropane derivatives, the observed reduction of DAT is usually more severe in PD and in MPTP-lesioned animals than the reduction of FDOPA accumulation [314, 315]. Thus, these tracers probably provide a more sensitive indicator of dopaminergic degeneration. In patients with early untreated PD, ^{18}F-2-β-carbomethoxy-3-β-(4-fluorophenyl)tropane (CFT) uptake (average of ipsilateral and contralateral side) in the total putamen was lower by 66 % than the mean control value. In all PD patients studies, that was more than 3 SD below the control mean [316]. Similar findings were observed with ^{11}C-WIN 35,428 [317], ^{11}C-RTI-32 [318] and (76)Br-FE-CBT, whereas in advanced PD, reductions were similar to those of FDOPA [319]. As in the case of FDOPA, the reduction of dopamine transporters is most severe in posterior putamen [320]. The DAT reduction in PD is in keeping with findings in monkeys lesioned chronically with MPTP [321]. Dopamine transporters are also reduced in the orbitofrontal cortex and in the amygdala, which may contribute to the mental and behavioral impairment observed in PD [322].

2.2.1.5   Vesicular Monoamine Transporter

Severe reductions in the vesicular monoamine transporter type 2 (VMAT2), which is also located presynaptically in dopaminergic neurons, are also seen in PD with ^{11}C-dihydrotetrabenazine (DTBZ) [323, 324]. VMAT2 loss in PD is more severe than the reduction in FDOPA uptake, but less severe than the reduction in dopamine transporters [325]. Although its reduction is not specific for the dopaminergic system, VMAT2 binding may be less markedly influenced by dopaminergic and anticholinergic drugs than is the case for other tracers, and therefore it may be more easily used in treated patients.

2.2.1.6   Dopamine Receptors

The binding capacity of striatal D2 receptors, as measured with RAC or with ^{11}C- or ^{18}F-labeled N-methylspiperone (NMSP) (see section 3.10.5), is increased in PD. The increase is most pronounced in posterior putamen, where the dopamine deficit is most severe [285, 326]. This effect has also been seen with 6-hydroxy-dopamine lesioning of the substantia nigra pars compacta in a rat hemiparkinsonian model, and saturation studies indicate that the 23 % increase measured in RAC-binding potential reflects a change in receptor affinity [327]. A similar effect was seen in monkeys after dopamine depletion by reserpine [328] and after lesioning with MPTP [329].

With PD progression, striatal D2 receptors return to normal or even fall somewhat below normal levels. Studies are difficult to perform and evaluate, however, because treatment with dopaminergic drugs may cause interaction with and competitive displacement of the tracer at the receptor [330], and patients with severe PD cannot usually be kept drug free for extended periods.

The binding capacity of D1 receptors, measured with ^{11}C-SCH 23390, is also preserved or even inversely related with FDOPA uptake and with DAT binding in the striatum in PD [331], whereas there is a parallel decline in normal controls.

2.2.1.7   Mesocortical Dopaminergic Innervation

There is also significant dopaminergic innervation of the cerebral cortex, in particular in the mesial frontal lobe, which probably is related mainly to cognitive rather than motor function. DAT binding is reduced there in early PD, whereas D1 receptors remain intact [322, 333]. With parametric imaging and statistical mapping, a bilateral decline of FDOPA accumulation is seen in the anterior cingulate area and ventral striatum and in the right caudate nucleus in patients with PD and dementia compared with PD patients without dementia [334]. Another study indicates that frontal monoaminergic activity is increased and that there is a sex difference in the prefrontal monoaminergic system in early PD [335].

2.2.1.8   Other Transmitters

Concurrently with the reduction in striatal DATs, a significant reduction in serotonin transporters (SERT) is also observed. This is consistent with the notion that the degeneration of aminergic cells in midbrain is not strictly limited to the dopamine system in PD [336].

Changes in the cholinergic system are also of clinical interest in PD. An increase in muscarinergic receptor-binding capacity in frontal cortex has been reported [337, 338]. It may be the consequence of cholinergic denervation, which has been inferred from the reduction of AChE activity [138, 339, 340].

2.2.1.9   Resting Glucose Metabolism

In most patients with PD, glucose metabolism as measured with FDG is generally normal. There may be a tendency for it to be elevated in putamen [285] and reduced in mesial frontal cortex [341], but these minor alterations rarely reach significance [342]. High putamen CMRglc is usually accompanied by low FDOPA uptake [341], and both are associated with the frequently observed hemispheric asymmetry. Parkinson tremor is associated with increased CMRglc in a metabolic network comprising the thalamus, pons, and premotor cortical regions [343]. The cerebellum also shows a tendency towards higher metabolism, which seems to be more closely correlated with akinesia and rigidity than with tremor [291].

With advancing PD, hypometabolism very similar to that seen in AD in frontal and temporo-parieto-occipital association cortices may develop. It is more frequently seen in patients who have already developed autonomic failure [344]. In contrast to AD, the reduction of CMRglc in association cortex is not generally associated with dementia or other cognitive impairment [111, 345–348]. Hemispheric asymmetry corresponds with asymmetry of putaminal FDOPA uptake and associated motor symptoms [349]. Metabolic deficits tend to be more severe and more focused in temporoparietal cortex in demented patients [350]. There is relatively little correlation with cognitive functions in nondemented patients [288, 351]. The only correlation that has been demonstrated is an association of parieto-occipital hypometabolism with mild impairment of memory and associative visual processing [352]. Impaired CMRglc in temporo-parieto-occipital association cortex in PD may be related to the occurrence of hallucinations and psychosis.

2.2.1.10   Activation Studies

Patients with PD usually show decreased activation of contralateral putamen, SMA, and motor cortex in motor tasks [353, 354], which can be partially reversed by dopaminergic drugs [355]. There is also a relative reduction in dorsolateral and mesial frontal activation during imagined movement [356, 357]. Nonetheless, there are also brain areas that may show increased activation in PD. Sequential finger movements in PD are associated with reduced activation in the mesial frontal and prefrontal areas, but increased activation in the lateral premotor and inferolateral parietal regions, suggesting a switch from the use of striatomesial frontal to parietal-lateral premotor circuits [358]. Cerebellar overactivation has also been reported [359]. Complex sequential finger movements may lead to more extensive recruitment of cortical regions to compensate for the dopaminergic striatal deficit [360]. Abnormally high activation is also found in ipsilateral cortex during motor tasks [356, 361]. Multiple parameters, such as voluntary or paced movement, rewards offered, and cognitive demands, may influence the results of such activation studies, as reviewed in more detail by Brooks [362].

Activation studies may also contribute to better understanding of the cognitive impairment associated with PD, such as bradyphrenia, memory deficits, and difficulty in decision making. The latter has been related to impaired frontal activation in PD [363].

2.2.1.11   Familial PD

The etiology and pathophysiology of PD may be elucidated by comparison with familial PD with known mutations. A substantial role for inheritance was also suggested for sporadic PD by a FDOPA study of twins, one of whom had the disease [364]. A higher concordance for subclinical striatal dopaminergic dysfunction was found in monozygotic than in dizygotic twin pairs (55% vs 18%, respectively) [365]. Mild parkinsonian symptoms, such as isolated resting tremor, are frequently found in otherwise asymptomatic relatives with abnormal FDOPA uptake [366]. A rapidly increasing number of mutations are being found in the α-*synuclein* and *parkin* genes. Occasionally, however, mutations in other genes, such as the *SCA3* gene, may also lead to symptomatic DOPA-responsive parkinsonism [274]. PET

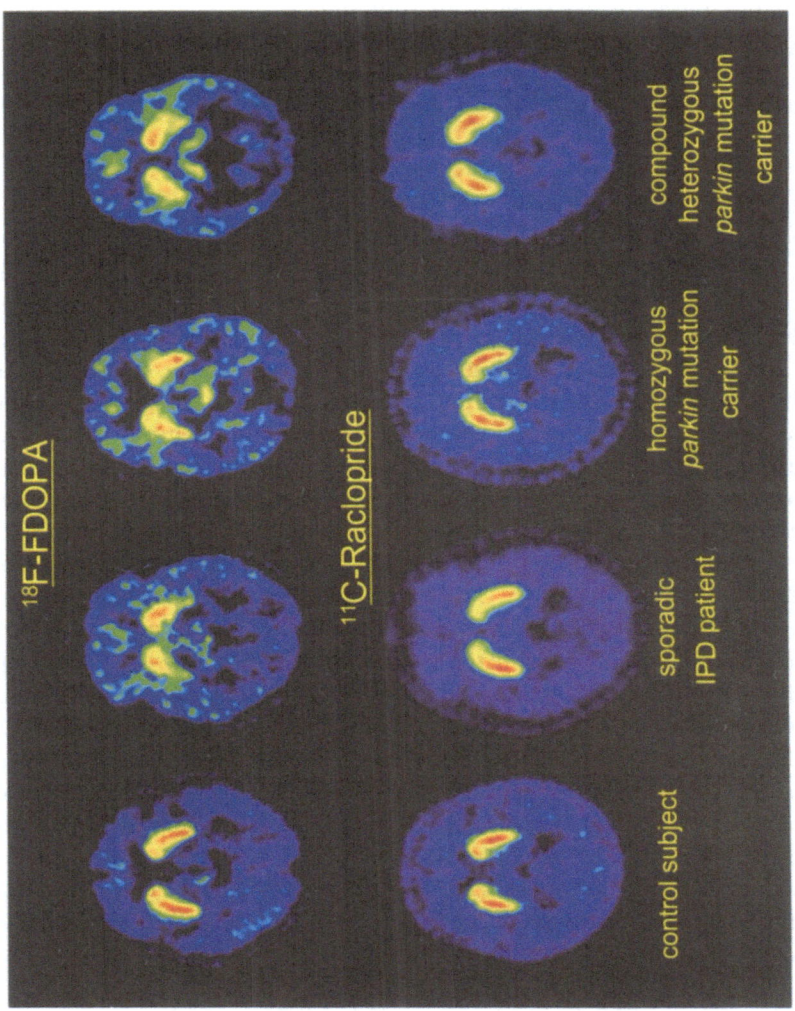

Fig. 2.16. Asymptomatic patients with *parkin* mutations show mild abnormalities of FDOPA uptake (*3rd and 4th images* in *top row*) and minor changes in RAC binding (not seen by visual image examination). (Adapted from [368], with permission)

studies to elucidate the metabolic phenotype of these genetic disorders with parkinsonism are only just beginning.

Familial forms of the disease tend to involve less pronounced asymmetry and less steep anterior–posterior gradients than PD, but otherwise similar reductions of FDOPA uptake. Bilateral presynaptic dopaminergic dysfunction without marked lateralization on ¹⁸F-FDOPA PET has been seen in a Chinese family [367]. Patients with mutations in the *parkin* gene show reduced FDOPA uptake similar to those seen in PD, but also have reduced receptor-binding capacity, which is a point of difference from PD [368] (Fig. 2.16). Clinically affected familial PD patients with a mutation at the *PARK6* locus on chromosome 1 have posterior dorsal putamen FDOPA uptake reduced by 85%, similar to idiopathic PD patients matched for clinical disease severity and duration. However, they show significantly greater involvement of the head of the caudate and anterior putamen, suggesting that *PARK6* pathology results in a more uniform loss of striatal dopamine terminal function than does PD [369].

2.2.1.12 Disease Progression

Disease progression in PD has been measured quantitatively by serial FDOPA studies. Putaminal FDOPA uptake declines annually by 4–5% relative to normal values [370], which corresponds to an annual reduction of 8–10% relative to baseline values in patients [284] and is significantly greater than the normal age-related decline [371]. The annual rate of decline is even higher for DATs. It is approximately 13% (relative to baseline) measured with ^{18}F-CFT [372]. Over the course of the disease, functional impairment of remaining neurons appears to proceed at a constant rate [373]. FDOPA was used in a study demonstrating that the dopamine agonist ropinirole may slow disease progression more than L-DOPA treatment [374].

The longitudinal studies give some insight into the presymptomatic duration of the disease by back-extrapolation. On the assumption that the constant progression rate seen during symptomatic disease is also present in asymptomatic disease, the preclinical phase is estimated to be no longer than 4–7 years [370, 375]. Subclinical impairment and progressive loss of dopaminergic function has also been observed in asymptomatic monozygotic twins of patients with sporadic PD, suggesting that there is also a strong genetic component in sporadic disease [364, 365]. On the other hand, progressive loss of dopaminergic function is also observed in drug abusers after transient exposure to MPTP, suggesting that a transient exogenous insult can trigger progressive disease [376].

With disease progression, there is also a decline of D2 receptor-binding capacity from its initially increased level. In a longitudinal study with RAC, binding was significantly lower in putamen and caudate nucleus after 3–5 years than when initially evaluated in a drug-naive state. Values were now in the control range in the putamen and lowered in the caudate nucleus [377]. There is also a substantial decline of cortical dopamine (D2/D3) receptors measured with ^{11}C-FLB 457 during disease progression, particularly in dorsolateral prefrontal and anterior cingulate cortex and in thalamus [378]. This loss of receptor-binding capacity may partially explain the reduced response to dopaminergic therapy in long-standing PD. It is not yet known whether these changes are induced by chronic dopaminergic therapy or occur independently of treatment, as a result of structural adaptation of the postsynaptic dopaminergic system to the progressive decline of nigrostriatal neurons.

2.2.1.13   Depression in PD

PD is frequently associated with depression, which may be due to impairment not only of dopamine but also of noradrenaline and serotonin synthesis. As in depression without PD (see section 2.7.1), this is associated with prefrontal and caudate hypometabolism [352, 379].

2.2.1.14   Improving and Understanding Treatment Effects

2.2.1.14.1 Drug Effects

The principal action of DOPA treatment is well studied and understood, but after several years DOPA treatment becomes less effective and side effects increase. This appears to be related to increased synaptic dopamine turnover [305, 380]. PD patients who suffer from levodopa-induced dyskinesias show less activation in left sensorimotor and left ventrolateral prefrontal cortex during a motor task than drug-naive patients and controls. This suggests that long-term levodopa treatment and disease severity affect the physiology of brain regions that depend on dopaminergic pathways [381].

Inhibition of catechol-O-methyl transferase (COMT) is used with the goal of achieving steadier brain DOPA levels [382]. FDOPA studies have indicated that peripheral COMT inhibition by entacapone mainly leads to increased FDOPA plasma levels and thus increasing its availability in the brain for decarboxylation [383, 384]. Central COMT inhibition has been demonstrated with FDOPA for tolcapone in experimental animals [385] and in patients [386].

Amantadine is moderately effective in PD, but the mechanism is not known. A PET study with RAC demonstrates that treatment increases the binding potential of D2 receptors [387]. In an FDOPA study, stimulation of DDC activity in striatum of healthy human brain was seen after amantadine in normal volunteers [388].

2.2.1.14.2 Surgical Lesioning

When drug treatment fails in PD, surgical intervention may be considered to improve symptoms. The theoretical framework for these interventions has largely been based on anatomical, electrophysiological, and animal studies, but the data on alteration of functional circuits in patients are limited and therefore substantial insight is expected from PET studies.

An FDG study of subthalamotomy indicates that it reduces basal ganglia output through internal globus pallidus/substantia nigra pars reticularis and also influences downstream neural activity in the pons and ventral thalamus [389]. Thalamotomy results in a reduction of FDG uptake predominantly in the lateral prefrontal and parietal cortex [390]. Thalamotomy relieves tremor but leads to decreased activation of the left sensorimotor cortex, lateral premotor cortex, and parietal area 7 on hand movement [391]. A case report indicates that a preoperative reduction of resting CBF and CMRO$_2$ in parietal cortex may normalize after thalamotomy [392].

Clinical outcome following pallidotomy suggests that patients with high preoperative CMRglc in this structure benefit most from this procedure [393]. Postoperatively, significant metabolic increases are noted in the primary motor cortex, lateral premotor cortex, and dorsolateral prefrontal cortex. Improvement in contralateral limb motor performance correlates significantly with decline in thalamic and increase in lateral frontal metabolism. These results suggest that pallidotomy reduces the preoperative overactivity of the inhibitory pallidothalamic projection [394]. Pallidotomy has also been associated with increased activation of premotor areas (supplementary motor area and dorsolateral prefrontal cortex) and reduced hyperactivity of the lentiform nucleus (augmented preoperatively) [395, 396]. There is also normalization of increased striatal D2-receptor binding potential after pallidotomy [397]. Nonetheless, there have also been observations of changes in personality, behavior, and executive and motor functions after pallidotomy [398], which require further study and argue for the use of nonlesional techniques, as described in the next section.

2.2.1.14.3 Deep Brain Stimulation

Surgical lesioning always involves a substantial risk of severe side effects, as paresis or cognitive impairment may be induced. Thus, the development of deep brain stimulation (DBS) to replace lesioning has dramatically improved the treatment options in advanced PD. However, the treatment is very expensive, and it is not effective in multiple-system atrophy or in some other diseases that can cause parkinsonism. Thus, PET may be used for ascertainment of diagnosis and, as with the lesioning techniques, to study the physiological effects of DBS.

The subthalamic nucleus (STN) is a target area in severe PD, primarily for the relief of akinesia and rigidity in the off-phases. In addition, STN DBS usually allows substantial lowering of DOPA dosage and there has been indirect evidence of increased striatal dopamine metabolism in experimental animals [399], but this has not been confirmed with RAC in human PD [400]. CBF measurements during motor activation show more activation of the anterior cingulate, rostral supplementary motor area, and premotor cortex during STN stimulation [401], corresponding to clinical improvement [402]. Abnormalities of resting CMRglc in association cortices are commonly found in advanced PD and may be reversed by DBS [403] (Fig. 2.17).

For treatment of severe levodopa-induced dyskinesias and severe generalized dystonia, the internal part of globus pallidus (GPi) is a commonly used target of DBS. GPi DBS improves UPDRS motor ratings (36%, $P<0.001$) and significantly increases resting CMRglc in the premotor cortex ipsilateral to stimulation and in the cerebellum bilaterally [404]. Abnormalities of resting CMRglc in association cortices are commonly found in advanced PD and may be reversed by DBS [404]. A motor activation study revealed a significantly enhanced increase in CBF by DBS in the left sensorimotor cortex (BA 4), bilaterally in the supplementary motor area

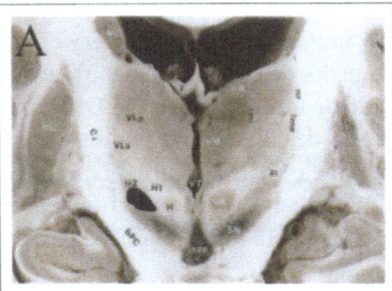
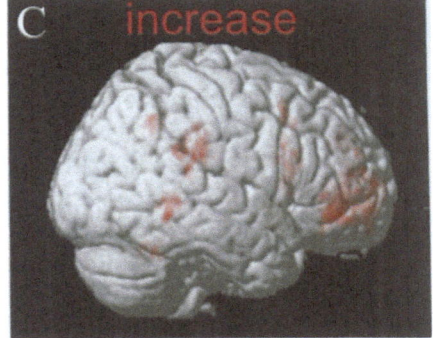
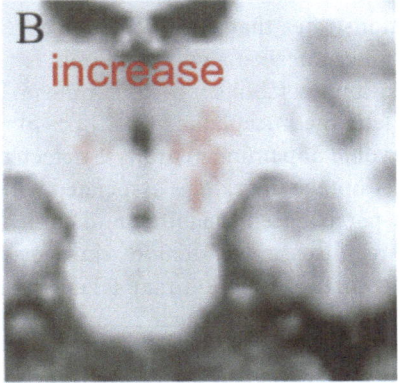
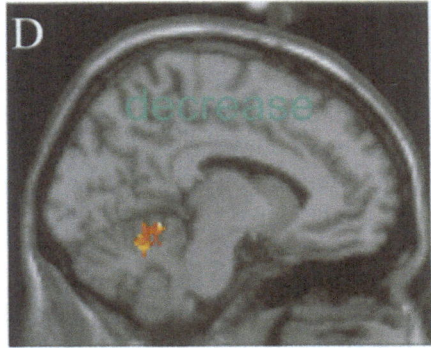

Fig. 2.17. Bilateral electrical stimulation of the subthalamic nucleus (A) increases CMRglc in adjacent nuclei (B) and in cortical association areas (C), but reduces abnormally high CMRglc in the upper part of the cerebellum (D) [403]

(BA 6), and in the right anterior cingulate cortex (BA 24/32) [405]. Another activation study suggested that internal globus pallidus DBS may also improve sequence learning by enhancement of the activity of prefrontal cortico-striato-pallidothalamic loops and related transcortical pathways [406]. The effect of GPi stimulation does not appear to be mediated by stimulation of the nigrostriatal dopaminergic pathway, and PET does not show evidence of increased striatal dopamine concentration with DBS in patients with PD [407]. In MPTP-lesioned monkeys an increase in CBF in ipsilateral premotor cortical areas during electrical GPi stimulation has been observed, which improved rigidity and bradykinesia. It suggests that disrupting the excessive inhibitory output of the basal ganglia reverses parkinsonism, via a thalamic relay, by activation of brain areas involved in the initiation of movement [408].

2.2.1.14.4 Gene Therapy

There could be a potential role for gene therapy in PD, to prenote recovery of lost neurons or at least transmitter function. To gauge the efficacy of such approaches, gene transfer and expression of key enzymes can be measured with PET. In a primate model of PD, lentiviral vector delivery of glial cell line-derived neurotrophic factor (GDNF) increased FDOPA uptake on the side of GDNF expression relative to preoperative levels [409]. Expression of AAAD after convection-enhanced delivery of AAV vector was demonstrated in monkeys with MPTP-induced parkinsonism using 6-^{18}F-fluoro-l-m-tyrosine [410].

2.2.1.14.5 Cell Transplantation

FDOPA-PET in MPTP-treated monkeys demonstrates that behavioral improvement after intrastriatal implantation of polymer encapsulated PC-12 cells is at least in part due to their highly specific uptake and metabolism of dopamine precursors [411]. In clinical studies, survival of fetal mesencephalic grafts [412–417] and fiber outgrowth from transplanted embryonic dopamine neurons is detected by an increase in FDOPA uptake [418]. FDOPA uptake in grafts is associated with clinical improvement [419]. In a case report, persistence of viability and clinical efficacy in spite of the underlying progressive disease was demonstrated over 10 years [420]. Recovery of movement-related cortical function may take longer than restoration of dopaminergic innervation of striatum. Such recovery was observed in a motor activation study 18 months after striatal embryonic dopaminergic grafts, and more than 12 months after improvement of striatal FDOPA uptake [421]. In a study that compared DAT ligand binding and FDOPA uptake in the grafted putamen, there was an increase in FDOPA only. This finding suggests that the clinical benefit induced by the graft is related more to increased dopaminergic activity than to improved dopaminergic innervation in the host striatum [422].

Re-establishment of normal FDOPA uptake and catecholamine reuptake sites (imaged by ^{11}C-NS-2214) was observed in MPTP-lesioned minipigs after intrastriatal grafts of mesencephalic neurons from fetal pigs [423]. In rats, a correlation between symptomatic efficacy of grafting with non-DA cells from dorsal mesencephalon and the degree of ^{11}C-CFT binding was found [424].

Direct brain infusion of glial cell line-derived neurotrophic factor in PD led to a 28% increase in putamen FDOPA storage after 18 months, corresponding to a 39% improvement in the off-medication motor subscore on the Unified Parkinson's Disease Rating Scale and a 61% improvement in the Activities of Daily Living subscore.

2.2.2   Multiple System Atrophy

Clinical signs of MSA are combinations of autonomic failure/urinary dysfunction, parkinsonism and cerebellar ataxia, and corticospinal dysfunction [262]. Hypokinetic parkinsonian symptoms are often unresponsive to L-DOPA. Clinical subtypes are striatonigral degeneration (SND, mainly parkinsonism without tremor, also known as MSA-P), olivo-ponto-cerebellar atrophy (OPCA, mainly with ataxia and corticospinal dysfunction, also known as MSA-C), and Shy-Drager syndrome (mostly with autonomic failure). Yet, as mentioned before, clinical features of MSA may also occasionally be present in PD. Frequent atypical presentations of PD are observed in patients of African-Caribbean and Indian origin [425]. A definitive diagnosis of MSA is possible only by histopathology, where glial cytoplasmatic inclusion bodies are characteristic. In vivo identification of the location and type of the pathophysiolgical disorder responsible is therefore of considerable scientific and clinical interest in these patients.

The distinction between PD and MSA with PET rests mainly on the fact that in PD there is a presynaptic dopaminergic deficit in the striatum but the glucose metabolism of striatal neurons is intact and their dopamine receptors are even upregulated, at least in the early stage of the disease. In contrast, both pre- and post-synaptic neurons of the dopaminergic system are impaired in MSA. These principal findings [426] have been replicated in many studies, but there is some overlap owing to reduced caudate (and sometimes also putaminal) glucose metabolism and D2 receptor binding, especially in advanced PD [267] and in the increasingly frequently recognized cases of PD with parkin mutations.

In contrast to PD, where there is little correlation between symptoms and CMR-glc, there is a clear correspondence between predominant symptoms and reduction of local CMRglc in MSA. Correlations have been found between (1) severity of parkinsonism and metabolic values of putamen and caudate; (2) severity of cerebellar signs and metabolism in the cerebellum; and (3) autonomic dysfunction and metabolic activity in the thalamus, frontal, and temporal regions, bilaterally [427].

The binding potential of benzodiazepine receptors is generally preserved in MSA in spite of atrophy and local reductions of CBF and glucose metabolism [428]. Binding of ^{11}C-diprenorphin to opiate receptors is slightly reduced in sporadic OPCA with autonomic failure [429].

2.2.2.1   Olivo-ponto-cerebellar Atrophy

OPCA is a mostly sporadic disease of old age without any known genetic cause. Dominantly inherited forms (dOPCA) actually most probably reflect spinocerebellar atrophy (SCA, see section 2.2.6.1), but were not recognized as such until the mid-1990s when current molecular examinations became available. Thus, the older PET studies do not use the current classification and therefore cannot contribute to the clarification of the relation between PET and molecular findings. At the beginning of the disease, cerebellar symptoms may be the only clinical abnormality, and it is not clear what proportion of patients will actually progress to the full clinical (and neuropathological) manifestation of MSA or will have symptoms restricted to cerebellar atrophy and dysfunction even during progression (Marie's disease).

The main finding observed with FDG-PET in OPCA is cerebellar and brain stem hypometabolism [427, 430, 431]. Hypometabolism may also involve frontal and temporal cortices, caudate, putamen, and thalamus [267, 431]. Brain stem and cerebellar hypometabolism appears to be present at an early stage of OPCA and to correlate with ataxia [432] and dysarthria [433], whereas thalamic and cortical metabolic impairment are common in patients with the full clinical picture of MSA including cognitive impairment [434] (Fig. 2.18). CBF and $CMRO_2$ are also reduced in brain stem and cerebellum in OPCA [435]. Metabolic impairment is not limited to the resting state, activated CMRglc in the pyramid of the vermis during treadmill walking is also reduced [436]. There is variably reduced FDOPA uptake in OPCA, usually involving caudate and putamen to similar degrees, whereas in PD the dorsal parts of the putamen are most affected [437].

The caudate and putamen contain a high density of opioidergic neurons and receptors, which can be imaged with ^{11}C-diprenorphine and have a close anatomical and physiological relationship with the dopaminergic system. There is a mild 10–15% reduction of striatal diprenorphine binding in OPCA [429].

2.2.2.2   Striatonigral Degeneration

Patients with the SND type of MCA have decreased CMRglc in the pallidum–putamen complex [427, 438–440], which clearly distinguishes the disease from PD and corresponds clinically to a lack of response to DOPA medication [441]. In contrast to OPCA, there is a relative sparing of the brain stem and cerebellar glucose metabolism [442].

There is a severe reduction in the density of striatal presynaptic monoaminergic terminals measured with ^{11}C-dihydrotetrabenazine, which is more pronounced than in OPCA and MSA patients with mainly cerebellar dysfunction [443]. Relatively small reductions in D2-receptor binding (approx. 10%) are noted in SND, which do not allow differentiation from advanced PD [444]. Significant reduction of D1 receptors is observed, mostly in posterior putamen (by approx. 30%), in contrast to their preservation in PD [445].

Striatal opioid receptor binding in PD, SND, and PSP was studied with ^{11}C-diprenorphine. Binding was normal in PD. Mean putamen, but not caudate, opioid receptor binding was significantly reduced in the SND group. In contrast, both caudate and putamen opioid receptor binding were significantly reduced in PSP [446].

2.2.2.3   Shy-Drager Syndrome

This clinical syndrome is a poorly characterized subtype of MSA with predominant autonomic failure. It should be distinguished from purely autonomic failure (also called idiopathic orthostatic hypotension), in which no neurological deficits other than autonomic dysfunction are present. Since autonomic failure is an essential symptom in all types of MSA, it has been suggested that separation of

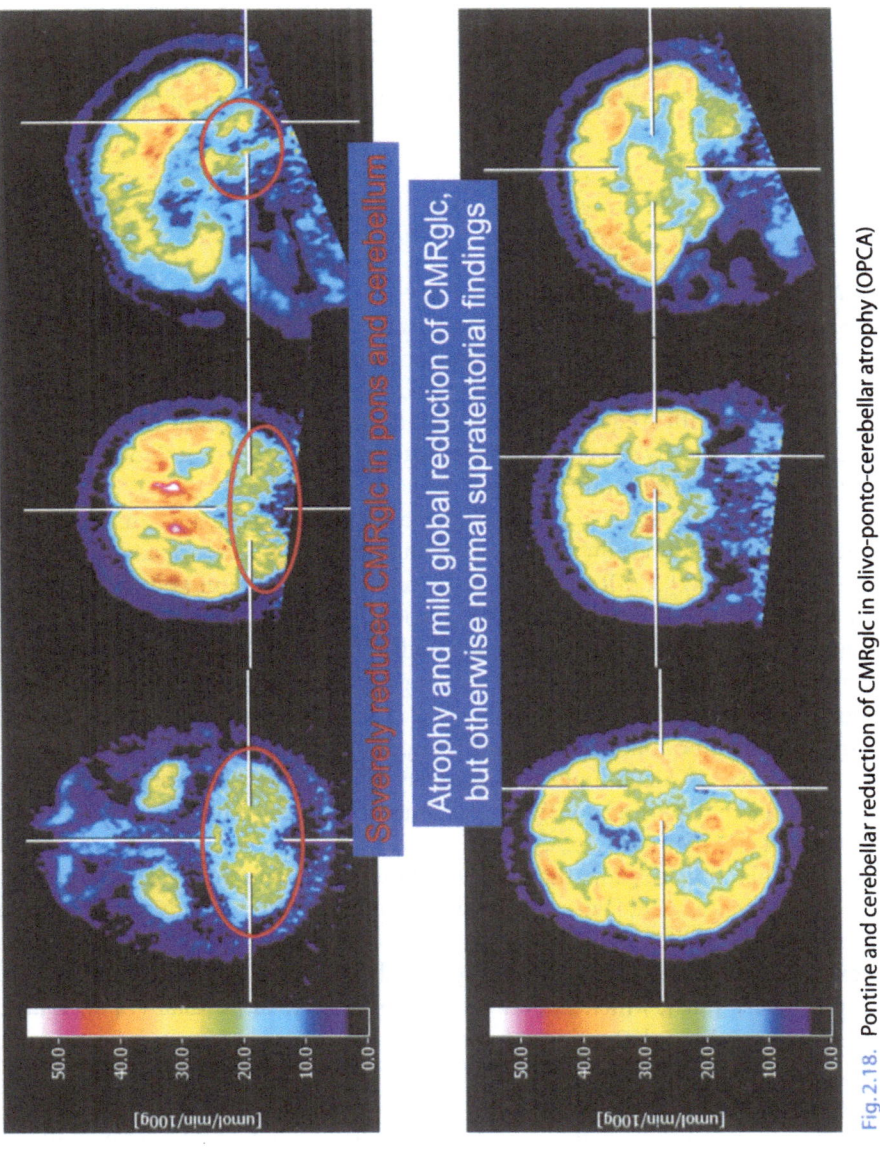

Severely reduced CMRglc in pons and cerebellum

Atrophy and mild global reduction of CMRglc, but otherwise normal supratentorial findings

Fig. 2.18. Pontine and cerebellar reduction of CMRglc in olivo-ponto-cerebellar atrophy (OPCA)

this subtype be entirely abandoned [262, 447]. As in other types of MSA, there is some reduction of striatal FDOPA uptake, at least in more severe disease with clinical signs of parkinsonism [442, 448]

A study of CBF regulation in Shy-Drager syndrome found that owing to the failure of autonomic vessel innervation, CBF depended on body position and passively followed blood pressure. An increase in PCO_2, however, still increased CBF, indicating intact metabolic regulation of CBF [449]. Increased cardiac uptake of 6-^{18}F-fluorodopamine in Shy-Drager syndrome indicated intact sympathetic terminals and absent nerve traffic, whereas other types of dysautonomia (with and without PD) were characterized by a loss of myocardial sympathetic nerve terminals [450].

2.2.3   Other Neurodegenerative Disorders

2.2.3.1   Progressive Supranuclear Palsy

PSP, also known as Steele-Richardson-Olszewski syndrome, is a mostly sporadic neurodegenerative disease with vertical gaze paresis (due to a supranuclear disorder in midbrain that gives the name to the disease), a parkinsonian movement and gait disorder with frequent and severe falls, apathy, and cognitive impairment. Neuropathologically it is characterized by tau deposits similar to those in FTD but mainly located in midbrain.

In PSP, CMRglc is reduced in prefrontal association cortex, most markedly in its mesial part, including anterior cingulate and orbitofrontal cortex, and in midbrain [210, 212, 451–457]. CMRglc may also be reduced in striatum and thalamus. The abnormality in midbrain is difficult to recognize on visual evaluation of FDG-PET images, because normal midbrain metabolism is lower than in cerebral cortex and basal ganglia. Therefore, quantitative image analysis is recommended for detection of this feature that distinguishes PSP from other disorders with frontal and striatal metabolic abnormalities [458] (Fig. 2.19).

Reduction of FDOPA uptake is relatively mild and may be absent in early PSP, suggesting that parkinsonism in PSP may relate to dysfunction distal to the dopaminergic neurons [272, 459–461]. Dopamine reuptake sites are also reduced, but the reduction is milder than in PD and there is no anterior–posterior gradient of striatal impairment [320, 437]. Reduction of the caudate dopaminergic innervation similar to that in the putamen is also noted and may contribute to frontal cortex dysfunction and apathy [462]. There is also a reduction of D2-receptor-binding capacity [463], which is even more pronounced in caudate (24%) than in putamen (9%) [444]. Opioid receptor binding is reduced in most patients in caudate and putamen, whereas such reductions are less frequently seen in SND and are absent in PD [446].

There is a reduction of thalamic AChE activity in PSP that is not seen in PD, but cortical AChE activity is largely preserved [339]. A frontal increase in muscarinergic binding such as is seen in PD is not present in PSP [338]. CMRglc was not altered by AChE inhibition with physostigmine [464]. Similar to CMRglc, cortical and striatal oxygen metabolism is reduced in PSP [465], but frontal accentuation of cortical impairment may not always be evident [466].

In rare cases PSP occurs as a familial disease. In the families affected, asymptomatic subjects at risk may show the typical changes with FDOPA- and FDG-PET

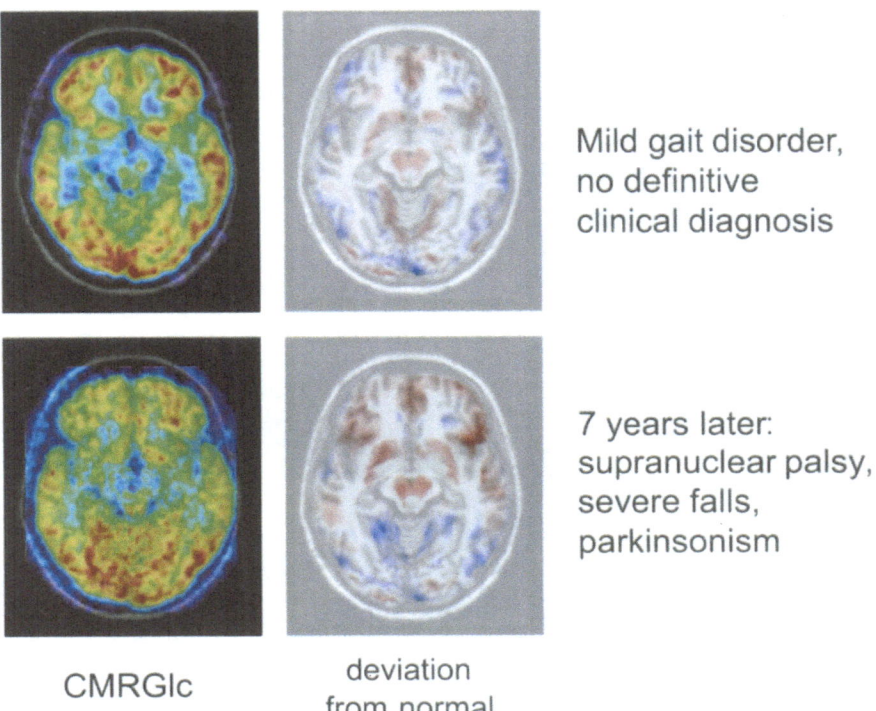

Mild gait disorder,
no definitive
clinical diagnosis

7 years later:
supranuclear palsy,
severe falls,
parkinsonism

CMRGlc   deviation
                from normal

Fig. 2.19. FDG-PET (fusion display with coregistered MRI) in progressive supranuclear palsy (PSP): 7-year follow-up of a patient who initially presented with mild unspecific symptoms but already had the typical PET findings (*top row*) with metabolic abnormalities in midbrain, basal ganglia, and frontal cortex (areas marked by *red color* on the *right-hand image*, which demonstrates the deviation from normal average). At follow-up he had the typical clinical symptoms and progression of the metabolic deficits (*bottom row*)

[467]. Occasionally, a clinical presentation and FDG-PET findings typical for PSP may be caused by other neuropathological processes, such as progressive subcortical gliosis [468].

2.2.3.2 Corticobasal Degeneration

Caudate FDOPA uptake in corticobasal degeneration (CBD) patients is lower contralateral to the clinical signs than in controls, but is higher than in PD of comparable clinical severity [469–471]. With FDG, CBD differs from PD by a metabolic decrease in premotor, primary motor, supplementary motor, primary sensory, and parietal associative cortices, and in caudate and thalamus [469, 471–473] (Fig. 2.20). Cortical and thalamic oxygen metabolism is also reduced [470]. There is usually prominent hemispheric asymmetry, with metabolic decreases on the

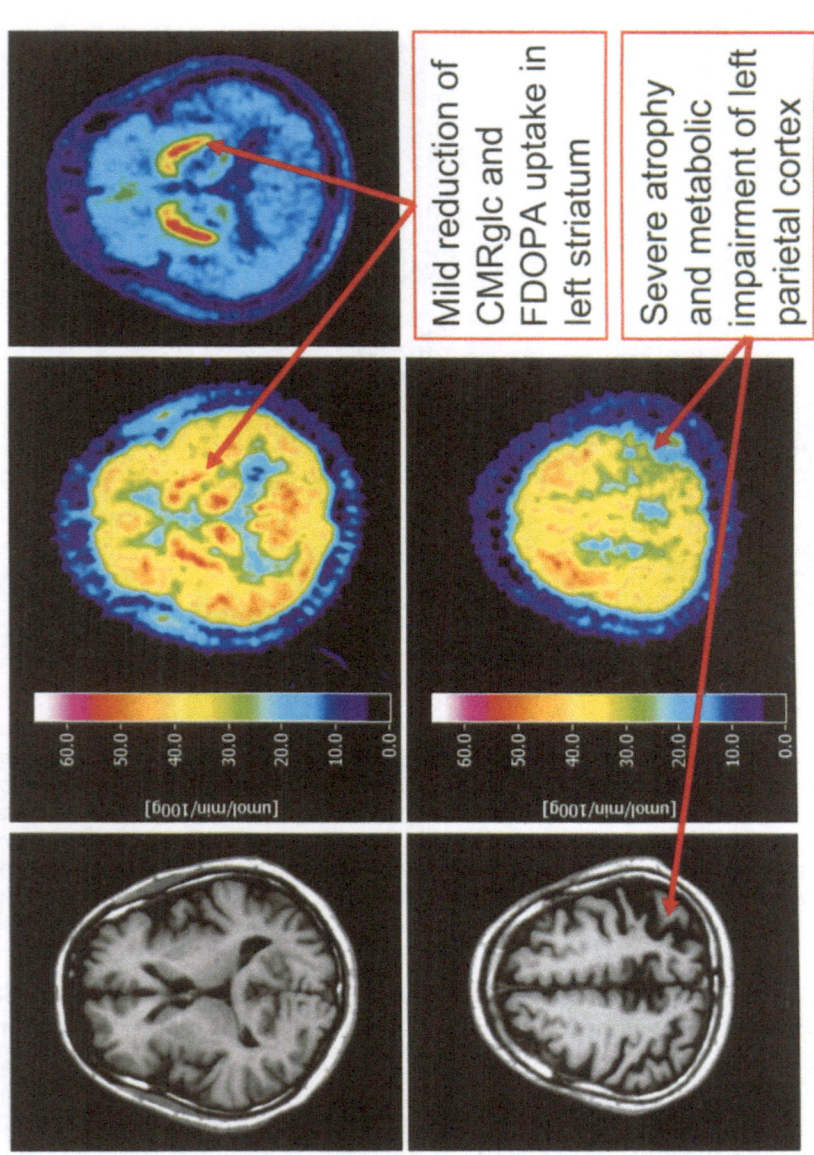

Fig. 2.20. FDG-PET in corticobasal degeneration (CBD), demonstrating typical unilateral decrease of parietal cortex and basal ganglia metabolism

side contralateral to the clinical signs [474]. The asymmetry could possibly be used to differentiate CBD from PSP, which is much more symmetrical with frontal accentuation of the metabolic abnormality [475].

2.2.3.3   Disorders with Abnormal Deposition of Metals

Hallervorden-Spatz syndrome is associated with abnormal deposits of iron, mainly in pallidum, that is visible on MRI and may be due to mutations in the gene encoding pantothenate kinase 2 (*PANK2*) [476]. FDOPA-PET may be normal, but CBF and CMRglc may be reduced in caudate nucleus, pons, and cerebellar vermis [477]. Cortical hypometabolism associated with dementia may also be present [478].

Wilson's disease leads to abnormal deposition of copper in the basal ganglia, which produces signal alterations on MRI and variable findings with pre- and postsynaptic dopaminergic PET markers [479–481]. CMRglc is reduced in putamen and in more severe cases also in thalamus and cortex [482, 483]. Dopamine-receptor-binding capacity and CMRglc may recover after successful treatment with D-penicillamine [484–487].

2.2.3.4   Parkinsonism Attributable to Toxic and Inflammatory Brain Damage

Since most cases of PD are sporadic, there is an ongoing search for possible environmental agents that might cause dopaminergic degeneration. A few toxins, such as manganese and MPTP, have been identified, but none is likely to be responsible for a large number of clinical cases. There has been some hope that functional neuroimaging might make it possible to detect sporadic subclinical cases or even identify a pattern of neuronal damage that would allow identification of causative agents [488], but findings reported so far do not allow such conclusions [489, 490].

Carbon monoxide intoxication, if survived, often leads to basal ganglia damage with symptomatic parkinsonism. In one case, the lesion was located in the pallidum, with intact and up-regulated dopamine striatal receptors [491]. Cyanide intoxication is rarely survived and can lead to pre- and postsynaptic striatal lesions [492].

Exposure to large doses of solvents may induce encephalopathy with parkinsonian symptoms. A few patients underwent investigation of dopaminergic functions with PET. Highly variable defects of pre- and postsynaptic systems were found, with some overlap but no exact correspondence with the typical pattern of PD [493–495]. In three subjects with solvent abuse and parkinsonism, normal FDOPA uptake and reduced raclopride binding were observed [496].

Manganese exposure can cause parkinsonian symptoms, and the primary site of neurological damage has been shown by pathological studies to be the globus pallidus [497]. Consistent with this observation, normal FDOPA uptake has been found in most patients studied so far who have had symptoms after exposure to manganese [498, 499], whereas striatal CMRglc and D2 receptor binding were moderately reduced. In an experimental study in monkeys, the most marked effect of chronic manganese intoxication was a severe reduction of dopamine transporter capacity [500].

Postencephalitic parkinsonism is a rare disorder today, and few patients have been studied with PET [501–503]. In all cases, a severe reduction of striatal FDOPA uptake was seen, affecting the putamen somewhat more than the caudate nucleus. Striatal CMRglc was normal or even elevated.

2.2.3.5   Amyotrophic Lateral Sclerosis

Amyotrophic lateral sclerosis (ALS) is a neurodegenerative disease that affects primarily upper and lower motor neurons and leads to severe weakness, disability and death. As a correlate to upper motor neuron degeneration, a reduction of CBF and energy metabolism is present in the primary sensorimotor cortex [504] and may extend to other cortical areas, putamen and cerebellum with progression of disease [505–507]. During motor activation, increased recruitment of nonprimary motor areas is observed, which may represent functional adaptation to a corticospinal tract lesion. In contrast, there is impaired activation of the medial prefrontal cortex and parahippocampal gyrus [508].

ALS is usually not associated with parkinsonian symptoms, although combinations have occasionally been found, predominantly on some Pacific islands. Subclinical nigrostriatal dopaminergic dysfunction has been demonstrated with FDOPA in some ALS patients due to copper/zinc superoxide dismutase mutations [509].

2.2.3.6   Other Rare Disorders

CMRglc is decreased in the cerebellum and brain stem, and FDOPA uptake is variable, in hereditary dentato-rubro-pallido-luysian atrophy [510], which is a genetic disorder with polyglutamine expansion in the gene for atrophin-1 that manifests itself with symptoms similar to those of MSA. Reduced striatal FDOPA uptake was also found in autosomal dominant parkinsonism and dementia with pallido-ponto-nigral degeneration [511].

Rapid-onset dystonia-parkinsonism is a genetic movement disorder characterized by an abrupt onset over hours to days of bradykinesia, postural instability, dysphagia, dysarthria, and severe dystonic spasms. In contrast to PD, in this condition PET does not show evidence of presynaptic dopaminergic impairment [512].

2.2.4   Hyperkinetic Syndromes

2.2.4.1   Huntington's Chorea (Huntington Disease)

Huntington disease (HD) is an autosomal dominant disorder with complete penetrance, which leads to severe hyperkinesias, disability, dementia, and death. It is associated with a severe and early reduction of CMRglc beginning in the caudate nucleus and rapidly also affecting the putamen. In the later stages, when demen-

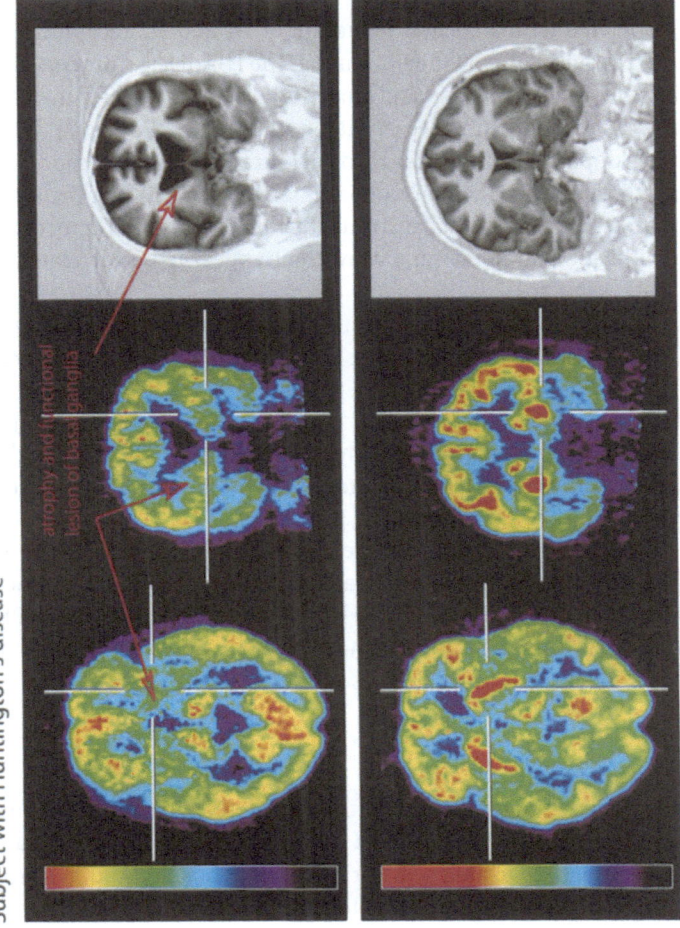

Huntington's disease vs. normal control
18F-FDG-PET and T1-weighted inversion recovery MRI

Subject with Huntington's disease

atrophy and functional lesion of basal ganglia

Normal control

Fig. 2.21. FDG-PET in Huntington disease (chorea; HD) with typical severe reduction of CMRglc in caudate and putamen

tia develops, hypometabolism also extends to thalamus and cerebral cortex [513–521] (Fig. 2.21). Thalamic hypometabolism is also noted as an early feature in the rigid juvenile form of HD [523], and it correlates with dystonia in adults [519]. Significantly reduced caudate CMRglc is observed in asymptomatic gene carriers [524–528], preceding the development of atrophy. Some researchers have noted subtle clinical abnormalities in subjects classified as at risk, with reduced caudate CMRglc [529]. With the development of reliable methods to detect the pathogenic trinucleotide expansion in the *huntington* gene on chromosome 4, however, interest in the use of PET for presymptomatic diagnosis has waned.

In keeping with the caudate atrophy and dysfunction, there is also impaired activation of frontal cortex in activation studies [530].

HD is associated with a striatal reduction of D1 and D2 receptors and of DAT [531–536]. PET results support the notion that the HD disease process is a function of trinucleotide length and age and that the development of clinical signs and symptoms is associated with CAG repeat lengths greater than 35.5 [537]. Both striatal D1 and D2 dopamine receptors are lost in parallel from both caudate and putamen in presymptomatic HD. Thus, dopamine receptor binding provides a sensitive means of detecting subclinical striatal dysfunction [538]. D2 receptor binding and glucose metabolism in the caudate nucleus appear to be correlated with some cognitive tests [539, 540].

There is also a reduction of ^{11}C-diprenorphine binding to striatal opioid receptors in HD [541]. Furthermore, reduced striatal VMAT2 binding is observed in HD, which is most severe in patients with muscular rigidity, suggesting additional nigrostriatal pathology similar to that of PD in these patients [542]. Benzodiazepine receptors, which are present in the striatum only in a low density, are reduced in caudate even in the early stages of the disease [543], but the decline does not parallel loss of D2 receptors [544].

So far no effective standard treatment is available for HD. There have been experimental trials with intrastriatal transplantation of fetal striatal neuroblasts, which have been shown by FDG-PET to be viable after one year [545]. PET studies have also been used to characterize longitudinal in vivo alterations in the development of animal models of HD [546–548] to facilitate drug studies.

2.2.4.2   Other Choreic Disorders

Other choreic diseases may be associated with striatal hypermetabolism instead of hypometabolism. This has been seen as a reversible finding in Sydenham's chorea [549] and in chorea associated with primary antiphospholipid syndrome [550], but not in some cases with chorea due to systemic lupus erythematosus [551]. On the other hand, hypoperfusion and hypometabolism in the striatum and other cerebral regions, more moderate than are seen in HD but otherwise similar, were observed in chorea-acanthocytosis [552, 553]. Benign hereditary chorea is associated with normal or moderately reduced striatal glucose consumption [554, 555].

2.2.4.3   Restless Legs Syndrome

Mildly reduced FDOPA uptake and reduced D2 receptor binding capacity in caudate and putamen are seen in the restless legs syndrome [556].

2.2.4.4 Essential and Orthostatic Tremor

This is a relatively benign, often familial, disease, which rarely leads to severe disability. Integrity of the dopaminergic system has been demonstrated with FDOPA uptake and D2 receptors [292] and also of DAT with ^{18}F-FE-CIT [292]. FDG has revealed mild glucose hypermetabolism in the medulla and thalami, but not in the cerebellar cortex [557].

The neurophysiology of essential tremor has been explored in ^{15}O-water studies. Increased bilateral cerebellar activity was described both at rest and during tremor [558], and a similar finding was observed in orthostatic tremor [559]. Further analysis has revealed involvement of connections of the red nucleus in essential tremor [560]. The tremor can be suppressed in many patients by alcohol ingestion. It has been demonstrated that this effect is probably mediated via a reduction of cerebellar synaptic overactivity [561]. Beneficial effects of stimulation of the thalamic nucleus ventralis intermedius in essential tremor are associated with increased activity in motor cortex and decreases in retroinsular regional CBF [562]. Variation of stimulation amplitude and frequency appears to exert a differential modulation of subcortical target and cortex during DBS [563].

2.2.4.5 Tourette Syndrome

In an FDG study, Tourette syndrome patients were characterized by decreased normalized metabolic rates in paralimbic and ventral prefrontal cortices and in some subcortical regions, including the ventral striatum [564]. Symptoms were associated with significant increases in metabolic activity in the orbitofrontal cortices [565]. In a study of metabolic covariations, Tourette syndrome was characterized by a nonspecific pattern of increased motor cortical activity that is also present in other hyperkinetic disorders, and by a reduction in the activity of limbic basal ganglia-thalamocortical projection systems [566]. These data are consistent with the hypothesis that altered limbic–motor interactions are a pathophysiological hallmark of this disease [567].

No major abnormality of D2 receptors or monoaminergic vesicles was found with NMSP [568] or ^{11}C-dihydrotetrabenazine (DTBZ) [569]. Greater FDOPA uptake [570], elevated striatal DAT [571], and increased dopamine release in putamen during pharmacological challenge are seen [572]. These results suggest that the underlying pathobiology in Tourette's syndrome is a phasic dysfunction of dopamine transmission.

In an activation study, aberrant activity in the interrelated sensorimotor, language, executive, and paralimbic circuits was identified. This may account for the initiation and execution of diverse motor and vocal behaviors that characterize tics in Tourette's syndrome and for the urges that often accompany them [573].

2.2.5   Dystonia and Related Disorders

Dystonia is a disorder of muscle tone that are found in a heterogeneous group of diseases. Most are probably due to genetic disorders (e.g., *DYT1* to *DYT7*) or secondary to other neurodegenerative diseases, but in many apparently sporadic "idiopathic" cases no cause has yet been identified. Neck muscles are very often severely involved in dystonia, leading to a condition called torticollis. Other frequent manifestations are blepharospasm and spasmodic dysphonia. Dystonia can also be the result of brain lesions (e.g., stroke or tumor) and then usually manifests itself as hemidystonia. PET studies have mostly been performed with the goal of better understanding the pathophysiology of dystonia and perhaps, as a result, improving classification and therapy of this disabling disorder.

2.2.5.1   Idiopathic Dystonia and Torticollis

Studies of blood flow, and glucose and oxygen metabolism in the resting state have not usually revealed large abnormalities. If present, abnormalities are most frequently observed in putamen or in thalamus, two brain structures well known for their physiological role in the extrapyramidal system. Increased CMRglc in putamen has been found in several studies [574–577], in some instances with extension into pons and midbrain or thalamus. The abnormality was present even after symptoms had been ameliorated by therapy with botulinus toxin. It is also seen in asymptomatic carriers of a *DYT1* mutation [578]. In a few studies, however, normal or even reduced CMRglc was found in the putamen and related structures in dystonia [579, 580], but probably these studies included more adults and more subjects with hemidystonic manifestations of the disease. It is also difficult to define a resting examination state in dystonia, because usually dystonic movements are present during the scans to a variable degree and associated neuronal activation may influence the results. To complicate matters even further, there also has been a report of focal uncoupling of oxygen metabolism and blood flow in dystonia in putamen [581].

Functional activation studies tend to point towards imbalances in regional cortical activity, which may be a consequence of dysfunction in the basal nuclei. During joystick movements, inappropriate overactivity of premotor and supplementary motor cortex but impaired activity of motor executive areas was observed [582, 583]. Reduction of frontal and motor cortex activity was observed under specific sensory stimulation that ameliorated symptoms [584]. Widespread reduction of activation was seen during electrical stimulation of globus pallidus internus for treatment of dystonia symptoms [585]. Activation during imagined movements did not differ from that in normal controls, suggesting that movement planning is intact [586].

Impaired dopaminergic transmission in dystonia was suggested by a decrease in dopamine D2-like binding in putamen that was found with ^{18}F-spiperone [587]. Reduction of striatal D2 binding was seen even in asymptomatic carriers of a *DYT1* mutation [588]. Nonetheless, no reduction of D2 receptors was found in another study of idiopathic dystonia [589] or in familial paroxysmal dystonic choreoathetosis [590]. Dopamine synthesis appears to be normal [591, 592] or increased in studies in a few patients [593]. Presynaptic markers have also been found to be intact in rapid-onset dystonia-parkinsonism [512]. Thus, PET studies currently do not allow firm conclusions with regard to dopaminergic dysfunction in these types of dystonia.

2.2.5.2   DOPA-responsive Dystonia

Molecular and biochemical studies indicate that most patients with DOPA-responsive dystonia (DRD, Segawa syndrome) have a defect of dopamine synthesis without degeneration of dopaminergic neurons. It may be due to impaired activity of tyrosine hydroxylase, which in turn may be caused by a defect in the gene for tetrahydrobiopterine, GTP cyclohydrolase I. In contrast, the activity of DOPA decarboxylase is normal. Correspondingly, normal FDOPA uptake was observed in DRD [592, 594–597]. However, an increased DTBZ-binding potential was seen in the striatum of DRD patients, which may indicate up-regulation of vesicular monoamine transporter type 2 (VMAT2) expression [598]. With RAC-PET, elevated striatal D2-receptor binding was seen in symptomatic and asymptomatic mutation carriers [599, 600], probably as a consequence of disturbed dopamine synthesis.

2.2.5.3   Dystonia Attributable to Localized Lesions or Degeneration

A lesion in the left substantia nigra and reduced FDOPA uptake in putamen were observed in a patient with hemiatrophy, juvenile-onset exertional alternating leg paresis, hypotonia, and hemidystonia and adult-onset hemiparkinsonism [601]. In another patient with hemidystonia-hemiatrophy, significant reductions in FDG uptake at the level of the basal ganglia and, to lesser extent, in the frontoparietal cortex contralateral to the clinically involved side were found but D2 receptors were normal [602]. Reduced striatal FDOPA uptake was seen in a group of patients with dystonia attributable to ischemic midbrain lesions involving the nigrostriatal pathway [603]. Reduced binding potential of dopamine transporters, suggesting a dopaminergic deficit, is also observed in Lesch-Nyhan disease, a genetic disorder that is characterized by hyperuricemia, choreoathetosis, dystonia, and compulsive self-injury [604].

2.2.6   Ataxia

The classification of neurodegenerative ataxia has often been revised in recent decades. With the recent rapid progress in molecular genetics, the hereditary ataxias are now usually classified according to the underlying mutations. The clinical diagnosis is usually made on the basis of clinical investigation of family history and symptoms and exclusion of some disorders with systemic metabolic abnormalities (such as hypolipoproteinemias, vitamin E deficiency, Lesch-Nyhan syndrome, Refsum disease, aminoacidurias) by laboratory tests. The role of PET in the diagnosis is not yet well established and is limited by negative findings in some types of ataxia.

2.2.6.1   Spinocerebellar Ataxia

More than 20 different types of autosomal dominant spinocerebellar ataxias (SCA) have so far been identified. Some types have been found to affect only a few families with a restricted geographical origin, while others (e.g., types 1, 3, and 6) are more common; few cases have been studied with PET. Within genetic types, there may be substantial phenotypical variation. Since FDG-PET is closely linked to neuronal function, it might be expected to be more closely related to the symptoms (phenotype) than to the genotype. SCA may be difficult to distinguish from multiple system atrophy (see section 2.2.2), and as yet PET data also do not provide a clear distinction.

In members of one family with type 1 SCA, widespread glucose hypometabolism was found. The metabolic pattern and histopathological alterations (including glial cytoplasmic inclusions) were similar to those in sporadic MSA [605]. There was normal binding of benzodiazepine receptors, which is also a point of similarity to MSA. In other patients with SCA1, the reduced CMRglc was limited to brain stem and cerebellum [606].

In type 3, also known as Machado-Joseph disease, decreased FDG utilization was found in the cerebellar hemispheres, brain stem, and occipital cortex. There was increased FDG metabolism in the parietal and temporal cortices in even asymptomatic carriers of the gene, suggesting preclinical disease activity [607]. Symptomatic patients showed severe and widespread brain glucose hypometabolism and also a reduction in FDOPA uptake in putamen [608], which is similar to the picture in sporadic OPCA. The only difference in terms of FDOPA uptake may be relative sparing of the caudate, which is different from MSA, where the caudate nucleus is usually severely impaired. D2 receptors were found to be preserved in SCA3 [609]. Impaired FDOPA uptake associated with DOPA-responsive parkinsonism has also been observed in SCA2 patients [610].

Widespread hypometabolism including cortical regions and basal ganglia as well as the cerebellar hemispheres and brain stem has been found in type 6, which is caused by an expanded CAG repeat sequence within the coding regions of the *CACNL1A4* gene [213].

2.2.6.2   Friedreich's Ataxia

Friedreich's ataxia is the most common autosomal recessive SCA and is caused by mutation of the *frataxin* gene on chromosome 9. It is a mitochondrial disorder, which explains why brain glucose metabolism (including that in the cerebellum) is increased in patients who are still ambulatory while it decreases later during further progression [611]. The highest local increase is seen in the basal ganglia, as in some other systemic metabolic disorders (see section 2.6.9). The cerebellum may appear less active metabolically by comparison. Thus, in this disease FDG-PET gives different results from those obtained in other ataxias, but absolute quantitation of CMRglc is required to interpret changes properly.

2.2.6.3   Other Ataxias

Widespread cerebral and cerebellar reduction of glucose metabolism is present not only in SCA, but also in other ataxias resulting from systemic and metabolic disorders, including paraneoplastic syndromes [612] and alcoholism [613].

2.3   Brain Tumors

Brain tumors are major structural lesions, and their diagnosis and clinical management are therefore dependent primarily on imaging with CT and MRI. These modalities are invaluable for delineating the anatomy and assessing the integrity of the BBB with contrast agents. It may be possible to improve clinical management, however, by using PET to provide physiological and biochemical information related to tumor metabolism, proliferation rate, and invasiveness. In addition, PET can demonstrate the interaction of a tumor with the surrounding and remote areas of the brain, including eloquent brain areas in which damage results in irreversible loss of language and motor function. The connection between tumor and brain function is particularly critical in malignant gliomas, because of their invasive growth and the lack of truly curative therapies. In these cases, therapeutic interventions are essentially palliative, which means they must preserve or improve brain function, rather than sacrifice it for a treatment with limited efficacy. On the other hand, treatment should be as aggressive and effective as possible, to restrict the degree of disability and prolong life. Therefore, treatment requires detailed information not only on tumor type, location, and extent, but also, ideally, on the functional status of the tumor and surrounding brain, which is information that can be provided by PET.

Initial studies with specific PET radiopharmaceuticals were used to develop methods for their use in brain tumor and to determine their ability to demonstrate abnormalities in different tumor types, especially malignant glioma. Subsequent research then focused on the utility of particular radiopharmaceuticals in the clinical management of individual patients. The PET radiopharmaceutical most widely used to image brain tumors is FDG, which provides an assessment of local glucose metabolism. FDG has clinical relevance because of the relationship between tracer uptake and the malignant behavior of tumors. A variety of radio-labeled amino acids have also been studied, and ^{11}C-methionine is in use clinically. Other promising but less well-established approaches include the use of tracers to image hypoxic brain tissue, methods to assess gene transfer and antiangiogenic therapy, and pharmacokinetic studies with radiolabeled chemotherapeutic agents.

A variety of clinical questions can be addressed with PET. These include: (1) determining the degree of malignancy and the prognosis of specific tumors; (2) selecting the site within a tumor for obtaining a biopsy specimen that best represents its pathology; (3) monitoring the response of a tumor to therapy and altering the course of therapy; (4) determining whether deterioration in a patient who has undergone treatment is due to tumor recurrence or radiation necrosis; (5) determining the functional status of tissue adjacent to tumor; and (6) assessing metastatic brain tumors and paraneoplastic syndromes. It should be stated, however, that the level of clinical research to support these applications varies, and that actual clinical practice differs among medical centers and geographic areas.

2.3.1   Biological Grading

Tumor grading relies primarily on histological criteria, such as degree of cellularity, nuclear and cellular atypia, number of mitotic figures, and presence or absence of necrosis and vascular endothelial proliferation. The World Health Organization (WHO) system classifies brain tumors into four grades, grade I being the most benign and grade IV, glioblastoma multiforme, the most malignant tumor with the worst prognosis [614]. Typically, for tumors that are felt to be of high grade on the basis of clinical presentation and structural imaging, a neurosurgical resection is performed to relieve symptoms and to provide tissue for pathological diagnosis. Histological grading is based on small tissue biopsies that may not be representative of the tumor, however, since gliomas are frequently heterogeneous [615]. Tissue may be more difficult to obtain in a recurrent tumor, which may of a higher grade than the original tumor and may require radiation or chemotherapy. Thus, PET indicators of tumor grade and prognosis may be useful as an adjunct to and/or even as a substitute for histological grading. This idea has been explored with PET tracers for glucose metabolism, amino acid uptake, and nucleoside uptake.

2.3.1.1   Glucose Consumption
It has been known for over 50 years that glucose consumption is increased in malignant tumors [616]. This principle led to the widespread use of whole-body PET imaging with FDG in patients with cancer [617]. The use of FDG to study brain tumors was a logical extension of its initial application for noninvasive

measurement of regional cerebral glucose metabolism, which was, in fact, the first use of PET in oncology. Considerable controversy soon arose, however, concerning the applicability of the FDG tracer model in brain tumors. Although correlations between malignant behavior and degree of FDG uptake were demonstrated, use of the FDG model with standard values for the lumped and rate constants to quantify the rate of tumor glucose consumption was felt to be incorrect [618, 619]. The use of ^{11}C-glucose shed some light on this issue, since it does not require the use of a lumped constant to estimate local glucose metabolism. Although there is evidence that high FDG uptake in gliomas actually reflects increased glucose consumption [620] (see also section 3.4), the lumped constant used to estimate glucose consumption from FDG uptake may be substantially higher in tumors than in normal brain [621, 622], leading to an overestimation of glucose consumption if the value for normal brain is used. Changes in the lumped constant could be due to the presence of hexokinase type II, which is expressed in some tumors but not in normal adult brain and which has a higher affinity for FDG than for glucose [623]. Changes in glucose transport may also have a role but appear to be relatively small compared with the changes in hexokinase activity or its affinity for FDG and glucose [624]. Thus, increased FDG uptake in tumors may be due to increases in both metabolism and the lumped constant; it may also reflect increased glycolytic rather than oxidative metabolism of glucose.

In clinical applications, FDG images are analyzed, visually or in a semiquantitative manner, with reference to the level of uptake in the contralateral, presumed normal, cerebral cortex and deep white matter [625, 626]. FDG uptake is related to histological tumor grade [627, 628] and to survival in both primary and recurrent gliomas [629, 630]. FDG uptake in low-grade gliomas (which are mostly grade II in adults) is usually close to that of normal white matter, whereas most grade III anaplastic gliomas have FDG uptake similar to or exceeding that in normal gray matter. Untreated glioblastomas, the most malignant of gliomas (grade IV) also show high uptake, which may be heterogeneous owing to the microscopic and macroscopic necroses that are typical for this tumor type (Fig. 2.22). After treatment and at a late stage, areas with low metabolism may prevail. Cutoff levels of 1.5 for the tumor-to-white matter FDG uptake ratio and 0.6 for the tumor-to-cortex ratio are useful in the differentiation of low-grade from high-grade gliomas [631]. The sensitivity and specificity of the use of these thresholds in detection of high-grade gliomas have been determined as 94% and 77%, respectively. PET can also show areas of high FDG uptake in low-grade tumors that have undergone malignant transformation. A clinical trial would be necessary, however, to demonstrate that regular FDG scans to detect malignant transformation, possibly before clinical deterioration, can improve clinical outcome.

High FDG uptake is also seen in other malignant brain tumors, in particular in primitive neuroectodermal tumors (PNET), medulloblastoma [632], and malignant lymphoma [633] (Fig. 2.23). High uptake in lymphoma can be used to differentiate it from toxoplasmosis in patients with acquired immune deficiency syndrome (AIDS) [634, 635]. Brain metastases from systemic cancers often show high

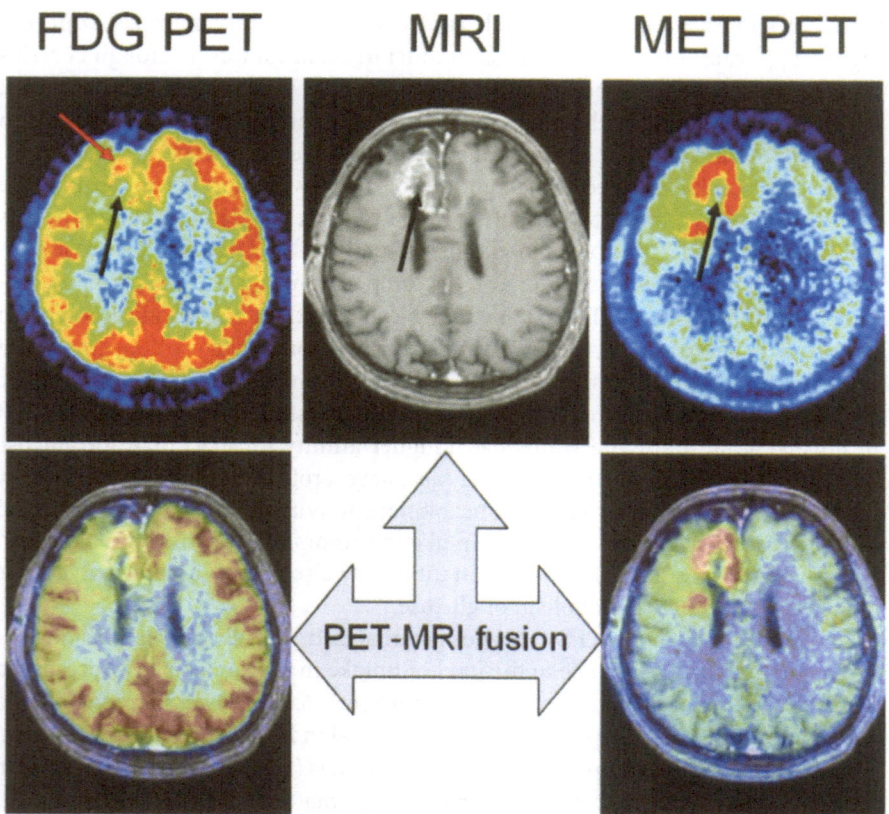

Fig. 2.22. Coregistered FDG-PET (*left*), contrast-enhanced T1-weighted MRI (*middle*), and ^{11}C-methionine (MET)-PET (*right*) demonstrate metabolic inhomogeneity of glioblastoma. A central necrosis (*black arrow*) is surrounded by tumor tissue with contrast enhancement. FDG uptake is similar to normal cortex in parts of this tumor (*red arrow*) and low in other parts. Distinction of metabolically active tumor from normal cortex is possible only by fusion with contrast enhanced MRI. MET uptake is high and exceeds normal brain uptake in all active tumor parts

FDG uptake, but PET is usually not indicated in such cases; the sensitivity of MRI is much greater, and systemic malignancy is often clinically evident [636]. Patients with systemic cancer frequently have whole-body FDG-PET scans for staging, but addition of a brain scan in such patients rarely yields clinically relevant additional information [636].

Meningiomas are generally found to be grade I tumors on histopathology, but FDG uptake is variable and can provide an index of tumor aggressivity and probability of recurrence [637]. Analogous observations have been made in a small series of cranial neuromas [638].

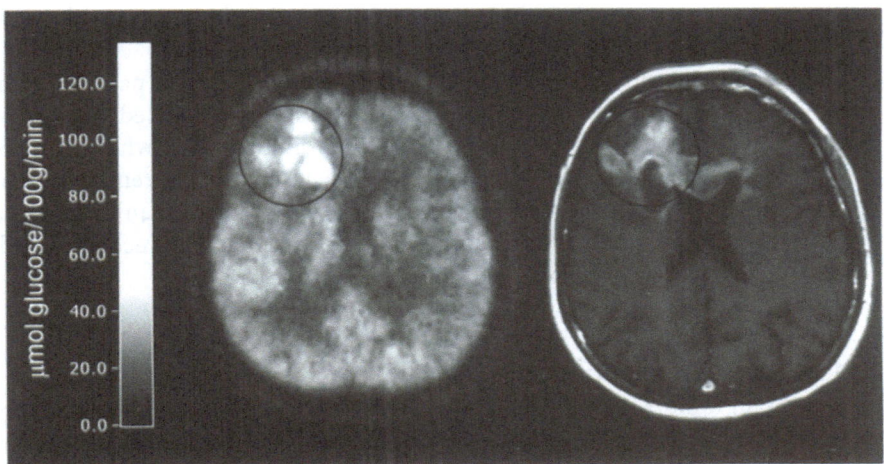

Fig. 2.23. Coregistered FDG-PET (*left*) and MRI (*right*) of lymphoma with exceedingly high FDG uptake, corresponding to peak glucose metabolic rates of more than 100 μmol/100 g per min, which is much higher than those in normal cortex

High FDG uptake has also been observed in benign pituitary adenomas, including microadenomas [639, 640]. Prolactinomas [641] and granular cell tumors of pituitary and hypothalamus [642] may appear hypometabolic on FDG.

Pilocytic astrocytomas are a rare type of astrocytoma classified histologically as grade I, which occur mainly in childhood. They are the most benign of gliomas and can be cured if complete resection is possible. However, they may exhibit radiological signs of malignancy without being malignant and also have variable glucose metabolism without a clear relation to prognosis [643, 644].

Several technical factors must be taken into account in performance and interpretation of FDG brain scans. The main limitation of FDG for clinical studies of brain tumors is the high glucose consumption of normal gray matter. Therefore, the high FDG uptake in malignant brain tumors is often similar to that in normal gray matter, and high-grade tumors may be missed if adjacent to or surrounded by intact gray matter. Small lesions are especially problematic in this respect. Also, because of the limited spatial resolution of PET and partial volume averaging, FDG uptake can be underestimated in small nodular lesions or in a thin tumor rim. The best assessment of FDG uptake in brain tumors is obtained if the location and gross structure of the tumor are accurately known. This information is best acquired by digital image coregistration with MRI. Transaxial T1-weighted MR images obtained with gadolinium enhancement are typically used for this purpose. Several techniques are available for coregistration of PET and MR images (see section 4.7). It should be borne in mind, however, that high FDG uptake is not specific for brain tumors, but may also be seen in florid inflammatory lesions (probably due to glycolysis in granulocytes) in patients with focal epilepsy during seizures (which may be subclinical), and in recent ischemic infarcts with non-oxidative glycolysis.

The glucose consumption in brain tissue is reduced in most patients with malignant brain tumors [645]. This reduction appears to be correlated with prognosis [646]. It may also be influenced by corticosteroid treatment, which is often used to treat edema in advanced glioma [647, 648]. FDG uptake may remain higher in brain tumors than in normal brain during hyperglycemia [649]. However, investigation of brain tumors during induced hyperglycemia is not recommended because there is a general reduction of the signal-to-noise ratio.

2.3.1.2 Amino Acid Uptake

In most brain tumors uptake of amino acids is elevated. This is probably due to increased carrier-mediated transport at the blood–brain barrier (BBB), rather than to gross BBB breakdown or increased incorporation of amino acids into tumor tissue. Thus, increased uptake is also seen in most low-grade gliomas in the absence of BBB damage; this is a substantial advantage of PET with amino acid tracers over CT, MRI, and FDG-PET [650–652]. ^{11}C-Methionine (MET) has been widely used for brain tumor imaging for several years, but other ^{11}C-labeled amino acids and ^{18}F-labeled analogues, such as ^{18}F-fluoro-tyrosine and O-(2-^{18}F-fluoroethyl)-L-tyrosine, have also been studied (see section 3.7). An advantage of tumor imaging with labeled amino acids such as MET is that compared with FDG, their uptake in normal gray matter is low. This improves the contrast between tumor and normal tissue. A relative disadvantage is that the 20-min half-life of the ^{11}C radiolabel requires an on-site medical cyclotron for tracer synthesis; FDG, with its 110-min half-life, can be provided from regional distribution centers.

In gliomas, MET uptake is greater in high-grade than in low-grade tumors [650, 653–655], and it also correlates with prognosis [656]. There is a correlation between MET uptake and microvessel density in gliomas classed as grades II–IV, indicating a relationship between angiogenesis and amino acid transport and uptake [657]. Some investigators have reported that MET is better than FDG for certain applications, e.g., definition of tumor margins. In contrast-enhancing gliomas, the spatial extent of increased MET uptake is larger than that of contrast enhancement [658, 659] and may include not only solid tumor but also the surrounding zone of tumor infiltration (Fig. 2.24). In a study of lesions that were hypo- or isometabolic with FDG in relation to normal gray matter there was increased MET uptake in 22 out of 24 gliomas, whereas in benign lesions uptake was decreased or normal [651]. MET uptake differs, however, with tumor type: in oligodendrogliomas uptake tends to be higher than in astrocytomas of the same histological grade, although they are clinically somewhat less aggressive than the latter [650, 655, 660]. MET uptake is increased in other malignant intracranial tumors, including lymphoma, metastases, leptomeningeal carcinomatosis, and hemangiopericytoma (Fig. 2.25), but benign meningiomas and hemangioblastomas also have high amino acid uptake (more than 2.5-fold that in normal tissue)

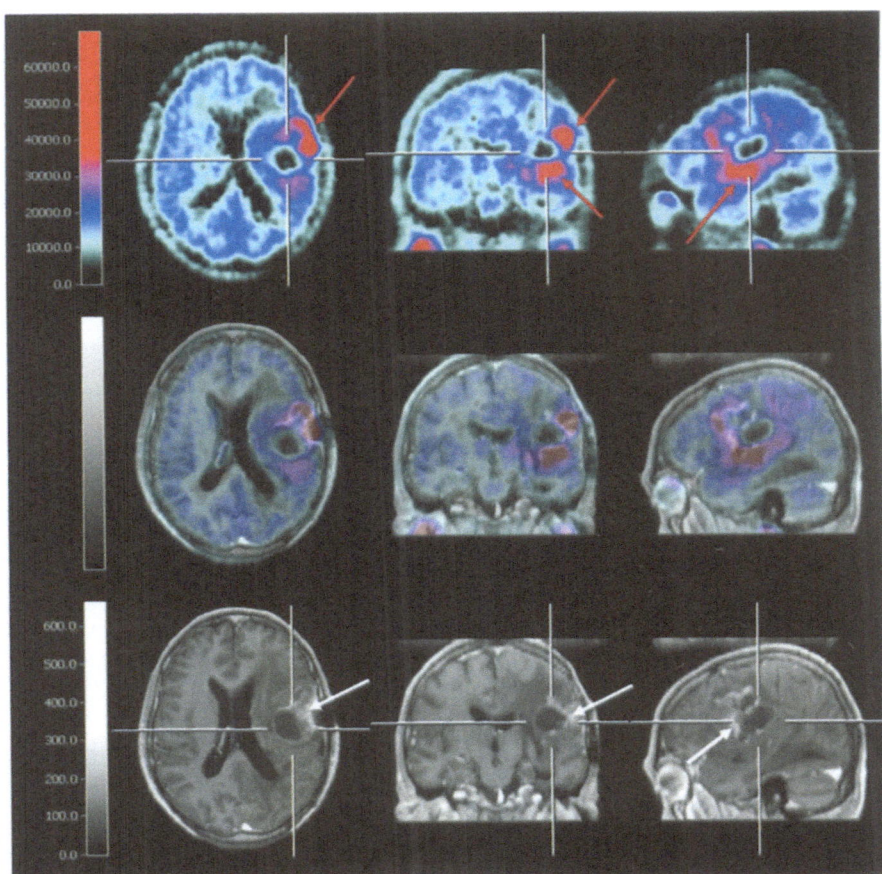

Fig. 2.24. ^{11}C-Methionine (MET) uptake in an anaplastic oligodendroglioma with a central cystic lesion is shown in three orthogonal slices (*top row*). Compared with contrast enhancement on MRI (*bottom row*, fusion image in *middle row*), the volume of significantly increased MET uptake is much larger and also includes low-grade tumor parts and tumor infiltration of brain tissue

[661–665], and moderately increased uptake (typically 1- to 2-fold) is also seen in neuromas [661–665]. Thus, tumor grading with MET requires knowledge of the histological tumor type.

A correlation between MET uptake and survival was observed in gliomas [666], even within groups of the same histological grade [656], and in childhood brain tumors [667]. In a series of 89 low-grade gliomas, MET uptake was a significant predictor of survival in patients with astrocytomas and oligodendrogliomas [668]. The results suggest that MET uptake could be useful for guiding therapy,

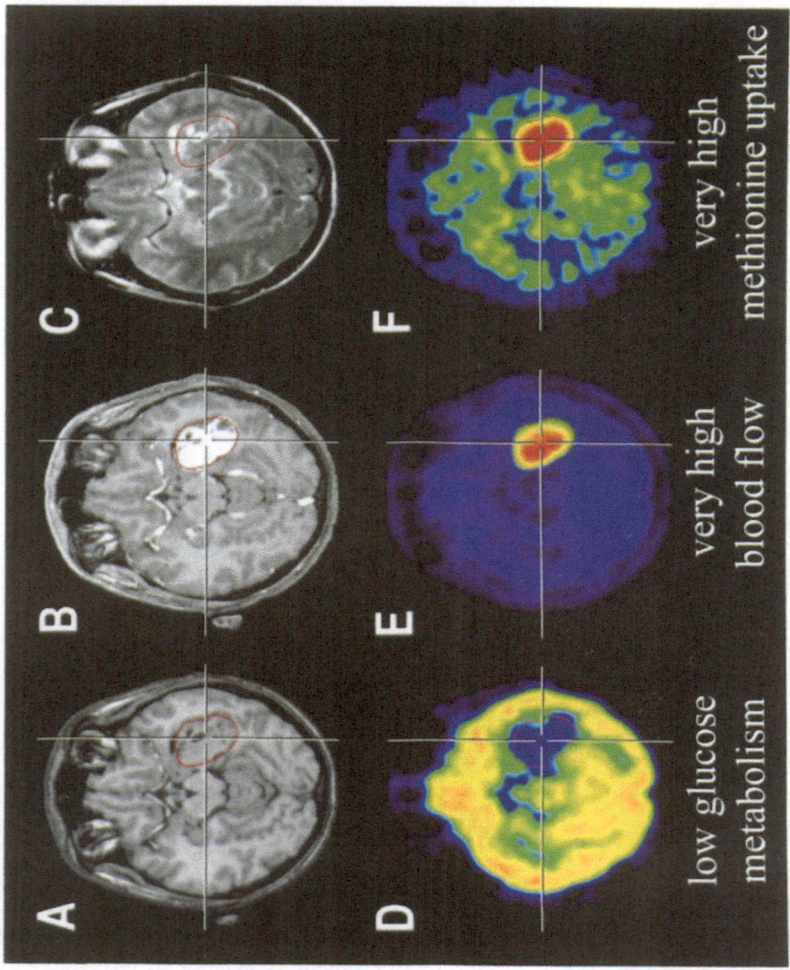

Fig. 2.25 A–C. Coregistered MRI (A T1-weighted, B with contrast enhancement, C T2-weighted), D low glucose metabolism, E very high blood flow, and F high MET uptake in hemangiopericytoma. (From [665], with permission)

because tumor resection is associated with a better prognosis in patients with high rather than low methionine uptake.

Amino acid PET may also help to differentiate between neoplastic and non-neoplastic lesions [651], even if tumor infiltration is diffuse without macroscopic solid tumor nodules, as is the case in gliomatosis cerebri [669, 670]. At a threshold of 1.47-fold uptake of ^{11}C-methionine in tumors compared with contralateral brain, this distinction was accurate in 79% of subjects in a large clinical series [655]. It must be borne in mind, however, that about 20% of low-grade gliomas remain below this threshold and that some uptake by acute ischemic infarcts, hematomas, and inflammatory lesions may be higher.

Amino acids also accumulate in the normal pituitary. This high normal uptake is difficult to distinguish from pituitary tumors with increased uptake [671, 672].

2.3.1.3 Nucleoside Uptake

In theory, labeled nucleosides used as indicators of cellular proliferation should provide the information that comes closest to histological grading (see section 3.8). Proliferation determines prognosis in tumor patients, and arresting tumor proliferation is the goal of therapy. In preliminary studies in brain tumor, 2-^{11}C-thymidine showed slightly higher uptake than in normal brain (tumor-to-cortex ratio ≥ 1.2), but no correlation was found between tracer uptake and tumor grade. There is a high background of radioactivity due to labeled metabolites, in particular ^{11}C-CO_2, and validation of the tracer has been hampered by its complex metabolism [673]. More recently, 3'deoxy-3'-^{18}F-fluorothymidine (FLT) has been introduced [674]. It has much higher tumor-to-brain ratios (in the order of 3–10) and fewer confounding metabolites. It has been validated as a proliferation marker in non-CNS tumors [675].

Another positron-emitting nucleoside is ^{124}I-iododeoxyuridine (IUdR), an analogue of thymidine. Its uptake is greater in higher grade malignancies, and the long half-life of ^{124}I (4.2 days) permits imaging to be delayed until labeled metabolites have been washed out [676].

2.3.2 PET-Guided Stereotactic Biopsy

There is some uncertainty about the best management of low-grade gliomas, because there has not yet been a convincing demonstration that treatment before malignant progression improves the prognosis. Therefore, some clinicians prefer to defer treatment with resection or radiation until malignant transformation is evident [677]. Stereotactic biopsy to obtain a histological diagnosis and thus help guide therapy in these and other situations may yield samples that are not representative of the most malignant part of the tumor [678]. This is a particular problem in higher grade lesions because of their heterogeneous histology [615]. It has been demonstrated that the most metabolically active tumor region on PET (FDG or MET) is the best location for removal of a stereotactic biopsy specimen [679–682].

2.3.3 Differentiation Between Recurrent Tumor and Radiation Necrosis

Detection of recurrent tumors is an important issue, because growth of a recurrent tumor will lead to worsening of symptoms and ultimately to the patient's death. FDG-PET has been used successfully for this purpose in high-grade tumors [632, 646, 683] (Fig. 2.26) and for the detection of malignant progression in low-

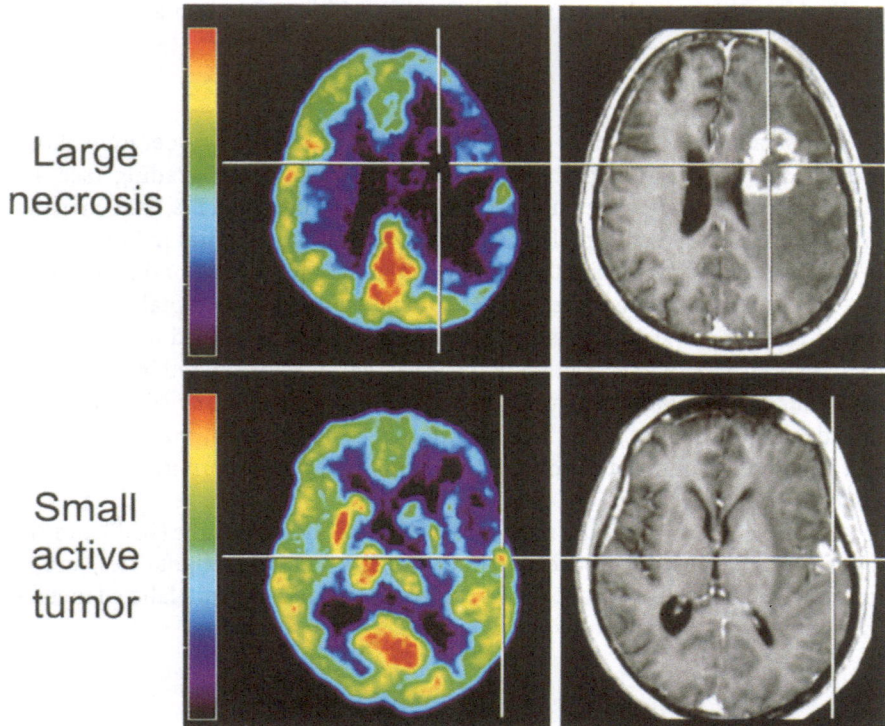

Fig. 2.26. FDG-PET and coregistered MRI in a patient with a large contrast-enhancing radiation necrosis (*top row*) and a small recurrent active carcinoma metastasis (*bottom row*)

grade gliomas [684]. However, differentiation from necrosis is difficult in gliomas, even with histological samples, because high-grade gliomas (II–IV) almost always contain residual tumor cells even in the absence of a solid tumor, and there is often partial necrosis, whether this has occurred spontaneously during tumor progression or is due to therapy. Frequently, these different conditions coexist in different parts of a tumor [685]. Therefore, the question "recurrent tumor or radiation necrosis" is an oversimplification [686, 687]. Nonetheless, it is important to decide whether active tumor growth or infiltration that could be halted by therapy is in train, as opposed to radiation necrosis. Both can present with clinical deterioration, and they are often impossible to tell apart by structural brain imaging. Metabolic imaging may be a more powerful tool for patient management than histological examination, which is the current gold standard. A definitive answer, however, would require a large, prospective clinical trial. The field currently has to rely on data comparing PET imaging with histopathological findings or clinical course. Initial optimistic reports noted high FDG uptake in tumor recurrences,

Table 2.4. Studies on differentiation between recurrent tumor and radionecrosis

Tracer	N	Sensitivity	Specificity	Lesion type	Remarks	Reference
FDG	47	75%	81%	Malignant tumor		[834]
FDG	15	43% (6/14)	100% (1/1)	Glioma	_a	[835]
FDG	84	73%	56%	Malignant tumor		[836]
FDG	38	88% (15/17)	81% (17/21)	Glioma		[837]
FDG	21	81% (13/16)	40% (2/5)	Tumor		[838]
FDG	9	80% (4/5)	100% (4/4)	Tumor	1	[695]
FDG	21	64% (9/14)	71% (5/7)	Metastases	1	[839]
FDG	54	83% (5/6)	96% (46/48)	Metastases		[840]
MET	12	100% (5/5)	86% (6/7)	Glioma	1	[841]

a All patients had histological verification of diagnosis

with areas of necrosis having low uptake [688–690]. More recent studies, however, have shown that there is considerable overlap [691] (see Table 2.4).

Although few data are available for MET compared with FDG, MET seems to have better sensitivity and specificity [692]. A recent study in 125 tissue samples obtained from 29 patients indicated that sensitivity for detection of recurrent tumors or active infiltration is 85% at a specificity of 93% [693]. Similar figures (78% sensitivity at 100% sensitivity) were observed in another series of 21 patients [694]. They also correspond to the results of previous studies, which indicated sensitivity and specificity of approx. 80% [655, 695]. Therefore, MET-PET is being used by several groups as a sensitive clinical tool to detect recurrent glioma (see Table 2.5 for a practical guide). MET uptake is substantially lower or absent in necrotic brain areas, but mildly increased uptake (up to 1.5-fold of that in contralateral brain) may be present (Fig. 2.27). It is probably due to passive diffusion of the tracer across the damaged BBB in necrosis. More studies are needed to establish specificity. Compared with coregistered contrast-enhanced MRI, the foci of highest methionine uptake are often found not to coincide with contrast-enhancing lesions but to be close to these. Case reports have shown that this may indicate an active tumor with high MET uptake in the vicinity of an area of necrosis with contrast enhancement [696].

2.3.4   Monitoring of Therapy

It is important to be able to monitor the effectiveness of treatment for brain tumors, particularly gliomas, because many tumors do not respond at all and the responses obtained are often incomplete. Continuation of ineffective chemotherapy has many side effects, including cumulative bone marrow toxicity, which could

Table 2.5. Practical guide for clinical use of MET-PET to detect recurrent glioma (based on report in [842] and continuing clinical experience)

Typical clinical situation: Patient with glioma is clinically stable or has mild symptoms, but MRI shows nonspecific signal changes that could indicate residual or recurrent tumor; therapeutic options (surgery, radiation, or chemotherapy) are available if active tumor can be identified

Patient preparation: Low-protein diet ("tea and biscuit") on the day of PET examination

PET: i.v. injection of approximately 740 MBq ^{11}C-methionine, data acquisition 20–60 min after injection

Image reconstruction with appropriate corrections for scatter and attenuation

Coregistration with MRI or CT whenever possible

Image evaluation:

- If increased uptake in tumor area is visible, obtain quantitative value (with an ROI of 8 mm diameter) in the hottest region and compare with the value in a mirror ROI (index = tumor ROI / contralateral ROI)
  - If index is ≥1.5, recurrent or residual tumor or active tumor infiltration is probable (except for situation described in next point). For accurate localization and guidance of neurosurgical intervention use coregistration with MRI (if not in exactly same place as contrast enhancement copies of coregistered images are necessary for intervention planning)
  - If index is only slightly above 1.5 and coincides with strong contrast enhancement, uptake is probably unspecific and due to passive tracer diffusion; recurrent or residual tumor is possible but not probable
  - If index is >1 but <1.5, this is a nonspecific increase and active tumor is unlikely (except for next point)
  - If index is <1.5 but close to 1.5 and contrast enhancement is absent, specific uptake is likely and the patient should be closely monitored for tumor progression (repeat study suggested in 3 months)
  - If there is no visibly increased uptake, active tumor is unlikely

be avoided by early detection of inefficacy. Similarly, in radiotherapy, dose escalation beyond standard tumor doses involves an increased risk of side effects and may be warranted only if viable tumor tissue is still present after standard therapy. This approach has successfully been implemented in a pilot study that used FDG-PET to determine the presence of residual tumor [697].

The probable gold standard for assessing therapy is the measurement of proliferation rates; preliminary results with FLT suggest that this is possible [698]. The role of FDG in monitoring therapy is still being defined after a few initial studies showed some correspondence between a change in FDG uptake during chemotherapy and outcome in the case of glioblastoma [646] and of medulloblastoma [632]. In a study of experimental brain tumor in the rat, reduced glucose consumption, but not reduced protein synthesis, was seen after 7 days of effective chemotherapy, in correlation with reduced DNA synthesis [699]. This finding corresponds with the results of some clinical studies in brain tumors [700]. However,

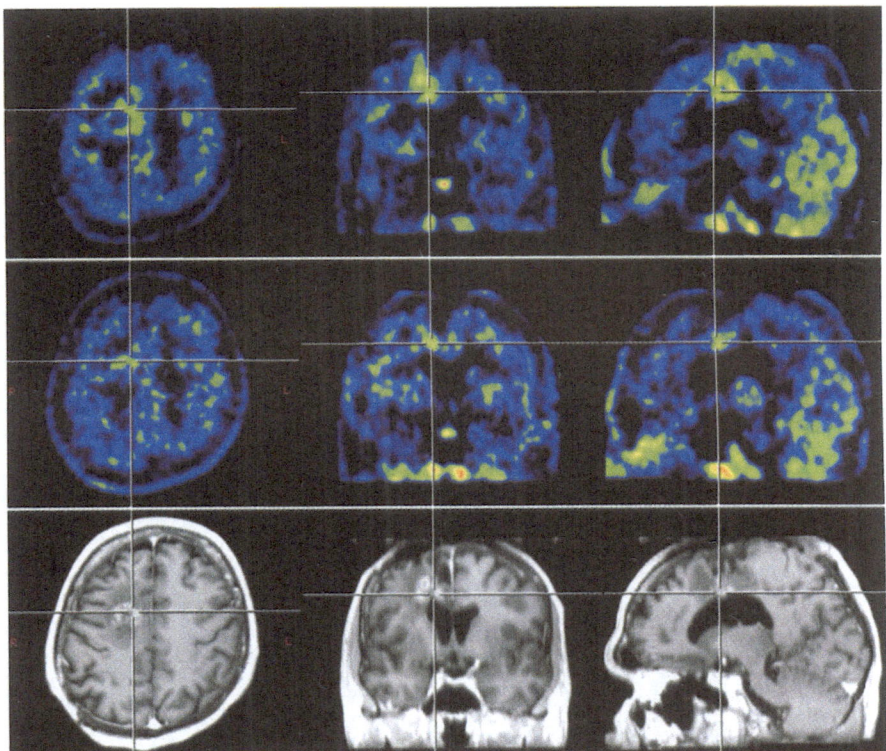

Fig. 2.27. Radiation necrosis after treatment of a low-grade astrocytoma with 125-iodine seeds with contrast enhancement on MRI (*bottom row*, lesion marked by *cross hairs* on orthogonal slices). There is some increase of MET uptake (*top row*), but it does not reach 1.5-fold the uptake in normal brain tissue and is therefore considered unspecific and probably due to damage to the blood–brain barrier (BBB). At follow-up 3 months later (*middle row*), MET uptake has decreased further

effective chemotherapy and radiation may lead to an initial transient increase of FDG uptake [701–703], so that the timing of PET is critical. Reduced FDG uptake has also been observed after radiotherapy [704], but in a study of experimental systemic tumors in the rat FDG uptake in a tumor remained high 6 days after irradiation, even with complete tumor regression [692]. This increase may be related to a local accumulation of metabolically active macrophages in areas of radiation-induced tumor necrosis.

In the same study, thymidine and MET uptake significantly decreased [692], suggesting that these tracers are better suited than FDG to monitoring therapy. In a clinical study of low-grade gliomas, MET uptake decreased after interstitial irradiation (brachytherapy) [705]. Another study showed a correlation between high pretreatment uptake of MET and reduction in MET uptake after radiotherapy

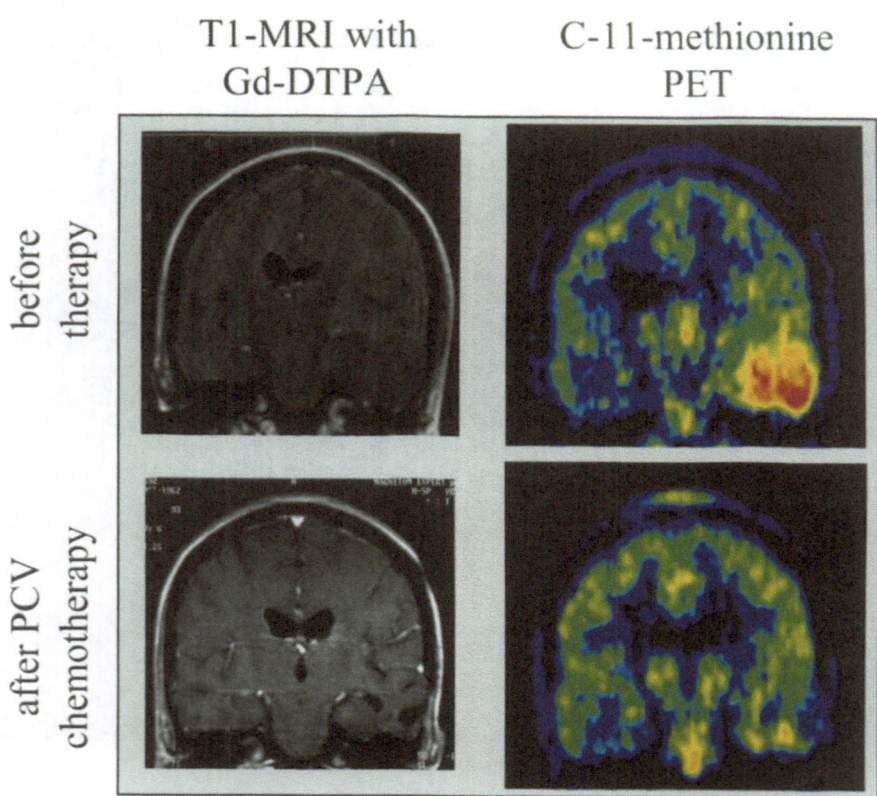

T1-MRI with
Gd-DTPA

C-11-methionine
PET

before therapy

after PCV chemotherapy

Fig. 2.28. Effect of chemotherapy on a recurrent oligoastrocytoma: Before therapy MET uptake was increased (in absence of contrast enhancement on MRI). It is reduced to values close to that in normal brain tissue after brain therapy with corresponding disappearance of the mass effect on MRI

[706]. An interesting observation was reported in lymphoma, where MET uptake persisted (corresponding to the usual poor clinical outcome) after radiation therapy, whereas CT contrast enhancement rapidly disappeared [662]. Data are fewer and less encouraging for ^{11}C-tyrosine; in a small series of brain tumors its uptake did not change after irradiation [707]. MET-PET has definite potential for monitoring the effectiveness of glioma therapy and should be studied in larger prospective series (Fig. 2.28).

Uptake of FDG, amino acids, and dopaminergic ligands in pituitary tumors decreases with effective therapy by irradiation or dopamine agonists, such as bromocriptine and octreotide [641, 708–710].

A special situation exists in the therapy of brain tumors with heavy ions, which may generate positron-emitting atoms in brain tissue by nuclear reactions [711,

712]. This can be used to monitor the effective field of radiation with PET with no need for radiotracer administration.

2.3.5   Identification of Hypoxic Tumor Tissue

Oxygen consumption is lower in most brain tumors than in normal brain [622, 713]. This is in contrast to their elevated glucose consumption, suggesting a substantial proportion of nonoxidative glycolysis with increased lactate production [620]. In site of this, tissue pH is usually alkalotic rather than acidotic in brain tumors [714,715]. Thus, there is no indication of general hypoxia in brain tumors, but owing to tumor inhomogeneity and irregularity of vessels there is reason to assume that certain areas within tumors may be hypoxic.

Hypoxic tissue in tumors is relatively resistant to radiation therapy. Therefore, it would be useful to identify hypoxic tissue to make it possible to predict the effectiveness of radiotherapy and to develop better treatment by improving tissue oxygenation. Tumor hypoxia is also a major stimulus for vascular proliferation and is therefore of interest in that it would help us to understand tumor growth as well as therapy. ^{18}F-Misonidazole and other ligands used to image hypoxic tissue are derived from misonidazole (MISO). MISO diffuses slowly into the brain with a uniform distribution and is selectively retained in hypoxic cells. MISO distribution in experimental tumors is variable, and tumor concentrations increase with time relative to that in brain [716]. ^{18}F-MISO has been used in patients with glioma [717], but its ability to predict radiation response has not yet been established.

^{18}F-4-Bromo-1-(3-fluoropropyl)-2-nitroimidazole has higher lipophilicity and brain uptake than ^{18}F-MISO, and may therefore be more useful than MISO for imaging hypoxia in both brain tumors and cerebrovascular disease [718]. Another candidate is the lipophilic molecule EF5, which has been used to measure tumor hypoxia in animals and humans using immunohistochemical methods; it has been labeled with ^{18}F [719].

2.3.6   Improving and Planning Therapy

Surgical resection should be as complete as possible and at the same time respect functionally intact brain tissue to avoid deterioration of neurological function. To achieve this goal, PET and MRI images need to be coregistered and presented in a format that facilitates planning of therapy (cf. Figs. 2.22–2.27). It has also been shown that integration of PET adds a new dimension to neuronavigation that may permit more effective surgical treatment than is possible with MRI and CT alone [720]. Demonstration of the clinical benefit of this comprehensive approach requires prospective clinical studies.

Drug delivery to brain tumors depends on blood flow, which can be measured with PET (see section 3.2) and is highly variable in malignant gliomas [721], with a tendency towards higher blood flow in oligodendroglioma than in astrocytoma [722].

Another important factor is drug transfer across the BBB, or better the blood–tumor barrier, which may be intact in low-grade gliomas but is usually disrupted in high-grade gliomas and absent in nonglial tumors, such as metastases and meningiomas. These barriers can be assessed with ^{68}Ga-EDTA [723, 724] or ^{82}Rb [725, 726] (see section 3.1). A reduction in permeability has been observed following corticosteroid treatment [727, 728], and increased permeability occurs after infusion of hyperosmotic agents [729–731]. Although similar results can also be obtained by analysis of contrast agent uptake with CT or MRI, the short physical half-life of ^{82}Rb (75 s) permits rapid repeat measurements of permeability.

For many drugs, including the nitrosoureas as commonly used chemotherapeutics for brain tumors, the BBB is not a major barrier but can be crossed by diffusion or by carrier-mediated transport. The actual transfer of chemotherapeutics to brain and brain tumors can be assessed by labeling of these drugs, as has been done for the nitrosoureas carmustine (BCNU), lomustine (CCNU) and SarCNU [732, 733], and for temozolomide [734]. Pharmacokinetics of superselective intra-arterially and intravenously administered ^{11}C-BCNU have been evaluated by PET [735]. On the basis of experimental studies, it has been postulated that temozolomide undergoes decarboxylation and ring opening in the 3–4 position to produce the highly reactive methyldiazonium ion that alkylates DNA. A PET study employing temozolomide labeled by ^{11}C in two different positions confirmed this mechanism in human gliomas [736].

Resistance to other chemotherapeutic drugs (which are not commonly used in brain tumors) may be caused by multiple drug resistance (MDR). This is mediated by P-glycoprotein (Pgp), which is encoded by the *MDR1* gene in humans, is highly expressed at the BBB, and can be studied by ^{11}C-verapamil [737]. Pgp in cell membranes pumps drugs out of cells and can be studied with positron-emitting substrates for Pgp, such as ^{11}C-verapamil [737]. Other tracers used to study MDR include ^{11}C-colchicine [738], N-^{11}C-acetyl-leukotriene E4 [739], and ^{18}F-paclitaxel [740].

The goals of radiotherapy include stopping the proliferative potential of low-grade tumor parts and invading tumor cells outside the solid malignant tumor core. With conventional techniques, including CT and MRI, it is impossible to identify these areas, and standard "safety margins" are often used to define the irradiation field. PET has the potential to contribute to a better definition of radiation treatment [697, 741–743]. There are also indications that pretreatment methionine uptake may be a marker for the radiosensitivity of low-grade gliomas [706]. Measurement of uptake of fluorine-18-labeled L-fluoroborono-phenylalanine (L-^{18}F-10B-FBPA) in high-grade gliomas was used for accurate neutron dosimetry during boron neutron capture therapy [744, 745]. The efficacy of integration of PET into radiation planning should be tested by appropriate studies.

2.3.7   New Therapies

Gene therapy involves local gene transfer to change genes or their expression in tumor cells [746]. One major issue is the effectiveness of this new therapeutic approach. As explained in section 3.9, the expression of herpes virus thymidine kinase (HSV-TK) and other nonmammalian enzymes can be imaged to assess gene transfer. This technique has been used in experimental studies of brain tumor treatment by adenovirally mediated gene transfer [747], and for replication-conditional, oncolytic herpes simplex virus type 1 mutant vectors [748]. In a preliminary study in humans with glioblastoma, the efficacy of liposomal vector-mediated HSV-TK transfer and treatment with gancyclovir was examined with ^{124}I-labeled 2'-fluoro-2'-deoxy-1-β-D-arabino-furanosyl-5-iodo-uracil (FIAU) and PET [749] (Fig. 2.29). The degree of HSV-TK expression, indicated by retention of phosphorylated radiotracer, correlated with therapeutic outcome.

The goal of antiangiogenic therapy is selective reduction of the blood supply to tumors. Angiogenesis in tumors due to up-regulation of vascular endothelial growth factor (VEGF) and hypoxia-inducible factors was imaged in experimental studies by means of dual reporter genes [750]. VEGF can be blocked by monoclonal antibodies. The in vivo distribution of such an antibody (HuMV833) has been demonstrated in vivo by labeling with ^{124}I [751]. The efficacy of antiangiogenic therapy can be monitored by PET measurements of blood flow and volume to assess the vascular system and of metabolism to assess cellular response. This has been done with ^{15}O-water and FDG in a phase I clinical trial with recombinant human endostatin (rh-Endo), a specific inhibitor of angiogenesis, in patients with systemic cancers [752]. No such trials have been performed to date in brain tumors.

Tumor growth is related to activation of several growth factor receptors that could be blocked by new therapeutic modalities. PET biomarkers targeting the epidermal growth factor receptor tyrosine kinase are being developed but have not yet entered the clinical arena [753]. In this context, expression of transferring

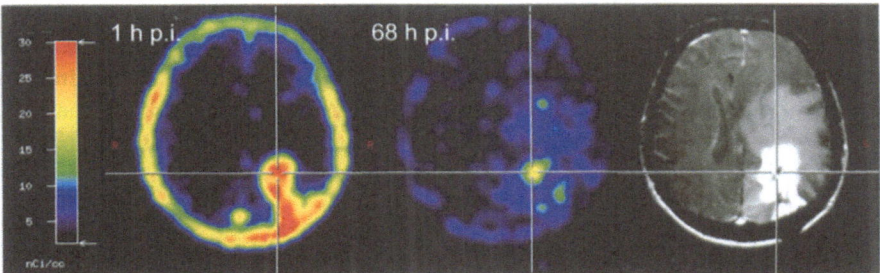

Fig. 2.29. Accumulation of ^{124}I-2'-fluoro-2'-deoxy-5-iodo-1-β-D-arabinofuranosyluracil (FIAU) in human glioblastoma center after liposomal gene transfer of HSV-TK. (From [749], with permission)

receptors at tumors is also of potential interest and can be imaged by Ga-68-trans-ferrin [754,755] and ^{18}F-holo-transferrin [756]. However, transferring receptors are difficult to assess in brain tumors because they are also expressed at the BBB [757].

2.3.8   Other Tracers for Brain Tumors

Many tumor cells exhibit increased choline transport and phosphorylation with subsequent trapping in the phospholipid pool. ^{11}C-Choline [758] and ^{18}F-fluoro-choline [759] have been used for brain tumor imaging, but target-to-background ratios are low and the potential usefulness of these tracers is unclear.

Somatostatin analogues bind to pituitary tumors and meningiomas, so that tracers labeled with ^{18}F, ^{86}Y, or ^{68}Ga could be used to image them [760–763]. Such tracers could also be used for planning isotope therapy of neuroendocrine tumors with yttrium-90-labeled octreotide [764]. Progesterone receptors on menin-giomas [765] could potentially be imaged by ^{18}F-labeled progestins that are being developed for breast cancer imaging [766]. Nonsecreting pituitary tumors (but not meningiomas) express high levels of MAO-B that can be imaged with deuterium-substituted ^{11}C-L-deprenyl [767]. Prolactinomas have more dopamine receptors than do normal pituitary and nonsecreting pituitary tumors. They have been imaged and measured with RAC [768], ^{18}F-fluoroethylspiperone (FESP) [769], and ^{11}C-N-methylspiperone (NMSP) [770].

2.3.9   Extracranial Tumors of the Nervous System

Schwannomas of the extremities have been imaged with ^{18}F-fluoro-methyl-tyro-sine [771] and with FDG [772, 773]. Malignant peripheral nerve sheath tumors, which occur most frequently in patients with neurofibromatosis, show high FDG uptake such as has not been seen in benign neurofibromas [774]. Pheochromocy-tomas can be localized with ^{11}C-hydroxyephedrine [775], a sympathomimetic amine, and 6-^{18}F-fluorodopamine [776–778], a substrate for the norepinephrine transporter. Potential alternatives are the norepinephrine analogue p-^{18}F-fluo-robenzylguanidine (PFBG) [779] and, though these are less specific, FDG [780, 781], FDOPA [782], and Rb-82 [783]. 4-^{18}F-Fluoro-3-iodobenzylguanidine, an iodobenzylguanidine analogue, has been explored for use in neuroblastoma [784, 785]. Some spinal cord tumors have been imaged with MET [786], but this has not generally been successful.

2.3.10   Detection of Primary Tumors

Neurological paraneoplastic syndromes (e.g., Lambert-Eaton syndrome, cerebellar degeneration, limbic encephalitis) often occur at a time when the tumor causing the syndrome has not yet been identified [787]. Whole-body FDG-PET offers an efficient way of detecting these tumors [788–790] even if standard clinical and radiological methods have failed. FDG-PET is also highly efficient in locating primary tumors in patients with brain metastases of unknown origin [791].

2.3.11   Activation Studies

Brain tumors close to eloquent brain areas present particular difficulties for surgical treatment. Intraoperative location of critical motor and language function can be achieved by electrophysiological methods, but in the case of language centers surgery must be performed with only local anesthesia, which is cumbersome. Exact presurgical location is therefore an important clinical goal. Infiltrative glioma growth poses a particular challenge for tailoring of resection, because infiltrated tissue may still be functional [792]. Since even total resection cannot prevent tumor recurrence in most cases of glioma, the emphasis must be on preservation of function. Compared with the more commonly used fMRI, PET provides a more physiologically specific and robust signal and is therefore a viable alternative to fMRI [793–796]. ^{15}O-Water is the most frequently used CBF tracer for this purpose, typically allowing up to 12 CBF measurements (e.g., 3 different conditions with 4 replications each; see Table 2.6 for a typical activation study protocol). Motor and language tasks can also be performed during the first 30 min after injection of FDG to record functional changes in local CMRglc. FDG activation studies are often performed in a separate session from the resting reference study, but there have also been suggestions for double-injection protocols to combine the activation and the resting condition in one session [797]. Coregistration and fusion image display with 3D-MRI are necessary for accurate anatomical localization [798–800] (Fig. 2.30), and integration into intraoperative neuronavigation is possible [801].

The locations of functionally activated areas may be altered in brain tumor patients as a result of several effects. First, there may be a mass effect that displaces the motor cortex, where the functional activation can be found at the anatomically expected location. Thus, it is important to have coregistered MRI scans and image fusion for direct comparison, although accurate anatomical localization may still be difficult if the sulci can no longer be identified because of the mass effect and edema. Direct effects of the tumor on motor or language cortex usually lead to reduced activation, often associated with impaired function. There have also been a few instances of false-positive activation, in particular in the vicinity of hyperperfused lesions [802, 803]. Functional activations may occur at atypical anatomical locations, apparently representing reorganization of functional networks [804,

Table 2.6. Practical guide for an ^{15}O-water activation study for localization of language and motor areas (based on [804] and related experience)

Typical clinical situation: The 3D MRI of a patient with a brain lesion considered for surgery (e.g., MPRAGE sequence with 1- to 1.5-mm-thick slices covering the entire brain) shows the extent of the lesion and defines cortical anatomy. If MRI is not sufficient to demonstrate lesion extent, an amino acid PET may also be necessary. An ^{15}O-water PET activation study is performed to localize motor and language areas and to compare them with lesion location and extent, which is useful in the planning of surgery. The patient's condition allows the performance of hand movements and a verb generation task.

No specific patient preparation necessary.
A set of twelve ^{15}O-water PET examinations is performed under three different conditions:
– A: Resting
– B: Verb generation task: the patient is asked to produce a semantically related verb for each item from a list of nouns with high association content that is presented to him/her at a rate of approx. 10 words per min
– C: Unilateral finger tapping

Each PET study is done in 3D mode and involves i.v. bolus injection of approx. 370 MBq ^{15}O-water and data acquisition over approx. 45 s. Conditions are performed in a balanced order (e.g., ABCBCACBABAC) with intervals of 6 min. between subsequent conditions (resulting in an approx. total study time of 90 min).

PET images of ^{15}O-water distribution are reconstructed (with appropriate corrections for attenuation and scatter) and coregistered to each other to adjust for minor head movements. Images are smoothed (e.g., by Gaussian filtering with FWHM 8 mm), brain masking is performed, and local blood flow increases during verb generation and during finger tapping (relative to the resting condition) are determined.

Activation images are coregistered to individual MRI and displayed in fusion mode to illustrate the topographical relation between anatomy, neuronal function, and lesion.

805]. The potential for reorganization is greater during early development in childhood than in later life [806]. The implications of these changes seen in activation studies for location of critical brain functions require further study. Not all areas that are activated during functional tasks are essential for task performance: a lesion would severely impair function, but might also include areas that are secondarily activated during task performance.

2.3.11.1 Motor Function
The main clinical issue is usually localization of motor cortex (arm and leg). Useful activation tasks are repetitive finger tapping and foot movements, which lead to reliable activations centered in the respective areas of contralateral motor cortex in healthy subjects [807, 808]. There are also regular activations of supplementary motor cortex and of cerebellum, more on the ipsilateral side [809, 810]. In patients with brain tumors, displacement of functionally activated areas has

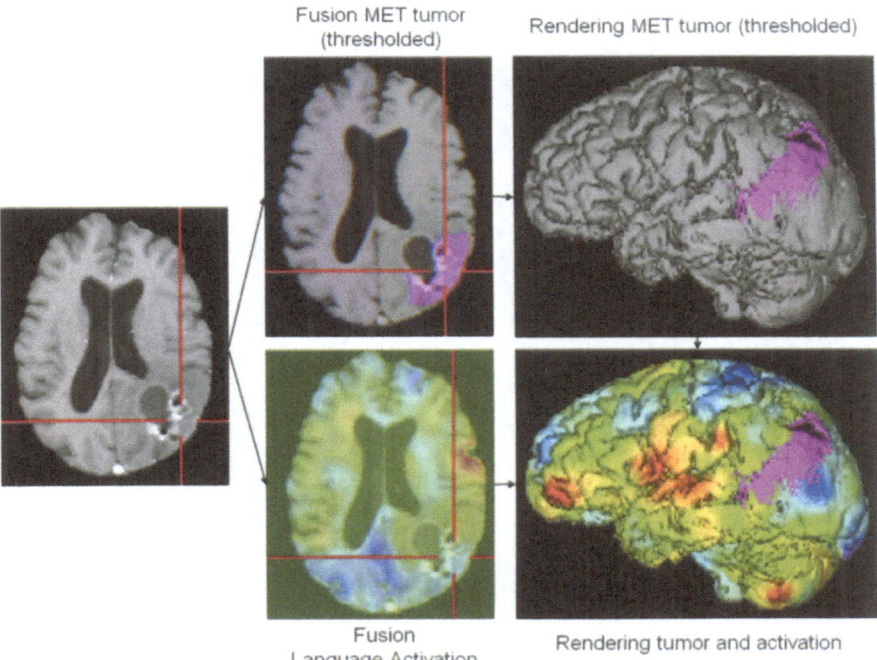

Fusion MET tumor (thresholded)

Rendering MET tumor (thresholded)

Fusion Language Activation

Rendering tumor and activation

Fig. 2.30. Multimodal image coregistration, fusion, and rendering to present comprehensive structural and functional preoperative findings in 3D mode. Image processing starts with removal of extracerebral structures from 3D MRI (*left*). Image fusion with MET-PET (thresholded at 1.5-fold normal brain uptake) shows the extent of glioma infiltration (*top middle*), which can then be displayed also by surface rendering (*top right*). PET activation data are also coregistered with MRI (fusion display, *bottom middle*). All three data sets (MRI, activation and thresholded MET-PET) are then displayed together (surface rendering, *bottom right*). The complete volume data of this patient and an animated video clip of the main findings are presented on the enclosed CD-ROM

been observed along the dorsoventral dimension of the precentral gyrus that exceeds anatomical displacement due to mass effect [811] (Fig. 2.31). In patients with cortical lesions that cause contralateral spastic paresis the activation of motor cortex may be largely abolished, often being replaced by more intense activation of secondary motor areas and motor cortex ipsilateral to the paretic limb. This has been studied more extensively in stroke than in brain tumors, and the functional implications are not yet entirely clear [812]. A comparison of PET findings with intraoperative electric stimulation and transcranial brain stimulation found overlapping results in 31 of 49 studies, and neighboring location of motor areas in 14 [793].

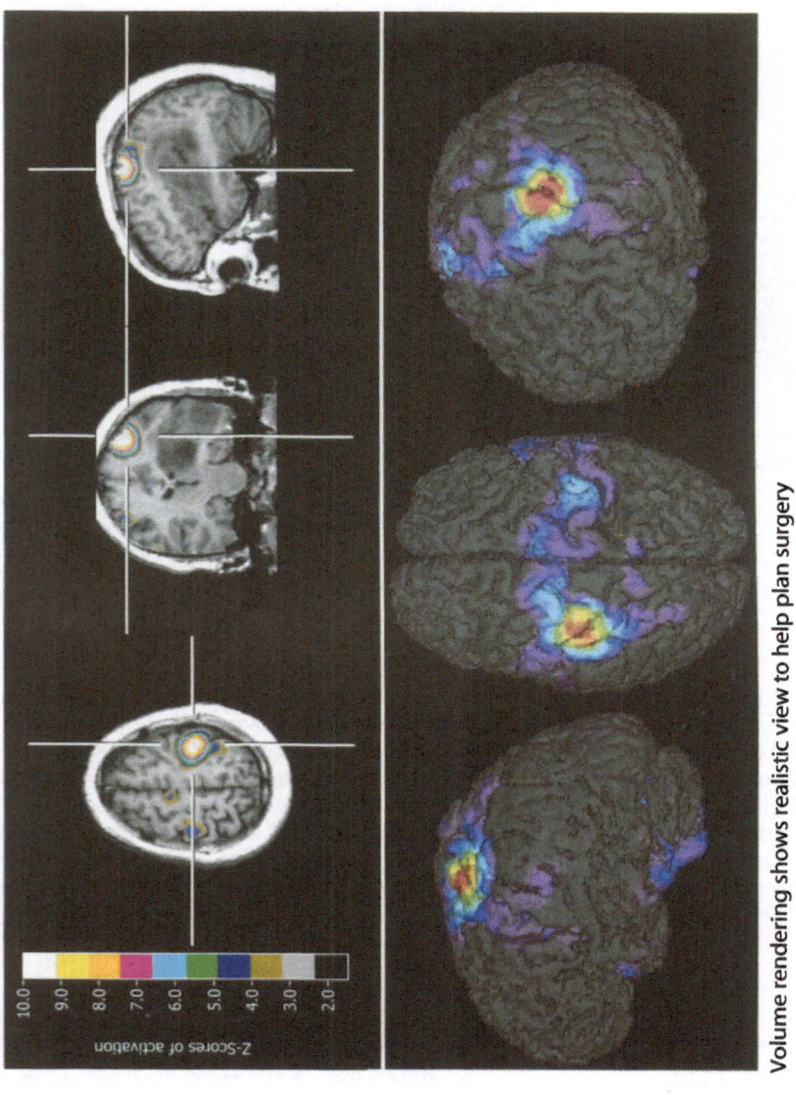

Volume rendering shows realistic view to help plan surgery

Fig. 2.31. Activation of the primary sensorimotor hand region by finger tapping in a patient with a large left fronto-temporo-parietal glioma. In the *top row*, orthogonal fusion images of a ^{15}O-water PET activation study with T1-weighted MRI are shown. In the *bottom row*, PET activations are shown on the 3D brain surface reconstructed from MRI

2.3.11.2 Language

The capacity to understand and to speak language is strictly lateralized in most subjects to the dominant hemisphere. With few exceptions this is the left hemisphere in right-handers, whereas in left-handers language may be represented in either hemisphere or even bilaterally [813]. In addition to language dominance, details of the anatomical localization of sensory and motor language areas (Wernicke's and Broca's), which may vary unpredictably even in normal individuals

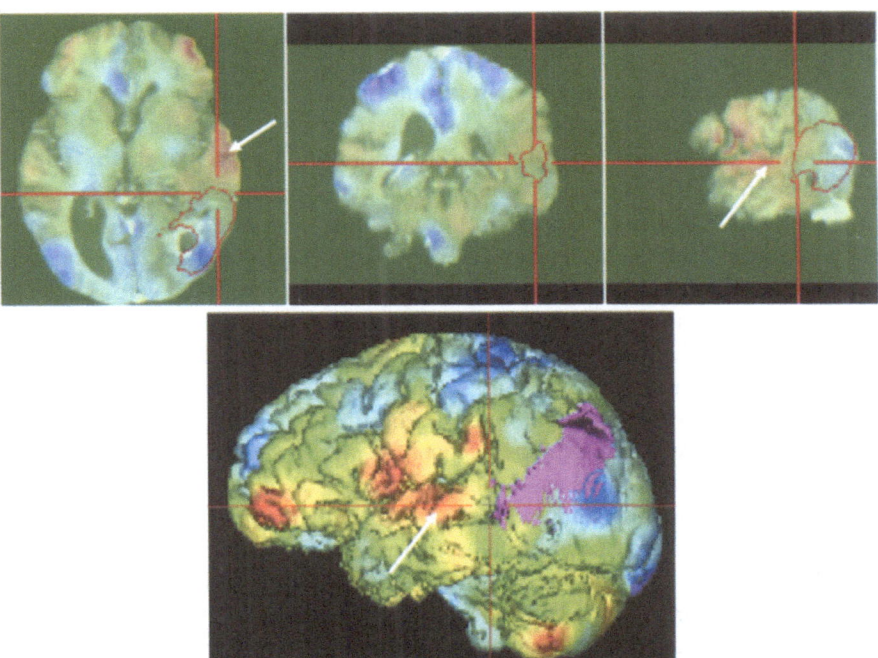

Fig. 2.32. Preoperative language activation by verb generation in left temporo-parietal malignant glioma, shown in orthogonal slices (*top*) and in a corresponding surface rendering (*bottom*) with point of correspondence marked by *red crosshairs*. The effects of language activation are shown by *rainbow color coding*, with distinct activation of Wernicke's area in superior temporal cortex (*white arrows*). Other major focal activations are located in the foot of motor cortex, inferior frontal cortex, and cerebellum. Superior temporal cortex activation is close to the anterior border of tumor infiltration (marked by *red contours* in the orthogonal slices and by *pink color* in the surface rendering), but tumor and functional brain still appear separated and tumor resection is possible

[814–816], are also of interest for surgical planning in patients with tumors in inferior frontal and temporoparietal association areas. Wada testing has been used as the gold standard for determination of language dominance, and PET activation studies have been validated against this gold standard in several studies [794, 817–819]. The more difficult question of whether the extent of language activation areas can provide guidance for tailoring resections has also been addressed [795, 820, 821]. Correspondence with intraoperative electrical stimulation is highly significant, although activated areas tends to include some sites where there is no language disturbance with electrical stimulation, and there are also a few sites with electrical language disturbance but no significant preoperative activation effect. Thus, language activation studies are helpful in the planning of surgery (Fig. 2.32), but cannot currently completely replace intraoperative monitoring.

A considerable variety of language activation paradigms have been tried for localization of language function [822–825]. Automated speech tasks are not very useful to localize language areas [826]. Passive listening activates temporal language areas in superior temporal cortex, but this is not always significant in individual subjects. More active semantic or language production tasks (e.g., the generation of semantically related verbs in response to presentation of nouns) provide more clearly lateralized activations, in particular in inferior frontal cortex of the dominant hemisphere [827] and in most instances also in superior temporal cortex, anterior cingulate cortex, and an adjacent supplementary language area and in cerebellum (predominantly contralateral to the dominant cerebral hemisphere). An essential advantage of PET in its clinical application of language activation studies is that active speaking during language production tasks does not induce technical artifacts (as is common with fMRI) and therefore direct monitoring of task performance is possible even in functionally impaired subjects.

Similar to motor function, there is considerable reorganization of language in patients with brain tumors in the dominant hemisphere [804]. This includes increased activation of secondary language sites and a shift to the right side, which may already be present in subjects with lesions but without aphasia. Lesion-induced plasticity is greater in childhood than in adults [828, 829], and even in adults the rightward shift of language activation tends to be stronger as a consequence of early as against late lesions [830]. It is not yet clear whether such apparent plasticity is sufficient to support language function to an extent that would allow surgical resections of primary language areas that are partially damaged following tumor infiltration.

Bilateral acoustic neuromas may induce deafness that can be improved by cochlear implants. These allow functional speech processing that is associated with activation of classic language areas [831, 832]. There is also evidence for altered functional specificity of the superior temporal cortex, flexible recruitment of brain regions located within and outside the classic language areas, and automatic contribution of visual regions to sound recognition [833].

2.4   Cerebrovascular Disease

Acute cerebrovascular disease, the cause of the clinical syndrome of stroke, is the most common neurological disorder, with an annual incidence of 150–200 per 100,000 in western industrialized countries. There are three main etiological categories: ischemic stroke (70–80%), spontaneous intracerebral hematoma (10–20%), and subarachnoidal hemorrhage (5–10%). The most direct functional measurements of oxygen supply and consumption, the most critical parameter in acute ischemic stroke, have been made with ^{15}O PET (see section 3.3 for methodological background). Functional effects on normal brain tissue have mostly been studied by measurement of CBF or CMRglc.

2.4.1   Ischemic Stroke

The cause of ischemic stroke is a severe and usually sudden decrease of CBF below a level of about 15 ml/100 g per min (a decline by about two thirds from its normal average level of about 50 ml/100 g per min in human brain). It is mostly due to atherothrombotic or embolic occlusion of a supplying artery. The severity and duration of this insult are the major determinants of tissue fate, whether infarction develops or whether recovery follows [843]. In most cases there is a central core of dense ischemia, in which residual flow is very low and the time to infarction consequently short, surrounded by an area of graded and less severe flow disturbance where function is impaired but morphology preserved for an ill-defined period. This area – the penumbra – has the potential for functional recovery providing that local blood flow can be re-established at a sufficient level and within a certain time window [844]. This is clinically possible by means of intravenous or intra-arterial thrombolysis [845], but the risk of intracranial bleeding is high if the complete infarct is already too large. Therefore, the main goal of functional imaging is currently the precise identification of the status and extent of penumbra to allow better informed selection of the therapeutic options and improvement of the treatment results.

Several studies have indicated that necrosis is present in tissue whose cerebral metabolic rate for oxygen ($CMRO_2$) is below 65 μmol/100 g per min (=1.5 ml/100 g per min) or whose CBF is below 12 ml/100 g per min [846, 847]. Voxels exhibiting initial flow rates between 10 and 22 ml/100 g per min were found in the area of the final cortical-subcortical infarcts [848]. On the other hand, voxels with the same penumbral criteria escaped infarction, and the minimum CBF of salvageable tissue was found to be 7 ml/100 g per min [849]. This means that these thresholds are variable, depending on the timing of measurement after the attack, and that brain regions with CBF between 12 and 22 ml/100 g per min can be considered as a penumbra zone [850]. In an experimental study in minipigs, brain tissue with CBF up to 30 ml/100 g per min deteriorated to below the threshold of viability within 3 hours [851].

Metabolically, acute ischemic stroke is characterized by a mismatch of relatively well-preserved oxygen metabolism but severely reduced CBF. This condition, which is often called misery perfusion, implies that blood flow is insufficient to meet the energy metabolic demand of still viable tissue for oxygen and substrate (Fig. 2.33) [847, 852–854]. It is characterized by an increase in the oxygen extraction fraction (OEF) to up to 50% above its normal values. It has also been observed as an immediate event after vessel occlusion in experimental stroke studies [850, 855]. It is indicative of the precarious condition of the tissue, but experimental studies show that it also holds promise for a full recovery if CBF can be restored [856]. Unfortunately, in most patients a slow continuous decline of $CMRO_2$ is seen in the border zone of an ischemic infarct [849, 854]. A few studies have suggested that it could be prevented by therapeutic intervention [852, 857].

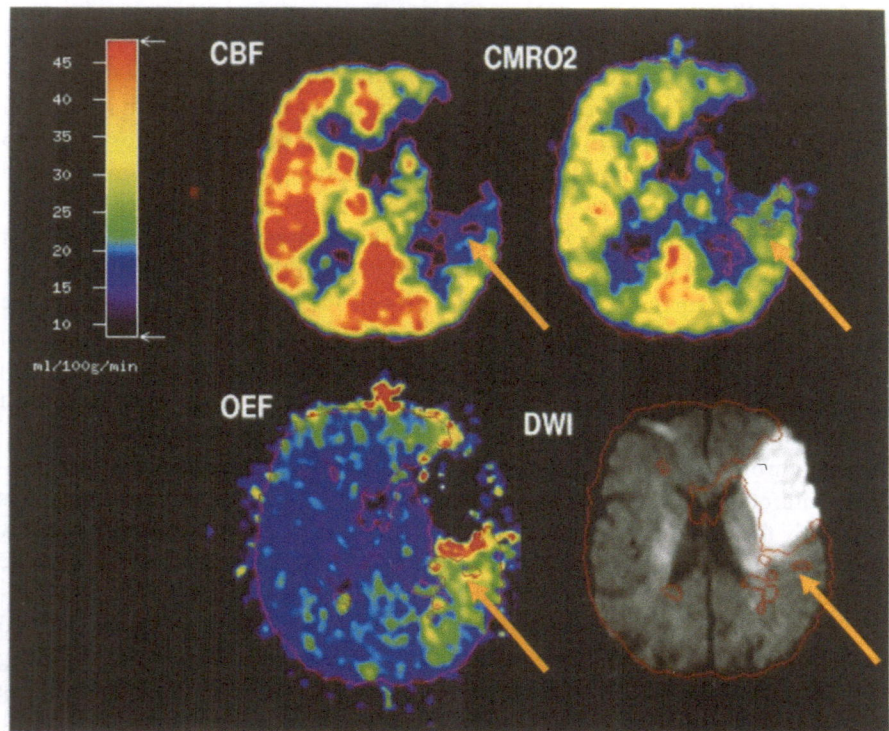

Fig. 2.33. Misery perfusion in acute stroke: images show a completed infarct in the left anterior middle cerebral artery (MCA) territory (*bright area* in diffusion-weighted MRI, *bottom right*) with severely reduced CBF and $CMRO_2$. Misery perfusion is located in the adjacent posterior MCA territory (*arrows*) with reduced CBF, near-normal $CMRO_2$, increased OEF, and normal diffusion-weighted MRI. A *red contour* (obtained by thresholding of the CBF image) was superimposed onto all images to facilitate regional comparison

Ligands to central benzodiazepine receptors, such as ^{11}C-flumazenil (FMZ), can be used as markers of neuronal integrity since they bind to the widely distributed γ-amino butyric acid (GABA) receptors of intact cortical neurons [858] (Fig. 2.34). The validity of this concept has been demonstrated in cat middle cerebral artery (MCA) occlusion, by comparing the FMZ binding to quantitative assessment of flow and energy metabolism [859]. It has also been applied to patients with acute ischemic stroke, demonstrating a close correlation with development of eventual infarction [860, 861].

Markers of hypoxic tissue were also tested with respect to their capacity to identify penumbral tissue. ^{18}F-Fluoro-misonidazole (FMISO) revealed increased uptake surrounding a zone of absent activity, and the area of high activity had disappeared by the time of follow-up, indicating that the hypoxic tissue had either infarcted or recovered [862–864].

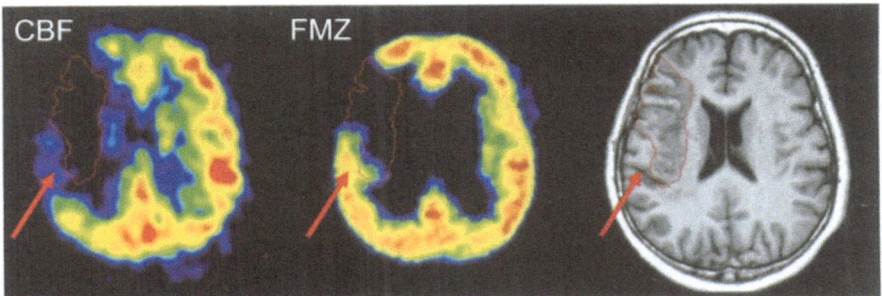

Fig. 2.34. Impaired CBF but intact flumazamil (*FMZ*) binding (*arrows*) in the posterior part of an acute MCA infarct indicates viable tissue in the ischemic penumbra that is still preserved on follow-up MRI (*right image*). A *red contour* (obtained at the viability threshold on the CBF image) outlines the completed infarct on all images

After an acute ischemic event arterial vessels may be damaged, and eventual reperfusion may then lead to postischemic hyperperfusion [865–867]. This hyperperfusion may develop in viable tissue, where it is often mild and does not prevent a good outcome [868, 869], but it may also occur within ischemic infarcts, when it is called "luxury perfusion" [870]. This phenomenon is a major reason for the inability of sole CBF measurements to predict tissue outcome reliably. In completed infarcts, a complex cascade of reactive changes is initiated, including microglial activation that can be imaged with ^{11}C-PK-11195 [871].

CMRglc is mostly reduced in acute ischemic infarcts [872–874], but to a lesser degree than $CMRO_2$. In a few cases – under 10 % of patients with acute and subacute infarcts – CMRglc is increased in some parts of the infarct to levels above that found in normal gray matter. This phenomenon is believed to indicate nonoxidative glycolysis, which may lead to intracellular accumulation of lactate [875]. Within the first few days of acute stroke, a decrease in tissue pH is observed [876]. In the subacute phase, intracellular pH in most infarcts shows an alkaline shift, which might be correlated with the occurrence of perfusion in excess of metabolic demand. It could also enhance the glycolysis rate or could represent mainly the pH_i of phagocytic cells, which use aerobic glycolysis to synthesize hydrogen peroxide [877, 878].

2.4.2   Intracerebral and Subdural Hemorrhage

In patients with intracerebral hemorrhage, PET can reveal the effects of the space-occupying lesion on CBF and metabolism and thereby provide information on the functional state of the surrounding tissue and on the prospects of recovery. The hematomas are manifest as defects of normal tracer uptake [879]. Large hematomas may lead to a larger oxygen extraction fraction (OEF) in the periphery, probably because of their mass effect, which demands evacuation [880–882]. Yet, in another study, a reduction in OEF in the periphery of intracerebral hematomas was seen [883]. The inflammatory tissue reaction during resorption of an intracerebral hematoma may lead to moderate focal increases of ^{11}C-methionine uptake [884, 885] (see section 2.3.1.2 for differentiation from brain tumors).

2.4.3   Subarachnoid Hemorrhage

Subarachnoid hemorrhage (SAH) may lead to severe cerebral vasospasm, which can in turn lead to ischemic infarcts. Good estimates of this risk are required for optimum timing of angiography and surgery (clipping of the aneurysm). In patients with postoperative vasospasm generalized impairment of oxygen metabolism with a reduced tissue oxygen supply has been observed, even in the apparently normal cortex, and additional impairment of regional perfusion in the territory of vasospasm [886] (Fig. 2.35). An experimental study has demonstrated that SAH is due to increased cerebrovascular resistance at the level of the small parenchymal vessels [887]. The generalized reduction of $CMRO_2$ is thought to be due to the initial aneurysm rupture [888, 889].

2.4.4   Remote Effects (Diaschisis)

Alterations of metabolism and blood flow in acute stroke are not limited to the infarct or hemorrhage and its immediate surroundings; there are also remote reductions, which are probably due largely to neuronal inactivation (diaschisis) [890]. The most obvious case is crossed (contralateral) cerebellar diaschisis (CCD), which is probably due to the inactivation of cortico-ponto-cerebellar pathways [891–899] (Fig. 2.36). It occurs within hours after supratentorial ischemic insults [900, 901], is most frequently seen after lesions to the basal ganglia and frontal and parietal lesions [902], and can persist over a long time with eventual cerebellar atrophy but usually lacks major correlates in neurological functional impairment [903, 904]. Preservation of CMRglc in the dentate nucleus has been noted, suggesting that cerebellar output activity (mostly originating in the Purkinje cells) is not impaired by diaschisis [897]. CCD was also observed with pontine lesions, involving specific cerebellar lobules to a different degree that is consistent with the pontine anatomy of the cortico-ponto-cerebellar pathway [893]. CCD associated with hemiataxia was demonstrated in few patients with thalamic lesions, and it was presumed to result from retrograde deactivation of the cerebellar hemisphere via the dentate-rubro-thalamic pathway [905]. The reverse, supratentorial diaschisis after cerebellar infarct, has been seen only rarely [906]. A study with ^{11}C-flumazenil suggested that reorganization of GABA-mediated mechanisms and glucose metabolism in cerebellum following cortical injury differs with size of lesion and age at the time of injury [907].

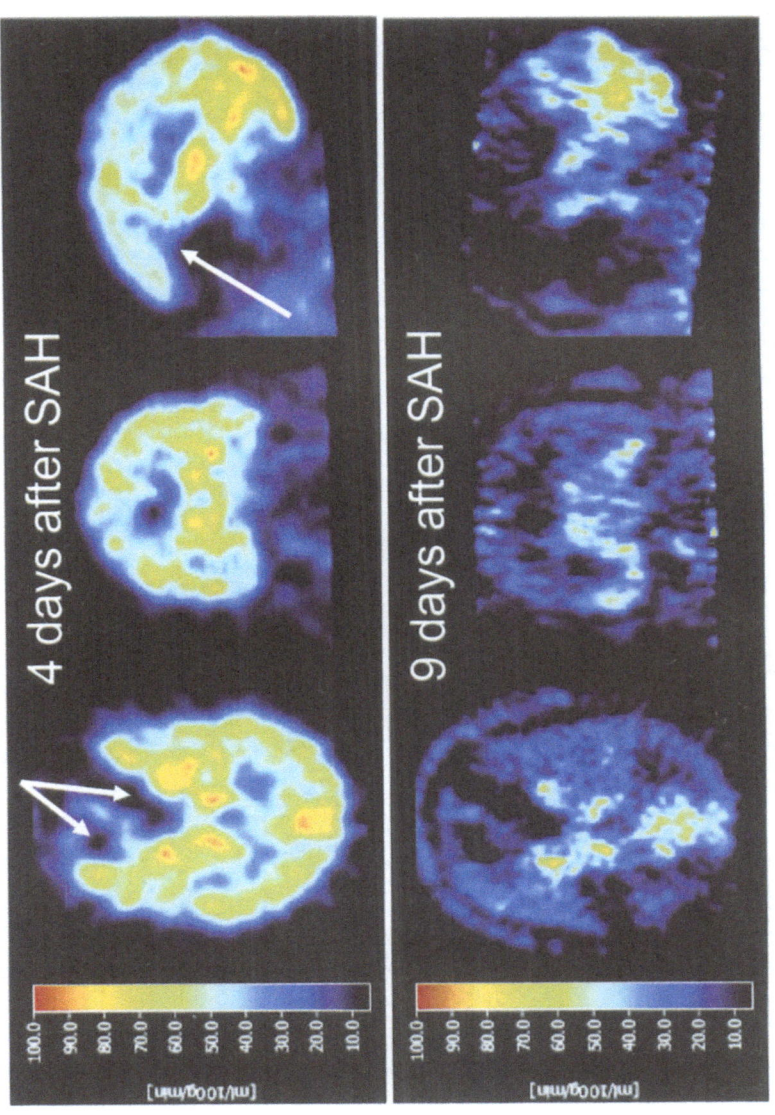

Fig. 2.35. ¹⁵O-Water CBF study (3 orthogonal slices) of the progress of cerebral vasospasm after subarachnoid hemorrhage (SAB): On the 4th day focal CBF reductions are seen in the frontal lobe (*arrows*), but CBF is still close to normal values in the rest of the brain. On day 9, severe generalized vasospasm is present in the whole brain, with cortical CBF values in the order of 20 ml/100 g per min

Cortical diaschisis is particularly prominent with thalamic infarcts, which often lead to pronounced thalamocortical diaschisis with corresponding cognitive deficits [247, 908–911] (Fig. 2.37). A distinction can be made between intrahemispheric diaschisis [912], in which noninfarcted gray matter structures in the same hemisphere are functionally impaired, and transhemispheric or transcortical diaschisis with impairment of the contralateral hemisphere [913, 914]. Intrahemispheric diaschisis is a very frequent finding that was noted in the very early

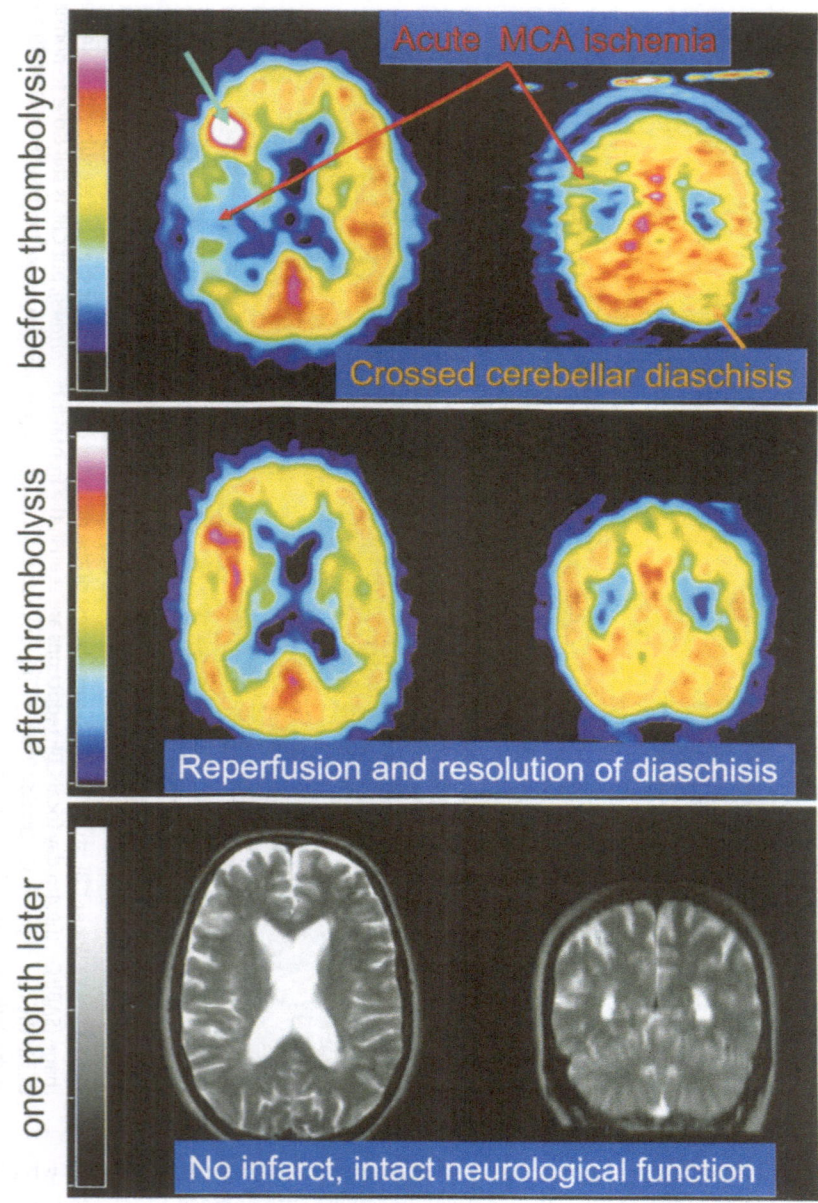

Fig. 2.36. Crossed cerebellar diaschisis in acute occlusion of the MCA, with resolution after successful thrombolysis. Images show CBF in a transaxial slice at a level just above the basal ganglia (*left column*) and in a coronal slice through the cerebellum (*right column*). There was already a focal area of postischemic hyperperfusion (*green arrow*) at the anterior border of the ischemic territory before the start of therapeutic thrombolysis. The outcome is documented by coregistered T2-weighted MRI

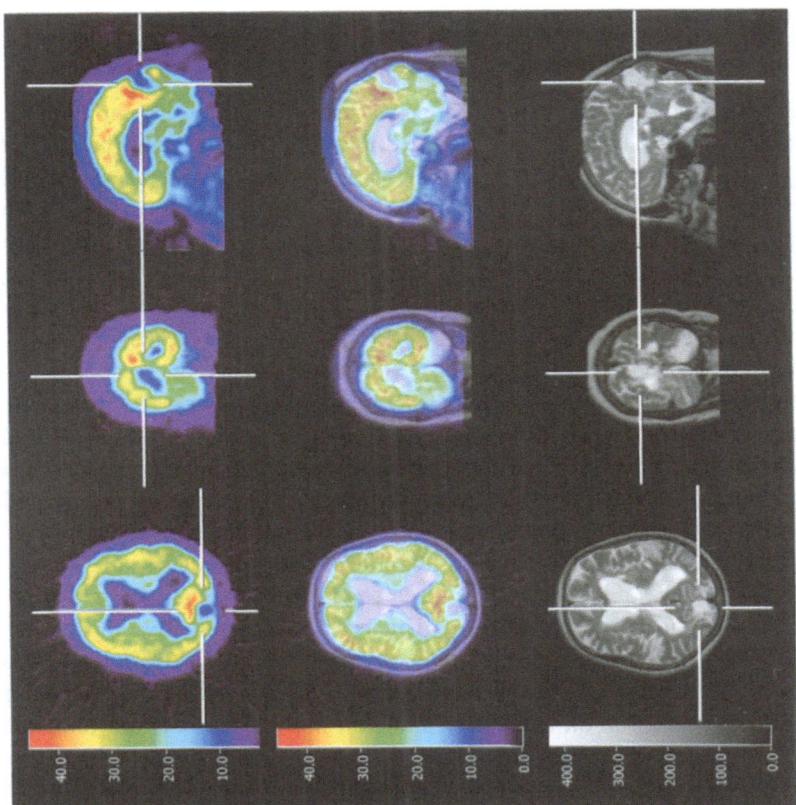

Fig. 2.37. Widespread impairment of CMRglc in structurally intact cortex in a patient with right occipital and left cerebellar infarcts and vascular dementia (orthogonal slices through occipital infarct with coregistered T2-weighted MRI and fusion images)

studies, can often can be related to functional deficits, and probably plays an important part in functional recovery [872, 912, 915–919] (see also section 2.4.8). It is most obvious with striatal infarcts, but in these cases, which often represent incomplete infarcts attributable to temporary occlusion of the MCA, there may also be a substantial contribution from diffuse ischemic cortical damage. Even small lacunar striatocapsular infarcts can be recognized with PET by their functional effects [920].

2.4.5   Chronic Hemodynamic Impairment

Chronic perfusion disturbance caused by arterial vascular disease is a precarious condition, necessitating vascular surgery in selected cases. Patients with a history of transient ischemic attacks (TIAs) but with normal clinical and CT examination usually have CBF values greater than 22 ml/100 g per min [921]. The distinct focal reductions in CBF and CMRglc found in many patients after TIAs often correspond to the location of the clinical deficits [879]. They are thought to be the result of small areas of embolic infarction not detected by morphological imaging or areas of selective neuronal loss without gross infarction.

Occlusion or high-grade stenoses of major arteries may reduce hemodynamic reserve capacity [922]. In its mildest stage, CBF is maintained within the normal range by peripheral vasodilatation, which is reflected in an increased cerebral blood volume (CBV). Thus, the CBF/CBV ratio is decreased, and its inverse, which is closely related to the mean transit time of intravascular indicators as used with dynamic CT and MRI, is increased. When the perfusion reserve is exhausted (i.e., at maximal vasodilatation) any decrease in arterial input pressure produces a proportional decrease in both CBF and CBF/CBV ratio. In this condition of hemodynamic decompensation the brain must draw on the oxygen carriage reserve to prevent energy failure and loss of function, as evidenced by an increase in the OEF from the normal 40–50 % up to 85 % [921, 922]. Patients with low CBF/CBV and submaximal elevations of OEF account for 10–15 % of all patients with cervical occlusive disease [922]; they exhibit the most advanced atherosclerotic lesions and are at high risk of recurrent stroke [923]. However, the relation between CBF/CBV ratio and OEF is rather variable, and both are needed to provide complete information on the severity of hemodynamic impairment and risk of stroke [924–926].

In some cases, the hemodynamic impairment with increased OEF, i.e. the condition of misery perfusion, has been treated successfully by extra-intracranial bypass surgery [852, 927–931]. Despite one study in which selection of patients for bypass surgery was based on PET and results did not confirm a better outcome after surgery [932], a new trial is now under way for which patients are selected on the basis of the hemodynamic compromise [933].

Vascular responsiveness as an indicator of hemodynamic reserve can also be assessed by comparing CBF at rest and during breathing of carbon dioxide or after administration of the carbonic anhydrase inhibitor acetazolamide or the adenosine uptake inhibitor dipyridamole. The dilatation of peripheral vessels and increase of CBF under this vasodilator challenge is taken as a measure of the residual reserve capacity and shows some correlation with CBF/CBV and OEF [934, 935]. This technique is most frequently used with perfusion measurements by SPECT, CT, or MRI when metabolic information cannot be obtained. It can of course also be used in connection with PET measurements of CBF [936–940].

Chronic cerebrovascular disease may lead to vascular dementia. In this condition, functional findings are dominated by the consequences of multifocal ischemic damage, which are described in section 2.1.5. There is some controversy about whether chronic hemodynamic impairment with reduced vascular responsiveness is characteristic for vascular dementia and could be used to differentiate it from degenerative dementia.

2.4.6   Moyamoya Disease

Moyamoya disease is most frequently observed in Japan. An often bilateral occlusion of the MCA (and in some cases also other major cerebral arteries) is accompanied by an extensive microvascular network that partially bypasses and compensates this occlusion. There is a high risk and frequency of cerebral ischemia and hemorrhage in moyamoya disease. The corresponding findings in ^{15}O PET studies are significant increases in CBV and cerebral transit time [941, 942]. In children and asymptomatic patients, CBF and $CMRO_2$ are usually maintained within the normal range [943] but reduced after ischemic or hemorrhagic events. The OEF is normal in most cases but may be increased, especially in children [944].

Extra-intracerebral bypass surgery is regarded as a standard therapeutic procedure to improve hemodynamics in moyamoya disease. Postoperative improvements of hemodynamic parameters have been observed predominantly near the cortex, where bypass surgery had been performed, and in the basal ganglia [945–947].

2.4.7   Genetic Disorders (MELAS, CADASIL)

MELAS is an acronym made up of the initial letters of the components myopathy, encephalopathy, lactic acidosis, and stroke-like episodes, and the MELAS syndrome is a mitochondrial disorder. As would be expected in a mitochondrial disease, $CMRO_2$ is reduced whereas CMRglc is usually within the normal range, indicating impaired oxidative and increased anaerobic glucose metabolism [948, 949]. CBF has often been found to be increased, especially after stroke-like episodes [948, 950, 951], which could indicate postischemic hyperperfusion but more probably demonstrates that the stroke-like episodes have a primary metabolic rather than a hemodynamic cause.

CADASIL is an autosomal dominant arteriopathy characterized by multiple brain infarcts, cognitive decline, and finally dementia, which is caused by mutations in the *Notch3* gene. With PET, a pronounced global reduction of CBF and CMRglc is observed [952, 953].

2.4.8   Brain Function and Recovery After Stroke

The potential for rehabilitation after stroke may be limited by several global factors, such as age and microvascular angiopathy, that affect basically the whole brain. These can be very strong factors that are evident from many clinical studies, and they are also reflected in reduced glucose metabolism even in the resting state [954]. The potential for recovery is high for lesions acquired during early life, when even hemispherectomy is often followed by good recovery of motor and language function [955]. Reorganization may be mediated by compensatory functional changes in corresponding cortical areas of the unaffected hemisphere, including the ipsilateral corticospinal pathway. Functional imaging studies suggest that metabolic recovery and functional plasticity in the remaining subcortical structures of the injured hemisphere may also play an important part [955, 956]. In these areas not only functional deactivation (diaschisis, see section 2.4.4), but also partial neuronal loss as the consequence of ischemia, may contribute to functional changes [957]. The potential for recovery is higher in childhood than in later life. Nonetheless, it is still not known whether there is a continuing influence of age beyond the limitations imposed by age-related multimorbidity on functional recovery after stroke [958].

2.4.8.1   Aphasia

Aphasia is a severely incapacitating symptom of stroke and is one of the main causes of disability. Most studies have been performed in right-handed individuals with language dominance in the left hemisphere. The left temporoparietal region, in particular the angular gyrus, supramarginal gyrus, and lateral and transverse superior temporal gyrus, are the most frequently and consistently impaired, and the degree of impairment is related to the severity of aphasia [959, 960]. In contrast, metabolic impairment of subcortical structures is related mainly to language fluency and other behavioral aspects, but not to aphasia severity [961]. In patients with aphasia attributable to purely subcortical strokes deactivation of temporoparietal cortex is regularly found, which is probably responsible for the aphasic symptoms [962].

Metabolic disturbance in the left temporoparietal cortex is related to the outcome of aphasia [963]. Investigations in the subacute state after stroke showed a highly significant correlation with language performance assessed at follow-up after 2 years. The receptive language disorder correlated with CMRglc in the left temporal cortex and word fluency correlated with CMRglc in the left prefrontal cortex [964]. In activation studies recovery was also associated with the ability to activate the left superior temporal gyrus [965–967]. Only the basic function of mere word repetition appears to be sufficiently supported by sole right hemispheric activation [968]. These results indicate that the functional disturbance as measured by rCMRglc in speech-relevant brain regions of the dominant hemisphere early after stroke is predictive of the eventual outcome of aphasia (Fig. 2.38). Facilitation of this activation has been demonstrated after treatment with piracetam [969]. The results also suggest that even limited salvage of peri-infarct tissue with acute stroke treatments will have an important impact on the rehabilitation of cognitive functions.

The wider language network is not confined to the dominant hemisphere. The role of the right hemisphere after ischemic infarcts of language areas in the left hemisphere has been addressed in several studies. Generally, more right hemispheric activations were seen in the subacute phase of an infarct with language

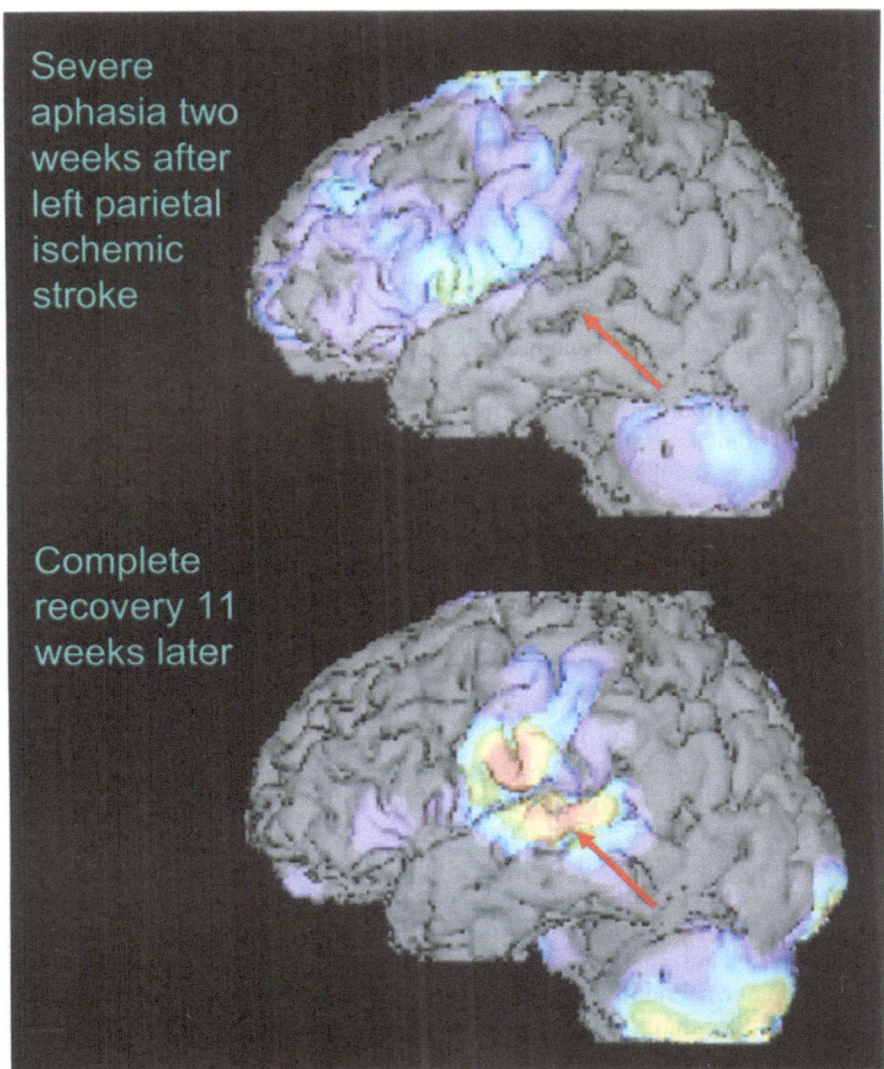

Fig. 2.38. Language activation by word repetition in a patient with a left parietal ischemic infarct and severe aphasia (47 errors in Token test) in the first 2 weeks (^{15}O-water PET study, shown as 3D fusion rendering with coregistered MRI). Activation of the left superior temporal gyrus (Wernicke's area, *red arrow*) was completely missing in the acute phase, but recovered corresponding with near-complete clinical recovery (2 errors in Token test) after rehabilitation

activation than in normals with the same tasks [970, 971]. Language recovery in the months immediately after aphasia onset was associated with regression of functional depression (diaschisis) in structurally unaffected regions, in particular in the right hemisphere [917]. Training-induced improvement in verbal comprehension in some aphasia patients was associated with activation of the posterior part of the right superior temporal gyrus and of the left precuneus [972]. Thus, there is ample evidence that the brain recruits right-hemispheric regions for speech processing when the left-hemispheric centers are impaired. However, outcome studies reveal that this strategy is significantly less effective than repair of the original speech-relevant network in adults [973]. The effectiveness of right-hemispheric compensation appears to be higher in childhood than later [828, 830].

Observations of crossed aphasia (i.e., aphasia in a right-hander due to a lesion of the right hemisphere) are of principal interest for the understanding of variants of the neuroanatomy of language. One study suggests that abnormal dominance for at least some language functions in the right hemisphere underlies the syndrome of crossed aphasia, but diaschisis of the anatomically normal left hemisphere is also seen [974].

The issue of language dominance and aphasia in left-handers has been studied with PET in normals. Activation studies show a greater participation of the right hemisphere in language processing in left-handers, but there is no clear relation between handedness and laterality of activation during language tasks [813, 975].

2.4.8.2   Motor Function

Capsular infarcts or hematomas are a common reason for lesions to motor fibers with contralateral hemiparesis but intact motor cortex. Patients with such lesions usually show the same degree of activation of the sensorimotor and premotor cortex as normal subjects [976]. Even imagining such movements can lead to substantial activation of motor cortex. Tactile exploration of shapes with the paretic hand after subcortical infarction leads to large activations in contralateral motor and sensory hand cortex [977]. These activations are also similar to those observed in normal subjects with the same task. In a study of six patients with capsular or pontine infarcts, passive movements of the paretic arm also led to activation of contralateral sensorimotor cortex [978]. With infarcts of motor cortex, the extent and intensity of activation obviously depend on the site and extent of the lesion.

Activations of ancillary motor areas, such as ipsilesional premotor cortex activation, are frequently observed [976, 979]. Secondary motor and frontoparietal nonmotor cortices are activated more in patients with lesion onset before age 4 than in those with onset after age 10, suggesting a greater potential for reorganization during early development than in later life [806]. Learning new movements also increases CBF in SMA and premotor cortex in normal subjects [980]. Learning-associated activations often also include sensorimotor and parietal cortex, basal ganglia, and cerebellum. Large interindividual variability has been noted

[981], which may be due to different learning strategies [982]. Thus, activation of ancillary motor areas after stroke could reflect a learning effect that could contribute to rehabilitation.

In normals and for the unaffected hand in stroke patients, little activation of ipsilateral motor cortex is seen during finger movement. Several studies have consistently shown that ipsilateral activation of motor cortex is stronger for movements of the paretic fingers after recovery from stroke [977, 983–986]. These data can be interpreted as evidence for stronger activity of the small noncrossing part of the corticospinal tract after a lesion to the crossing fibers or neurons.

It is not clear whether increased activity in the ipsilateral motor cortex contributes to functional recovery in adults [979, 987, 988]. Nonrecovered hemiplegic patients show stronger activations, both ipsilateral and contralateral, during passive movements than do normal controls [978]. In a small series, ipsilateral activation was only seen in the most severely impaired subject [989]. In a study of resting $CMRO_2$ there was even a slight decline in the unaffected hemisphere during the early recovery phase after stroke [990]. It was not related to neurological recovery and was interpreted as a possible consequence of transcallosal fiber degeneration.

The thalamus is an important relay station for sensory afferences to the cortex, and it is also involved in the extrapyramidal motor system. A reduction of ipsilateral thalamic CMRglc at rest is often seen after hemispheric stroke [902] and is associated with poor motor recovery [991]. By multiple regression and discriminant analysis, Azari et al. [992] found a close correlation of CMRglc in bilateral SMA, ipsilateral thalamus, and contralateral cerebellum in recovered patients, suggesting a stronger functional association of these structures than in normals or in nonrecovered patients. In a related study, a statistically significant association was found between activation of a similar network including bilateral occipital and bifrontal cortex and cerebellum, contralesional cingulate, hippocampus and thalamus and recovery of motor function [918].

Cerebellar activation is seen in most activation studies. In patients after stroke it has been variously reported as normal [976] or reduced, especially in ipsilateral cerebellum [977, 993]. There is less cerebellar activation in patients with lesions received in early childhood than in those who have sustained lesions in later life [806]. Reduced activation may be a consequence of damage to cortico-ponto-cerebellar pathways by stroke, which is probably also the cause of the crossed cerebellar diaschisis seen even in the resting state [904].

2.4.8.3   Neglect

Neglect is a disorder of spatial attention usually on one side of the body and in the related field of action and vision. Even though it is not due to hemianopia or hemiparesis, it can substantially impair rehabilitation. Clinical observations indicate an important role of the subdominant right hemisphere and involvement of the parietal cortex as the association cortex that is most important for visuospatial orientation. There have been a few PET studies, which generally support the clinical evidence [994]. They also indicate that functional diaschisis may have an important role [995]. A report of two cases suggested that the remission of unilateral neglect after left-hemispheric stroke might be associated with a functional metabolic recovery in both the undamaged left hemisphere and the unaffected regions of the right hemisphere [996]. In an activation study with a visuospatial task, regions that were notably more active after recovery from neglect were found almost exclusively in right-sided cortical areas and largely overlapped with those observed in a group of normal subjects performing the same visuospatial task [997].

2.5   Epilepsy

Epilepsy is a heterogeneous group of conditions, which vary according to seizure type, age of onset, and underlying pathophysiology. Recurrent seizures are broadly classified into generalized epilepsies and localization-related or focal epilepsies. Generalized epilepsy, including absence seizures and generalized tonic-clonic seizures, often begin in childhood. Their pathophysiology is not well understood, but such functional abnormalities as abnormal ion channels [998] have been implicated, rather than focal brain lesions. Simple focal seizures, with transient motor or sensory manifestations but no impairment of consciousness, are typically associated with a focal brain lesion that can be localized by CT or MRI, e.g., posttraumatic scar, tumor, inflammatory lesion, or hemorrhage. The most difficult clinical problems are posed by complex partial seizures. These consist of episodes of impaired consciousness, typically accompanied by automatisms and often progressing to a generalized tonic-clonic seizure. These seizures have a focal origin, most frequently in hippocampus or frontal lobes. Only 50–70 % of patients thus affected become seizure free with antiepileptic medication; the remainder may benefit from surgical treatment. This requires accurate preoperative localization of the epileptogenic focus. Prior to resection of the focus, it is often necessary to define adjacent eloquent cortex, that is, cortex in which an injury results in a motor or cognitive deficit. Special diagnostic and therapeutic considerations arise in childhood with its specific epileptic syndromes and in patients with cortical malformations.

There are several clinical and research applications of PET in epilepsy. In patients with medically intractable focal seizures who are being assessed for surgical therapy, PET can assist in localization of the epileptogenic focus. In fact, the finding of interictal hypometabolism in the temporal lobe harboring the focus in patients with complex partial seizures was one of the first clinical observations to be made with PET [999]. Preoperative PET activation studies with ^{15}O-water can be used to define eloquent cortex involved in memory and language [1000]. The role of PET in these applications is currently being redefined and becoming more limited as advances are made in anatomical and functional MR [1001–1003]. Current clinical indications include discordant findings between surface EEG and MRI prior to placement of intracranial electrodes and suspected focal epilepsy if less complex methods of seeking for localized abnormalities have failed [1004]. PET may also provide information on the prognosis of patients who have been surgically treated.

PET has been widely used to study pathophysiological aspects of epilepsy. Studies with FDG and ^{15}O-water have examined changes in CMRglc at sites distant from the focus, the effect of a seizure on blood flow and glucose metabolism, and the effect of anticonvulsant medications on cerebral metabolism.

A wide variety of radioligands have been used to study neuroreceptor systems in epilepsy [1005, 1006]. There are several possible interpretations when an abnormality is detected. The affinity (K_D) of the receptor for the radiotracer may be abnormal. The amount of receptor (B_{max}) per unit volume of tissue may be changed; this could be due to a change in the number of receptors per neuron, or a change in the number of neurons. Lastly, the local concentration of endogenous ligand may be increased or decreased, resulting in altered competition for binding with the radiotracer being imaged (see section 4.9.3 for methodological details). Such changes could potentially be involved in the pathophysiology of epilepsy. It is also possible to observe changes in radioligand uptake because of nonspecific abnormalities, such as alterations in regional CBF or in the BBB. Tracer-kinetic modeling techniques can be used to account for these changes.

2.5.1   Localization of Epileptogenic Foci

The seminal observation with FDG-PET in focal epilepsy was made in 1980 by Kuhl and Engel and their colleagues [999]. In the interictal state, there is a reduction of local CMRGlu in the epileptic focus. Glucose metabolism is focally increased during focal seizures, and globally increased during generalized absence seizures [1007]. These basic findings have been replicated many times and have been compared with clinical, electrophysiological and other imaging data.

2.5.1.1   Temporal Lobe Epilepsy

Most patients who undergo surgical therapy for epilepsy have complex partial seizures that arise in the mesial temporal lobe (temporal lobe epilepsy, TLE). The most common underlying pathology is hippocampal or mesial temporal sclerosis. This consists of atrophy of the hippocampus with microscopic neuronal loss and gliosis. Heterotopias or low-grade tumors are less commonly implicated. The main finding yielded by FDG-PET, which has been replicated in numerous studies (see [1008] for review), is interictal focal hypometabolism in the mesial temporal lobe (Figs. 2.39, 2.40). This is found in approximately 80 % of patients. The hypometabolic zone is generally much larger than the area of pathological involvement and extends to involve temporopolar and temporolateral neocortex [1009–1011]. This

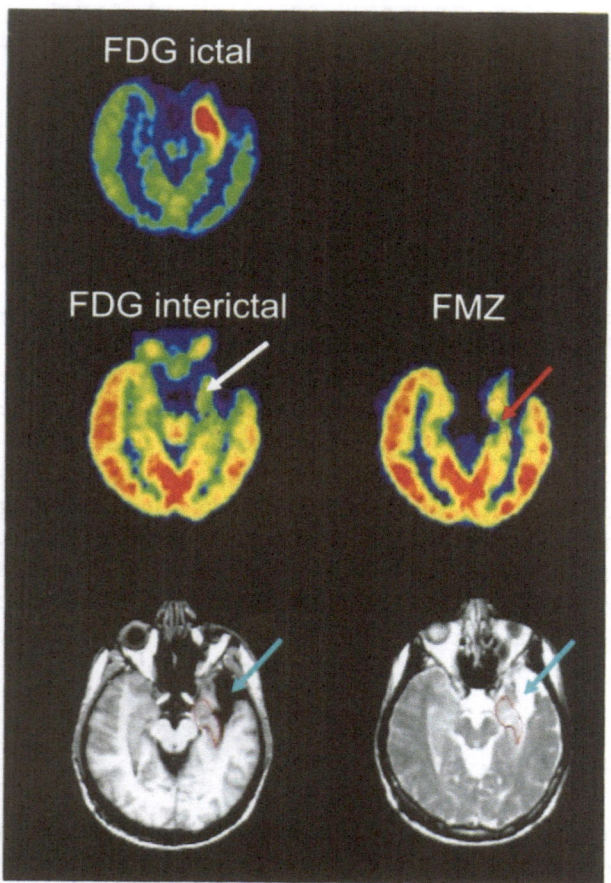

Fig. 2.39. Ictal hyper-metabolism (*top image, red area*), interictal hypometabolism in same area (*white arrow*), and small region with reduced FMZ binding (*red arrow*) in posttraumatic temporal lobe epilepsy (TLE). The ictal hypermetabolic region is outlined in *red* on coregistered MRIs, which also show the adjacent posttraumatic scar (*blue arrows*)

may be due to deafferentiation of temporal cortex distant from the hippocampal lesion. In the case of patients with temporolateral impairment only and absence of temporomesial hypometabolism, however, the seizure origin is usually in the lateral temporal lobe [1012]. In some cases, severe neocortical metabolic impairment may be due to microscopic cortical dysplasia [1013].

A pathophysiological correlate of temporomesial hypometabolism may be a reduction in the rate of glucose oxidation in the CA3 pyramidal subfields [1014]. A coupled reduction of glucose metabolism and glutamate cycling, which may be linked by common metabolic pathways, as explained in more detail in section 3.4, was also demonstrated in epileptic lesions with FDG-PET and proton MRS [1015] and could also play a part in the pathophysiology of focal epilepsy. Pharmacological activation of GABA-A receptors may partially reverse glucose hypometabolism [1016]. An alteration of BBB GLUT1 glucose transporter activity in epileptogenic cortex could also contribute to reduced FDG uptake [1017].

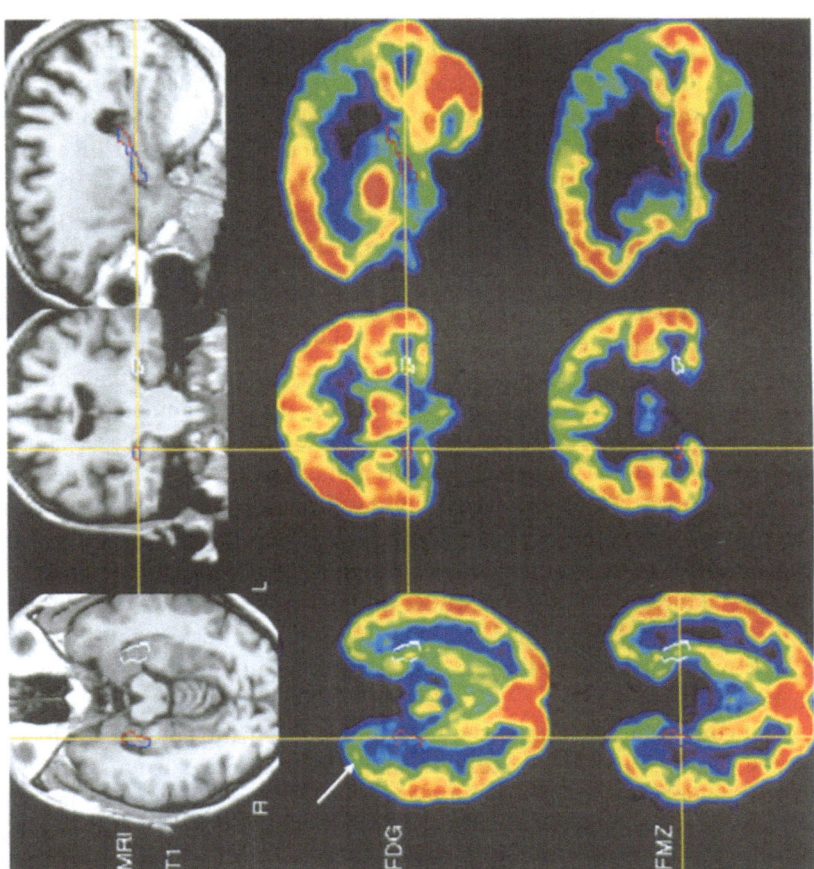

Fig. 2.40 Extended reduction of CMRglc (*middle row, white arrow*) in the right temporal lobe and more focal reduction of FMZ binding (*bottom row*) in hippocampal formation in TLE

The reduction in mesial temporal CMRglc is not simply an artifact due to partial volume averaging and volume loss in the hippocampus. Knowlton et al. [1018] reported that after partial volume correction, hippocampal metabolism per unit volume of tissue persists although hippocampal metabolism is directly correlated with hippocampal volume, suggesting that neuronal loss is an important underlying factor. However, several other authors have noted that the degree of hypometabolism is not related to the severity of hippocampal damage assessed with either quantitative MRI or histopathology [1019–1021]. In addition, reductions in FDG are not correlated with those of N-acetyl-aspartate (NAA), an MRS marker for the local number of neurons [1022]. Therefore, hypometabolism may be due to a combination of neuronal loss and suppression of glucose metabolism in any remaining neurons. This is consistent with data suggesting that hypometabolism is an independent predictor of surgical outcome [1019].

Initial studies of temporal lobe epilepsy (TLE) suggested that focal interictal hypometabolism is related to the severity of the pathological lesion [1009] and the duration of epilepsy [1023]. The degree of left temporal hypometabolism has been correlated with impairment of verbal memory [1024]. The frequency of interictal spikes on EEG does not generally show a relation to focal CMRGlc changes, although there has been a case report of an increase in CMRGlc and CBF associated with interictal spiking [1025].

Hypometabolism often extends to brain regions distant from the involved temporal lobe, most commonly ipsilateral thalamus, but also basal ganglia, and frontal and parietal cortex [1010, 1026]. Hypometabolism may extent also to the insular cortex; its anterior part may be involved in emotional symptoms and the posterior insular cortex, in somesthetic symptoms [1027]. Ictal dystonic posturing in TLE has been associated with striatal hypometabolism [1028]. In a study that found a correlation between hippocampal cell loss and subcortical hypometabolism the hypothesis that decreased efferent synaptic activity to thalamus and basal ganglia causes decreased neuronal activity in these remote structures and hypometabolism was set up [1029]. Functional activation has been suggested as a method of differentiating between areas with local synaptic dysfunction that are potentially epileptogenic and remote hypometabolism due to deafferentiation [1030–1032].

The interictal focal reduction of CMRGlc in TLE is associated with a reduction of CBF that is usually less marked and less reliable [1023, 1033, 1034]. PET-CBF studies have probably also not been as widely used because of the higher image noise than with FDG.

EEG is, of course, the most important method of localizing an epileptic focus. With partial seizures, and in particular in TLE, EEG recordings from scalp electrodes are often not conclusive. Before the introduction of modern imaging technologies, invasive sphenoidal or depth electrode recordings were the next step for accurate localization. Early PET studies suggested that combined use of FDG-PET and surface EEG may obviate the need for invasive recordings when surface EEG confirms that a corresponding hypometabolic zone is epileptogenic [1035]. If PET is concordant with surface EEG, it is a reliable indicator of the side where the abnormality is sited and provides important information for use in the planning of surgery [1036, 1037]. Partial complex seizures may not be clinically apparent. Therefore it is necessary to obtain an EEG recording during FDG-PET to avoid false localization due to potential confusion of ictal, postictal, and interictal findings. In 2001, FDG-PET was approved by the Centers for Medicare & Medicaid Services of the United States for presurgical evaluation of a focus of refractory seizures.

In many instances, potentially epileptogenic lesions are associated with discrete macroscopic morphological changes, and because of their high sensitivity in the detection of such changes, modern MRI techniques have replaced functional imaging in presurgical work-up to some extent [1038–1040]. Still, FDG-PET appears to be more sensitive than MRI [1021, 1041–1043]. In a recent series of 113 TLE patients who had surgically and pathologically confirmed lesions and a good surgical outcome, sensitivity was found to be 89 % and the specificity was 91 % [1012]. In general, if a patient has a localized EEG focus and a concordant neuroimaging study (PET, ictal SPECT, or MRI), invasive electrophysiological studies are not required. PET may be valuable for focus localization in patients with a negative MRI.

2.5.1.2   Neocortical Focal Epilepsy

FDG-PET can also be useful to detect abnormalities in patients with neocortical (mainly frontal) rather than temporomesial partial seizures [1044–1046]. However, sensitivity appears to be lower (approximately 50–75 %, than in TLE [1041, 1047–1049]. Sensitivity may be improved by standardized quantitative evaluation [1050]. Hypometabolic regions are often associated with structural imaging abnormalities. As in TLE, regional hypometabolism often includes but is not specific for the epileptogenic region [1051]. Thus, the clinical relevance of FDG-PET appears to be primarily in directing placement of intracranial electrodes for presurgical evaluation of refractory neocortical seizures.

2.5.1.3   Ictal Studies

Ictal studies performed with ^{15}O-water or FDG are logistically difficult. They require the presence of very frequent or continuous seizures, reflex seizures, or pharmacological seizure induction [1052, 1053]. In some patients, epileptic discharges occur regularly during sleep and can then be studied with PET, as in one patient with continuous spike-wave discharges during sleep accompanied by partial motor and atypical absence seizures who had a corresponding focal increase in CMRglc [1054]. The interpretation of ictal FDG scans can be difficult, because glucose metabolism is not at steady state and because the value of the lumped constant for FDG changes during the extreme hypermetabolism of a seizure (see section 4.9.2.2 for methodological aspects). The FDG images may reflect a combination of ictal hypermetabolism, postictal depression of metabolism, and baseline interictal hypometabolism. In general, though, these studies have shown increases in both flow and metabolism in the area of the focus that are often large. Changes can also occur in subcortical structures and in contralateral brain regions. Changes in local metabolism, in relation to the interictal state, can persist for 48 h or longer after a localized seizure [1055]. As in functional activation, there appears to be coupled focal increase in CBF and CMRglc in focal epilepsy but a smaller increase in $CMRO_2$ [1056].

Although ictal PET is rarely practical [1057], ictal blood flow imaging with SPECT (and a rapidly trapped tracer such as Tc-99m-HMPAO or ECD) is more commonly used to identify the site of seizure onset [1058–1060]. Very early and rapid tracer injection after seizure onset is crucial to avoid bias from late ictal and postictal activity distributions [1061].

The match between electrical discharges and blood flow increases may not be perfect. A study using SPECT in two patients reported a focal CBF increase has prior to electrographic seizure onset [1062]. A PET study of focal epileptic discharges induced by intracerebral electrical stimulation demonstrated that areas of increased CBF always indicated an underlying epileptic discharge but some discharging regions showed no CBF change [1063]. The coupling of CBF and CMRglc may not be preserved in the postictal phase, for which persistent hyperperfusion [1064] and hypometabolism [1055] has been described. Focal epileptic status (epilepsia partialis continua) may be associated with focal hypometabolic or hypermetabolic alterations [1056, 1065, 1066]. Focally increased CMRglc was observed during periodic lateralized epileptiform discharges [1067].

2.5.1.4   GABA-A and Benzodiazepine Receptors

The most widely-used PET radioligand to study epilepsy has been ^{11}C-flumazenil (FMZ), which binds to benzodiazepine receptor (BZR), as part of the central GABA-A receptor complex. A consistent observation in TLE has been reduced binding of FMZ in epileptogenic foci [1068]. A region of decreased FMZ binding may be found with a sensitivity as high as 94% [1036, 1049]. This finding is even more consistent after correction for partial volume effects [1069]. Reduced FMZ binding can be present even in the absence of abnormalities on MRI [1070-1073]. Comparative studies have shown that the area of reduced FMZ binding is much more focally restricted than the area of reduced FDG uptake [1074-1076]. Correlations with invasive electrophysiological methods done so far in few patients indicate that the area of reduced FMZ receptor binding corresponds to the seizure onset zone [1045, 1059]. Minor reductions of FMZ binding have also been observed in projection areas of epileptogenic foci with normalization after successful focus resection [1077, 1078] and in the cerebellum [1079].

Several investigators have studied the underlying basis for the reduction in FMZ binding in TLE. Neuronal loss has been observed in pathological studies of resected hippocampi. The fact that the reduction persists after correction for partial volume effects and occurs in the absence of MRI abnormalities, however, indicates that it is not due to hippocampal atrophy alone. In a group of patients with hippocampal sclerosis, there was excellent agreement between the partial-volume corrected reduction in FMZ binding and ex vivo autoradiographic measurements of H-3-FMZ receptor density [1080]. A detailed autoradiographic study of receptor density using ^{3}H-FMZ showed B_{max} (receptor density) to be decreased in all components of resected sclerotic hippocampus. In the CA1 region, B_{max} was reduced greater than could be attributed to neuronal loss, while elsewhere, B_{max} and neuronal reductions were matched [1081]. Receptor affinity was increased in subiculum, hilus, and dentate gyrus. In vivo PET measurements have also suggested decreased B_{max}. One study, however, reported that B_{max} was increased in resected hippocampi, with decreased FMZ binding attributed to altered affinity, which was decreased to an extent greater than the B_{max} increase [1082]. Overall, it appears that altered PET FMZ binding reflects neuronal loss, with possible additional reduction in B_{max}. Interestingly, increased FMZ binding in temporal lobe white matter has been noted in TLE and may indicate microdysgenesis [1083].

In spite of the extensive positive findings with FMZ in TLE, its current clinical role is limited by advances in quantitative MRI. It may have a role, however, in localizing the seizure focus in patients with normal MRI.

In neocortical focal epilepsy, extensive cortical abnormalities on FMZ-PET predict poor outcome in epilepsy surgery, whereas resection of focally restricted FMZ abnormalities in the lobe of seizure onset is associated with excellent outcome even in the absence of a structural lesion [1084, 1085]. FMZ-PET is probably also useful for identifying epileptogenic cortex in the vicinity of structural lesions [1086]. For example, epilepsy associated with dysembryoplastic neuroepithelial tumors or cavernomas may be related to reduced benzodiazepine receptor density in the vicinity of these lesions [1076, 1087] (Fig. 2.41).

2.5.1.5 Other Transmitters and Receptors

An increase of μ-opioid receptor binding capacity in the ipsilateral temporal neocortex in patients with TLE is observed with ^{11}C-carfentanil [1088, 1089]. It has been hypothesized that this increase may represent a tonic anticonvulsant system that limits the spread of electrical activity from other temporal lobe structures. Increased binding has also been observed with the delta opioid receptor ligand ^{11}C-methylnaltrindole [1090]. In the same patients, increased μ-receptor binding was confined to the middle aspect of the inferior temporal cortex, whereas binding of delta receptors was increased in the mid-inferior temporal cortex and anterior middle and superior temporal cortex. These changes are apparently specific for these receptor subtypes, because no increase in receptor binding was found with ^{11}C-diprenorphine [1089, 1091] and ^{18}F-cyclofoxy [1092], which do not have opiate receptor subtype specificity.

Competition of endogenous opioids was measured during reading-induced seizures. ^{11}C-diprenorphine binding to opioid receptors was lower in left parieto-temporo-occipital cortex. These findings suggested that opioid-like substances are involved in the termination of reading-induced seizures [1093]. Similar ictal studies with other radioligands would be of great interest.

Uptake of i.v.-administered ^{11}C-ketamine was found to be reduced in the affected temporal lobes of patients with TLE [1094]. This could be due to several factors, including reduced NMDA-receptor density, reduced perfusion, and focal atrophy. A reduction in 5-HT_{1A} serotonin binding potential was observed in temporal lobe epileptogenic foci [1095]. It is noteworthy that activation of 5-HT_{1A} receptors in experimental models inhibits neuronal firing. There is a report of increased histamine receptor binding assessed with ^{11}C-doxepin in epileptic foci [1096], but the specificity of this finding remains to be determined.

Increased levels of serotonin and quinolinic acid have been observed in epileptogenic lesions, raising the possibility that MTrp may accumulate these lesions.

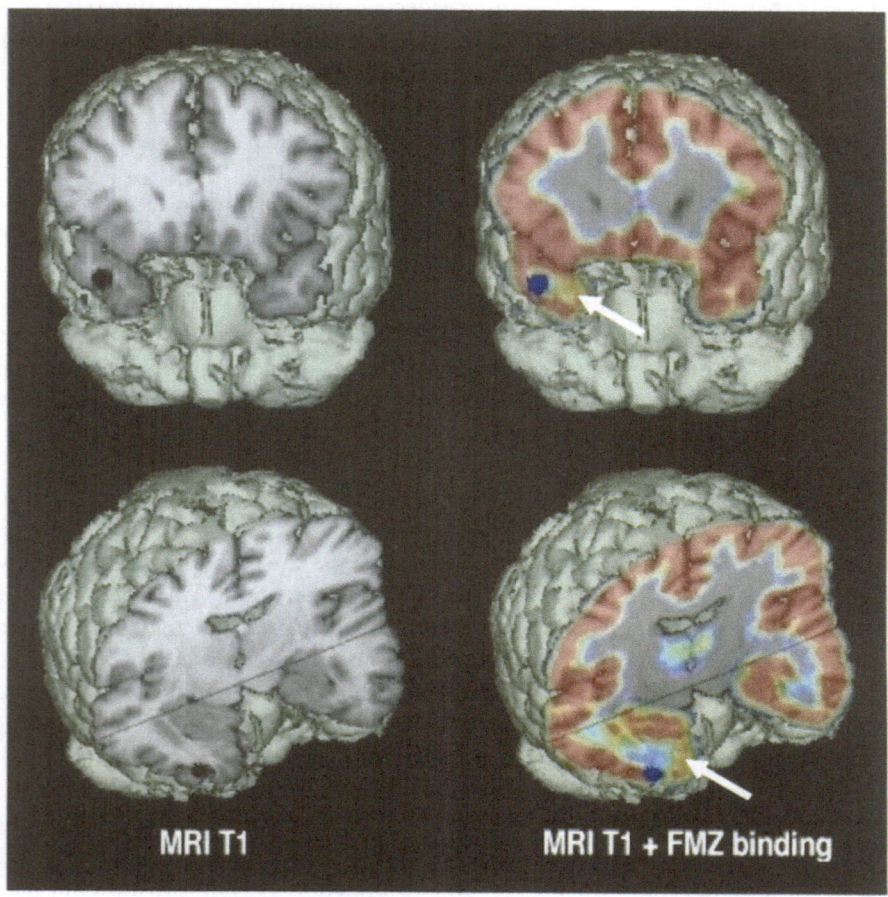

Fig. 2.41. Reduced FMZ uptake (*white arrow*) in the vicinity of a right temporal cavernoma, which is marked by *blue color* on 3D fusion rendering (*right column*) with coregistered MRI. (From [1076], with permission)

This has in fact been demonstrated in patients with focal epilepsy due to tuberous sclerosis. There was increased uptake of MTrp in those tubers that were subsequently demonstrated to be epileptogenic [1097, 1098]. In contrast, FDG uptake is typically reduced in these lesions [1099].

2.5.1.6 Inflammatory Lesions and Glial Reaction

Epileptic lesions may be associated with focal inflammation. The peripheral benzodiazepine receptor ligand ^{11}C-PK11195 can be used to image microglial activation. Positive findings have been demonstrated in Rasmussen's encephalitis [1100] and in vasculitis [1101]. Increased hippocampal gliosis in TLE may be imaged by ^{11}C-deuterium-deprenyl, which binds to monoamine oxidase B (MAO-B) expressed by glial cells [1102, 1103]. Interictal hippocampal hypermetabolism was observed with FDG-PET in patients with limbic encephalitis [1104].

2.5.2   Progression of Epileptic Lesions

There is an ongoing controversy about the progression of tissue damage in epileptic foci [1105]. Cross-sectional studies with PET provide some evidence for progression. There is a low incidence of abnormal FDG-PET in children with new-onset partial epilepsy [1106], and the severity of focal hypometabolism is related to epilepsy duration [1023, 1107]. These PET findings concur with MRI studies, which indicate a relation between hippocampal atrophy and disease duration. Also, patients with recent-onset seizures have less hippocampal atrophy than those with chronic TLE [1105].

2.5.3   Prediction of Surgical Outcome

PET can provide information on the prognosis of patients with focal epilepsy who have been surgically treated. A better outcome is associated with hypometabolism only in the affected temporal lobe or if there is a greater degree of asymmetry of temporal lobe metabolism as well as the presence of hypometabolism in the lateral temporal neocortex [1108–1110]. Extratemporal cortical hypometabolism outside the seizure focus, in particular hypometabolism in the contralateral cerebral cortex, may be associated with a poorer postoperative seizure outcome in TLE and may represent underlying pathology that is potentially epileptogenic [1111]. Bilateral temporal hypometabolism is present in approximately 10 % of patients with TLE. It is associated with a higher percentage of generalized seizures and worse prognosis for seizure remission after surgery [1112, 1113]. Bilateral thalamic hypometabolism, and in particular thalamic metabolic asymmetry in the reverse direction to that of the temporal lobe asymmetry, is also associated with poor outcome [1114].

2.5.4   Malformations of Cortical Development

Increased FMZ binding was observed in ectopic neurons of band heterotopia and in cortex adjacent to cortical dysplasias (Figs. 2.42, 2.43). A high incidence of focally increased FMZ binding in gray or white matter was observed in neocortical epilepsy with normal MRI, indicating that in many of these subjects migrational disturbances may be a cause of epilepsy that cannot be readily detected by MRI [1115]. In general, reduced FMZ binding is seen in subependymal nodular het-

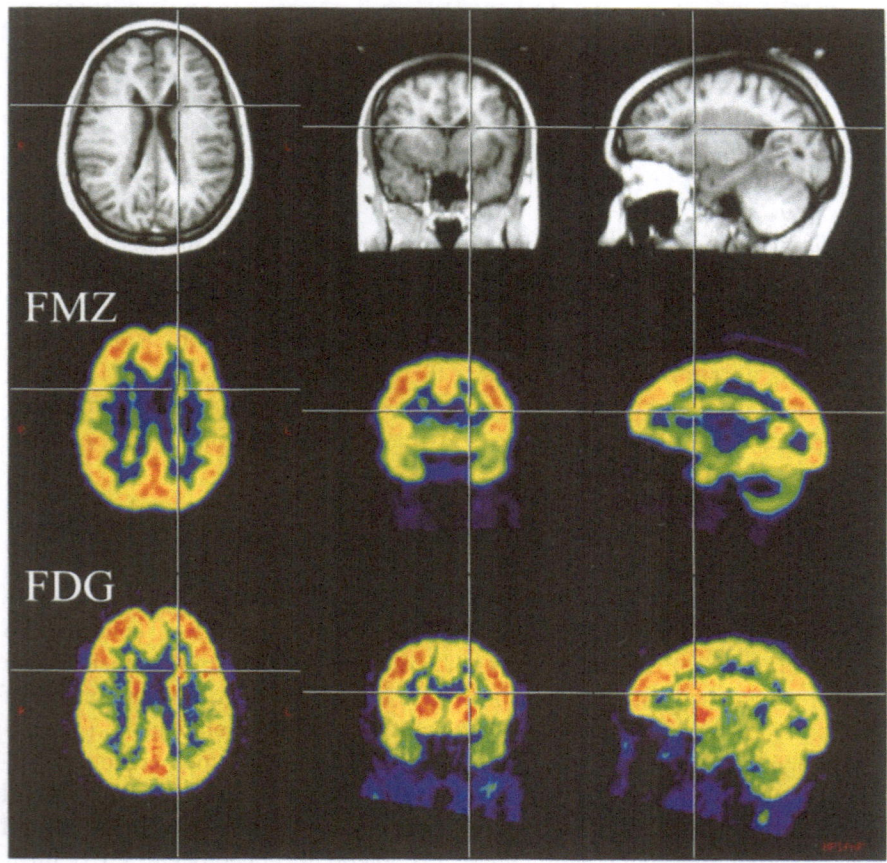

Fig. 2.42. Abnormal FDG and FMZ uptake in ectopic periventricular cortex (marked by *crosshairs* on coregistered orthogonal slices)

erotopia, focal cortical dysplasia and polymicrogyria [1116], and dysembryoplastic neuroepithelial tumors [1087].

Active glucose consumption suggesting synaptic activity has been detected in band heterotopia [1117, 1118] and in heterotopic nodules and displaced gray matter [1119, 1120]. Imaging rCBF with ^{15}O-water showed that regions of malformations of cortical development (MCD) can be activated during cognitive tasks, suggesting that heterotopic neurons synapse with neurons in other brain regions [955]. Abnormalities in the activation pattern of apparently normal cerebral cortex can also occur, indicating more widespread abnormalities of cortical organization [1121].

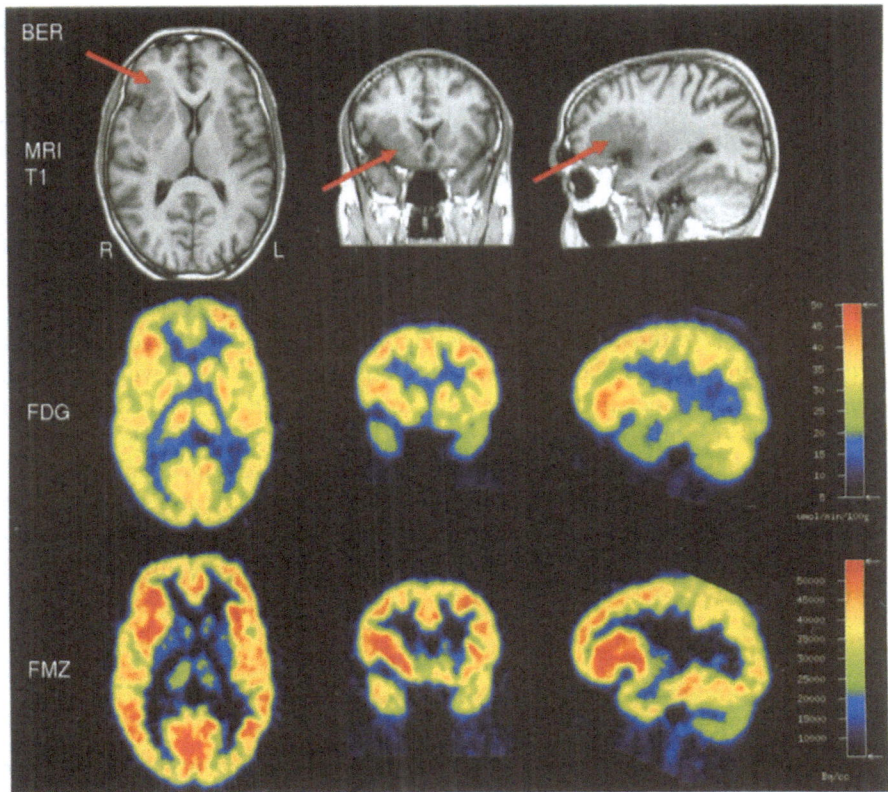

Fig. 2.43. Abnormal FDG and FMZ uptake in cortical malformation (*arrows*)

2.5.5   Childhood Epileptic Syndromes

There are several encephalopathies of childhood that are characterized by developmental delay and epileptic seizures. In patients with infantile spasms and West syndrome, FDG-PET studies suggest that the spasms are the result of secondary generalization from cortical foci [1122]. Other PET tracers, such as FMZ and (α-^{11}C-methyl-L-tryptophan, *MTrp*), are being used to investigate developmental abnormalities of serotonergic and GABAergic neurotransmitter systems in this syndrome [1123]. Patients with the Lennox–Gastaut syndrome have a variety of cerebral metabolic patterns, including normal, focal and diffuse (unilateral and bilateral) hypometabolism [1124, 1125]. Diffuse cortical dysfunction is common in the epileptic encephalopathies and may reflect the underlying cause of the condition or arise as a consequence of uncontrolled seizures. Altered thalamic glucose metabolism is further evidence of subcortical involvement in these conditions [1126].

As in the other epileptic disorders, hypometabolism most likely reflects impaired synaptic activity, but it often also includes the epileptogenic cortex [1127]. Hypometabolism was prominent in the temporal lobes in Landau–Kleffner syndrome [1128], an acquired aphasia that begins in childhood and is thought to arise from an epileptic disorder within the auditory speech cortex. In a few cases, however, focally increased CMRGlu in left superior temporal cortex has been noted [1129]; this may have been a consequence of epileptic discharges during the study. In scans performed during slow-wave sleep there is also a marked bilateral increase in glucose metabolism in these areas [1130]. Even after cessation of epilepsy in adulthood, impaired verbal short-term memory may persist; this is reported to correspond to reduced activation of superior temporal cortex [1131]. Similar findings to those in Landau-Kleffner syndrome are observed in the syndrome of continuous spike-and-wave discharges during slow wave sleep, with abnormalities in association cortices that are probably related to a disturbance of neuronal maturation [1132].

In few instances hemispherectomy is used to treat severe focal epileptic syndromes in childhood. Functional changes related to that treatment have been studied with PET and recorded in a case report suggesting extensive disease-related alterations and some reorganization of language and motor function [1133].

2.5.6   Language Dominance

Determination of language dominance may be relevant prior to temporal resections. It can be done by language activation studies with ^{15}O-water [817, 820, 1134, 1135], which are described in more detail in section 2.3.11.2 with reference to brain tumors. More recently, activation studies have mostly been performed with fMRI [1136–1142] rather than PET. Interpretation of functional activation studies should take into account the potential for functional plasticity in epileptic patients. They often acquire their brain lesions early in life, and language dominance may therefore not be as clearly lateralized as in normal subjects [1143].

Functional activation of dysplastic motor cortex associated with epilepsia partialis continua has been observed [1144]. Abnormal activations of uncertain functional significance may occur in association with hyperperfused lesions.

2.5.7   Effects of Surgical Intervention and Medical Treatment

Remote reduction of CMRglc in TLE in ipsilateral inferior frontal cortex and bilateral thalamus was reversible after resection of the epileptogenic lesion [1145–1147]. In a case report on childhood epileptic encephalopathy (Lennox-Gastaut syndrome) interictal temporal lobe unilateral hypometabolism disappeared corresponding to seizure control after callosotomy [1148].

Most antiepileptic drugs reduce CMRglc, and these effects are very similar in normals and in patients with epilepsy (see also section 3.5.2). Barbiturates, including phenobarbital and primidone, cause the greatest decrease, by approximately 37 %, in global CMRglc. Valproate reduces global CMRglc to a greater extent (22 %) than either phenytoin (13 %) or carbamazepine (12 %), while vigabatrin causes only an 8 % reduction. In general, the decreases in CMRglc attributable to these medications parallel their neuropsychological effects, with barbiturates having the greatest effect on cognition. Valproate has been found decrease global CBF by 15 %, vigabatrin by 13 % [1149–1154].

The regional pharmacokinetics of antiepileptic drugs can also be studied with PET. This has been done for ^{11}C-labeled diphenylhydantoin, demonstrating that normal brain regions of medically resistant epileptic patients bind diphenylhydantoin as effectively as in nonepileptic patients [1155]. Clinical efficacy of antiepileptic drugs may be reduced by activation of P-glycoprotein, which transports drugs back from brain to blood. The activity of P-glycoprotein, which may also be responsible for multiple drug resistance in tumors, can be studied with specific PET tracers (see section 2.3.6).

2.5.8   Generalized Epilepsy

Petit mal absences are a frequent manifestation of generalized epilepsy in childhood. Patterns of CMRglc are normal and identical for ictal and interictal scans but a 2.5- to 3.5-fold diffuse ictal increase in global CMRglc is evident when ictal studies are compared with hyperventilation control studies in which no seizures occurred [1156]. FMZ binding was reported to be reduced in thalamus, but increased in cerebellum in generalized epilepsy [1157].

After provocation of serial absence seizures, there was increased ^{11}C-diprenorphine elimination from the association cortex, but not from the thalamus. This suggests that endogenous opioids are released in the association cortex at the time of serial absences and lead to increased receptor occupancy [1158]. There is no interictal overall abnormality of opioid receptors in patients with childhood and juvenile absence epilepsy [1159].

2.6   Other Neurological Disorders

2.6.1   Traumatic Brain Injury

CMRglc is often reduced in patients with acute brain trauma and, as with other lesions, the extent of abnormalities is usually larger than that seen in CT and MRI [1160, 1161], although this may be due, in part, to partial volume averaging with unaffected tissue. In accordance with experimental findings [1162], FDG-PET in the subacute phase is correlated with outcome, but is judged not to improve prediction beyond that possible on the basis of clinical and CT/MRI data, and correlation between changes of CMRglc and clinical recovery is poor [1163–1165].

Studies of oxygen metabolism often demonstrate highly inhomogeneous findings, including focal evidence of misery perfusion and hyperglycolysis in brain contusion [1166–1168]. Low oxygen metabolism is associated with impaired cerebrovascular autoregulation and poor prognosis [1169]. Hyperventilation may be used for treatment of increased intracranial pressure, but has been controversial because of associated reduction of CBF. A PET study demonstrated that hyperventilation indeed produced large reductions in CBF but not energy failure, even in regions in which CBF fell below the threshold for energy failure defined in acute ischemia. Oxygen metabolism was preserved due to the low baseline metabolic rate and compensatory increases in OEF; thus, it was felt that these reductions in CBF were unlikely to cause further brain injury [1170]. An increase in CBF would be expected as the consequence of therapeutic elevation of cerebral perfusion pressure if autoregulation is impaired. In brain trauma patients, pericontusional CBF increased modestly with increased CPP; greater CBF increases in distant brain tissue suggested a more widespread disturbance of autoregulation [1171].

There is some evidence from activation studies that patients in the subacute phase of traumatic brain injury perform memory tasks using altered functional neuroanatomical networks (e.g., less thalamic and parietal and more frontal and occipital activation). These changes may be the result of diffuse axonal injury and may reflect either cortical disinhibition attributable to disconnection or compensation for inefficient mnemonic processes [1172–1174]. Bilateral hypometabolism in hippocampus and anterior cingulate was observed in a patient with normal MRI and persistent posttraumatic amnesia [1175].

Some patients with whiplash injury experience persistent impairment of cognitive function of uncertain organic cause. Case reports suggest that PET could clarify the functional disturbance in some of these patients [1176], but a systematic study has not established any significant relation between symptoms and PET findings [1177, 1178].

2.6.2   Persistent Vegetative State

Resuscitation and intensive care after hypoxic brain damage are often associated with uncertain outcome. Hypoxic brain damage may result in a vegetative state, which imposes a significant burden on physicians and relatives of patients and can involve enormous costs. Thus, one would wish to have better tools that could be applied early after hypoxic brain damage to predict prognosis.

On the first day after severe hypoxic brain injury, cortical CMRglc metabolism is reduced by approximately 50%, but it has not been possible to establish a relation between the severity of metabolic impairment and outcome [242] (see also section 2.1.7.1). Even if coma persists for several weeks, substantial recovery is still possible, albeit rare. A method that could predict the best possible outcome would be highly valuable, as it would help determine the intensity of therapeutic measures to be used. Attempts have been made to establish whether PET could deliver such information, but measurements are difficult to perform in these critically ill patients and data interpretation is usually limited by small numbers. After hypoxic brain damage, global CMRglc is usually drastically decreased (by approx. 50%) in patients in a vegetative state and to a lesser degree (by approx. 25%) in conscious subjects. Most frequently, the parieto-occipital cortex, which is the "frontier" between vertebral and carotid arterial territories, frontomesial junction, striatum (in association with dystonia), and visual cortex are affected most severely, whereas the cerebellum, and in particular the vermis cerebelli, appears to be relatively spared [1179–1182]. In some vegetative patients with behavioral fragments, segregated corticothalamic networks that retain connectivity and partial functional integrity have been observed [1183].

Neuronal loss has been studied with ^{11}C-flumazenil PET. In the chronic state a considerable reduction of BZR-binding sites below normal values was documented in all cortical regions [1181] (Fig. 2.44). The reduction of FMZ binding grossly corresponds to the extent of reduction of cerebral glucose metabolism assessed with FDG-PET, whilst the cerebellum is spared from neuronal loss [1184].

Reactivity to external stimuli can be determined by activation studies. Increased regional activity (CBF) was found mainly in the posterior insula and in parietal cortex with electrical nociceptive stimulation in long-term postanoxic vegetative state, although some of these patients did not show electrophysiological responses to these stimuli and had severe impairment of resting CMRglc [1185]. This suggests that a residual pain-related cerebral network remains active at the cortical level, although this probably does not imply any involvement of conscious processing. Significant cortical activation of CBF was also elicited in vegetative patients by auditory click stimuli. However, this activation was limited to primary cortices and dissociated from higher order associative cortices, thought to be necessary for conscious perception [1186]. Cognitive stimulation paradigms may provide prognostically more relevant information. In two apparently unconscious patients clear regional CBF responses were observed during well-documented activation paradigms (face recognition and speech perception), which was followed by significant clinical recovery after a few months [1187].

2.6.3   Perinatal Brain Damage

Global brain CMRglc is very low in preterm infants (8.8 μmol/100 g per min) [1188] and infants with hypoxic brain damage [1189]. Higher values (18–55 μmol/100 g per min) that correlated with outcome were observed in another study [1190]. In infants with perinatal hypoxic-ischemic brain damage, FDG accumulated most actively in the subcortical areas (thalami, brain stem and cerebellum) and the sensorimotor cortex during the neonatal period. Follow-up PET studies showed that the uptake of FDG was high and increased in all brain areas of infants with

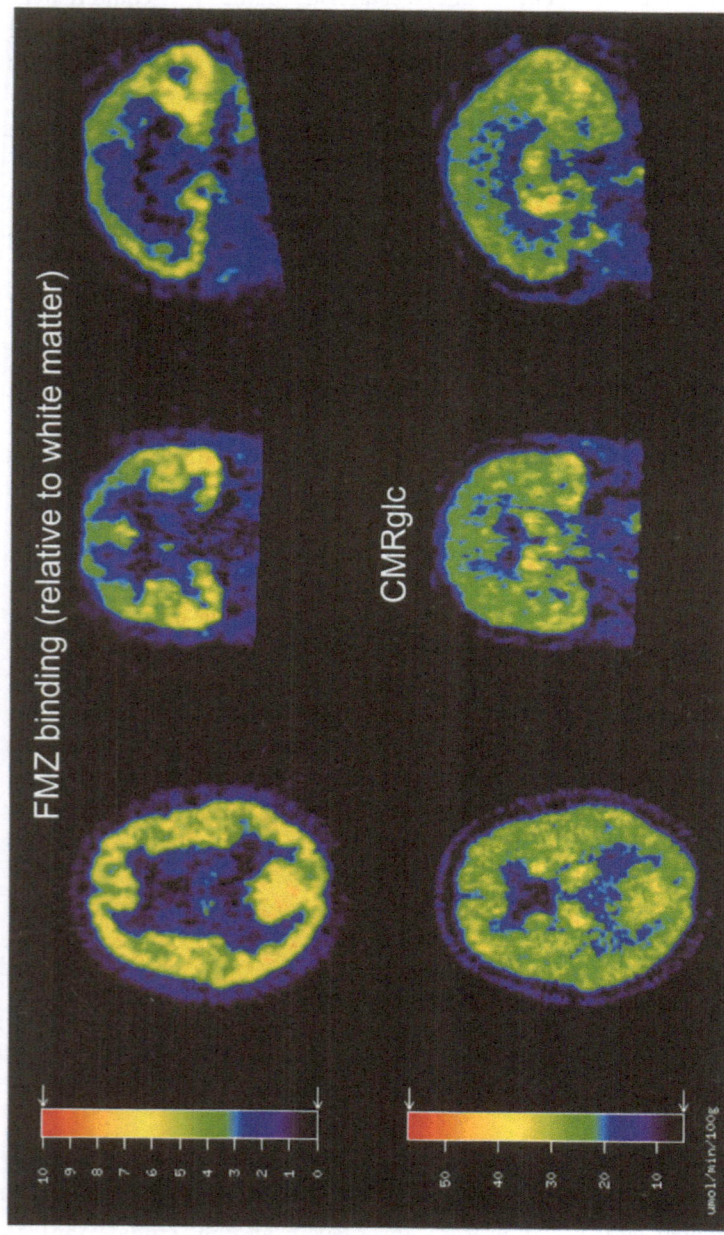

Fig. 2.44. FDG- and FMZ-PET in a patient with persistent vegetative state (orthogonal slices through left thalamus). FMZ binding is significantly reduced in most cortical regions (*top row*, values are scaled with reference to white matter, the normal range in cortex is 5–8; see [1184]). Global CMRglc is also reduced (*bottom row*)

normal development, whereas those with delayed development had significantly lower values [1191].

The most frequent finding in infants with perinatal asphyxia was a relative deficit in CBF in parasagittal regions, i.e., the cerebrovascular watershed regions, suggesting that the brain injury is basically ischemic [1192] and that the pathogenesis relates to impaired cerebral perfusion, perhaps secondary to systemic hypotension occurring in association with perinatal asphyxia. This mechanism may also be responsible for intraventricular hemorrhage with hemorrhagic intracerebral involvement, because extensive areas of severely reduced CBF were also found in these infants [1193, 1194]. Preterm infants tend to have lower CBF than infants born at term. Thresholds of CBF and $CMRO_2$ for irreversible brain damage appear to be lower in infants than in adults, and CBF values of less than 10 ml/100 g per min can be survived without major neurological deficits [1195, 1196].

2.6.4   Inflammatory Disease

Studies of CMRglc in encephalitis demonstrate major differences depending on how acute the inflammation is. In the acute phase, focal increase of CMRglc and CBF has been observed in herpes simplex encephalitis [1197, 1198], and limbic encephalitis [1104, 1199, 1200], probably reflecting the metabolic needs of inflammation rather than neuronal function. Acute brain abscesses may also be accompanied by high glucose metabolism (similar to that seen with malignant brain tumors), probably reflecting anaerobic glycolysis by macrophages [1201–1203].

In the chronic phase, when there is no longer acute tissue inflammation, focal and remote areas of hypometabolism are usually seen (Fig. 2.45). A pattern of inhomogeneous areas of hyper- and hypometabolism that changes with progression of the disease has also been observed in subacute sclerosing panencephalitis [1204–1206].

Rasmussen's encephalitis, a progressive encephalitis, has similar findings. It is characterized by glucose hypometabolism in the affected hemisphere [1207, 1208]. Within this hemisphere, there may be discrete foci of hypermetabolism that could be related to seizure activity [1209]. Focal and diffuse increase in binding of ^{11}C-PK11195 has been observed throughout the affected hemisphere, corresponding to activated microglia [1100]. Hemispherectomy may be needed to control the severe seizures that accompany this disease. Postsurgical reversibility of hypometabolism in the nonresected prefrontal cortex and cognitive improvement was observed in one case, illustrating recovery of functionally impaired tissue after removal of the primary lesion [1210].

The activation of microglia in herpes simplex encephalitis has been demonstrated by extensive increased binding of ^{11}C-PK11195. In contrast to a more transient increase of CMRglc in inflammatory lesions, this increased binding may persist many months (>12) after successful antiviral treatment [1211].

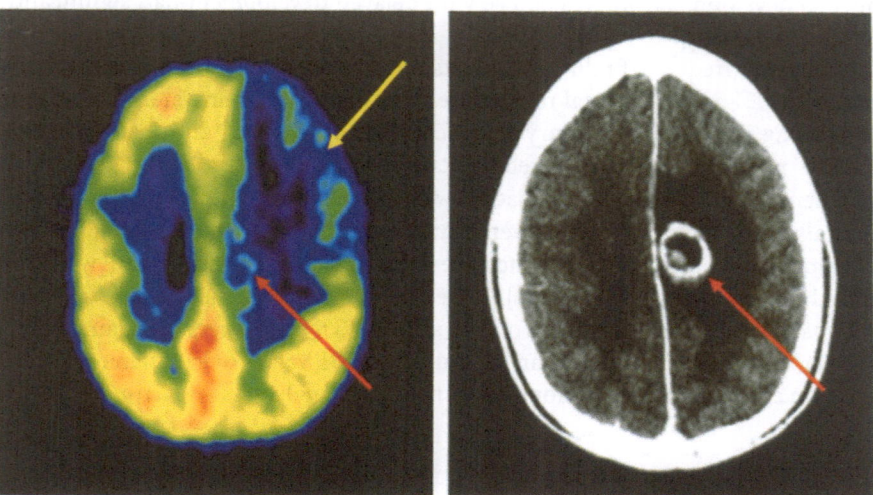

Fig. 2.45. FDG-PET (*left*) and corresponding CT (*right*) of cerebral toxoplasmosis, showing low FDG uptake in contrast-enhancing lesion (*red arrow*) accompanied by a severe reduction of CMRglc in structurally intact frontal cortex (*yellow arrow*)

Multiple sclerosis (MS) may be associated not only with demyelinating lesions in white matter, which are detected with high sensitivity by MRI, but also with neuronal dysfunction and eventual neuronal loss, which can be detected by PET with progression of disease [1212]. There is no general alteration of cortical metabolism in MS, but asymmetry may be increased [1213]. $CMRO_2$ is reduced in patients with cerebral atrophy, cognitive impairment or a high number of relapses [1214, 1215]. Fatigue is associated with frontal cortex and basal ganglia CMRglc reduction [1216].

Active MS lesions show increased uptake of ^{68}Ga-EDTA [1217] and of ^{55}Co [1218], probably as a consequence of BBB damage. An increase of peripheral benzodiazepine receptor-binding capacity, which is typical for activated microglia, is found during relapse, even in white matter that has appeared normal on MRI [1219, 1220]. Compared with normal white matter, lesions may show increased CMRglc [1221].

Predominantly global reduction of CMRglc is found in Lyme disease, which may produce symptoms similar to MS [1222].

Reductions of cortical CMRglc and CBF have also been found in cerebral lupus erythematosus, which may produce MS-like symptoms, extrapyramidal movement disorders and psychiatric symptoms [1223–1225]. Reductions are typically found in patients with neuropsychiatric symptoms, even when MRI findings are normal [1226]. Parieto-occipital regions are most commonly affected, followed by parietal regions [1227]. Increased binding of ^{11}C-PK-11195 has been observed in cerebral vasculitis with focal epilepsy [1101].

2.6.5   Pain

Pain accompanies many diseases as a symptom, but can also occur as a disease in its own right in chronic pain syndromes. Effective treatment of pain is one of the primary goals of clinical medicine, and a better understanding of its pathophysiology probably is essential for progress, especially in the chronic pain disorders.

The functional neuroanatomy of cerebral pain processing has been studied with measurements of CBF and CMRglc during various kinds of painful stimulation in normals [1228–1234, 1234–1245]. Relative regional activations were accompanied by a global decrease of CBF (by approx. 22%) not due to hyperventilation [1246]. Relative CBF increases to noxious stimuli are almost always observed in second somatic (SII) and insular regions, and in the anterior cingulate cortex (ACC), and with slightly less consistency in the contralateral thalamus and the primary somatic area (SI). Activation of the lateral thalamus, SI, SII and insula are thought to be related to the sensory-discriminative aspects of pain processing. There appears to be a gender difference, with greater activation of the contralateral insula and thalamus in female subjects [1247]. ACC does not seem to be involved in coding stimulus intensity or location, but appears to participate in both the affective and attentional concomitants of pain sensation, as well as in response selection. Subdivisions of the ACC and their relevance for pain processing have been studied in detail [1248, 1249]. ACC can also be activated by an illusion of pain [1250]. Activation of the insula was associated with allodynia in mononeuropathic pain [1251] and with muscle pain [1252] in one study, whereas in another, a medial thalamic pathway to the frontal lobe was activated by heat allodynia [1253]. Activation by phantom limb pain was found to be very similar to that by actual painful stimuli [1254]. Although some small differences were observed between patients with nonspecific low back pain and controls during painful stimulation, these were not considered sufficiently striking to suggest abnormal nociceptive processing [1255].

The key role of the thalamus in processing pain was seen in patients with a chronic pain syndrome who had reduced CBF in contralateral thalamus that normalized with effective therapy by percutaneous high cervical cordotomy [1233, 1256]. PET during thalamic stimulation for pain treatment revealed significant activation of the thalamus in the region of the stimulating electrodes, as well as activation of the insular cortex ipsilateral to the electrodes [1257]. Differences between acute and chronic pain were observed in another patient with thalamic stimulation. Chronic pain was associated with CBF increases in the prefrontal and anterior insular cortices, hypothalamus and periaqueductal gray, whereas short-term thalamic stimulation increased CBF in the amygdala and anterior insular cortex [1258].

The relevance of affective processing in ACC for chronic pain syndromes has been demonstrated in patients with atypical facial pain, who had higher blood flow (than controls) in the ACC and decreased blood flow in the prefrontal cortex during painful stimulation [1259]. In contrast to patients with neuropathic pain, patients with chronic rheumatoid arthritis or acute post-dental extraction pain showed reduced ACC activation [1260, 1261]. Relief from chronic neuropathic pain has also demonstrated the involvement of bilateral anterior insula, posterior parietal, lateral inferior prefrontal, and posterior cingulate cortex [1262]. With motor cortex stimulation for pain control, CBF in ACC increased in patients with good analgesic efficacy, while it decreased in those with poor clinical outcome [1263, 1264]. Similar effects were also seen with thalamic stimulation [1265], with electrostimulation of the trigeminal ganglion [1266], and with both opioid and placebo analgesia [1267].

A related but more complex pattern emerged in a study of the analgesic effect of fentanyl. Fentanyl increased CBF in the ACC and contralateral motor cortices, and decreased rCBF in the thalamus (bilaterally) and posterior cingulate during both noxious and nonnoxious thermal stimuli. During combined pain stimulation and fentanyl administration, fentanyl significantly augmented pain-related rCBF increases in the supplementary motor area and prefrontal cortex. This activation pattern was associated with decreased pain perception, as measured on a visual analogue scale, suggesting that ACC activation alone cannot explain the analgesic effect [1268]. Hemispheric asymmetry was observed in a study of electrical extradural precentral gyrus stimulation in trigeminal neuropathy. Increased regional CBF was detected in the right caudal ACC and anterior limbic thalamus during the chronic pain state, in comparison with the pain-alleviated state, regardless of the side where the neuropathy occurred [1269].

The relevance of endogenous opioids and their receptors for pain perception is obvious. In a study of healthy volunteers with ^{11}C-carfentanil, sustained pain induced the regional release of endogenous opioids interacting with μ-opioid receptors in a number of cortical and subcortical brain regions. The activation of the μ-opioid receptor system was associated with reductions in the sensory and affective ratings of the pain experience, with distinct neuroanatomical localization [1270]. Similar findings were observed in the contralateral thalamus, consistent with competitive binding between ^{11}C-carfentanil and acutely released endogenous opioid peptides in response to a painful stimulus [1271]. In another study, men demonstrated larger magnitudes of μ-opioid system activation than women in the anterior thalamus, ventral basal ganglia, and amygdala. Conversely, women demonstrated reductions in the basal state of activation of the μ-opioid system during pain in the nucleus accumbens, an area that was previously associated with hyperalgesic responses to the blockade of opioid receptors in experimental animals [1272]. Alfentanil (a μ-opioid receptor agonist) reduced subjective pain intensity and the associated striatal dopamine release (measured with ^{11}C-raclopride) that was associated with a mechanical pain stimulus [1273].

In patients with trigeminal neuralgia, the volume of distribution of ^{11}C-diprenorphine binding was significantly increased after thermocoagulation of the relevant trigeminal division, suggesting an increased occupancy by endogenous opioid peptides or down-regulation of binding sites during trigeminal pain [1274]. In a patient with central poststroke pain that had developed after a small pontine hemorrhagic infarction, reduced ^{11}C-diprenorphine binding was more accentuated than glucose hypometabolism on the lateral cortical surface contralateral to the symptoms and a differential abnormal distribution between the tracers was seen in pain-related central structures [1275].

An alteration of the dopaminergic system with reduced FDOPA uptake and increased binding of ^{11}C-raclopride in putamen was observed in a unique pain syndrome, the burning mouth syndrome, which is probably not comparable to other pain syndromes [1276, 1277].

The sympathoneural imaging agent 6-^{18}F-fluorodopamine was used to visualize sympathetic innervation, and ^{13}N-ammonia to visualize local perfusion of the extremities, in patients with painful diabetic neuropathy. The results provided evidence for regionally selective sympathetic denervation [1278].

2.6.6   Migraine and Cluster Headache

Migraine is a very common type of headache with unknown etiology, associated with a functional disorder of the brain [1279]. There are familial forms that have been associated with neuronal channelopathies [1280], and there is evidence for release of vasoactive neuropeptides from perivascular nerve terminals. Serotonin (5-HT$_{1B}$) receptor agonists, the tryptans, provide effective treatment in most patients. Classic migraine attacks have a sequence of symptoms, beginning with neuronal dysfunction (aura, typically with visual field abnormalities, more rarely with other neurological deficits, such as hemiparesis), which are followed by pulsating headache and nausea. Rather diverse functional imaging findings have been observed [1281, 1282], and it is very likely that the inconsistency of findings is related to timing of measurement during the rapid sequence of functional events in this disease.

PET studies with α-^{11}C-methyl-L-tryptophan, show increased uptake in patients with migraine, and provide some evidence of increased regional serotonergic activity [1283, 1284]. No abnormalities of 5-HT$_2$ receptors were found in cerebral cortex of migraineurs studied using ^{18}F-fluorosetoperone [1285].

PET studies of CBF and CMRO$_2$ at baseline, during aura and headache, and after treatment with sumatryptan showed reduced CBF and CMRO$_2$ in primary visual cortex during the aura phase only [1286]. A reduction in CMRglc was observed in reserpine-induced headache and visual disturbance in patients with migraine [1287]. Functional abnormalities in familial hemiplegic migraine can persist over several days, and a corresponding reduction of CMRglc (without associated changes of CBF) in the affected hemisphere has been observed [1288]. The findings support the concept of a primary neuronal dysfunction that causes neurological deficits in migraine and that may trigger subsequent vascular dysregulation.

In one study, increased CBF in the brain stem was observed during migraine attacks [1289]. Brain stem activation was not observed in an acute trigeminal pain state that was elicited in normal volunteers by a small amount of capsaicin administered subcutaneously in the right forehead [1238].

Cluster headache is characterized by short but very severe attacks of unilateral headache that are accompanied by vegetative symptoms. It is regarded as a vascular pain syndrome that has similarities with migraine, but also clear differences [1290]. In the acute pain state, activation is seen in the ipsilateral inferior hypothalamic gray matter, the contralateral ventroposterior thalamus, the anterior cingulate cortex, and bilaterally in the insula, but not in the brain stem [1291, 1292].

2.6.7   Narcolepsy

Narcolepsy is characterized by abnormal daytime sleepiness and intrusion of rapid eye movement sleep phenomena into waking. Recent advances provide compelling evidence that narcolepsy may be a neurodegenerative or autoimmune disorder resulting in a loss of hypothalamic neurons containing the neuropeptide orexin (also known as hypocretin) [1293]. There are no known positron-emitting tracers for this transmitter system. Earlier studies investigated D_1 and D_2 receptors [1294–1297] and muscarinic receptors [1298], but did not reveal any abnormalities.

2.6.8   Hypoparathyroidism (Fahr's Disease)

The disease is characterized by progressive calcification and degeneration of the basal ganglia, cerebellar dentate nucleus, and other brain structures, leading to movement disorders and cognitive impairment. Although calcifications are rather localized, reduction of CMRglc is much more widespread and corresponds to clinical symptoms [1299, 1300].

2.6.9   Systemic Inherited Metabolic Disorders

Many brain metabolic disorders are associated with reduced energy metabolism, which is associated in turn with impaired brain function, irrespective of the primary metabolic cause. For instance, global reduction of cortical CMRglc was found in type I alpha-N-acetylgalactosaminidase deficiency [1301]. Even in Duchenne muscular dystrophy, in which the primary manifestation is in muscle, CMRglc is decreased [1302]. Global reduction of CMRglc is also found in myotonic dystrophy [1303–1305], which is an autosomal dominant disorder characterized by myotonia, muscular dystrophy, cataracts, and hypogonadism caused by an amplified trinucleotide repeat in the untranslated region of a protein kinase gene on chromosome 19. Uncoupling of blood flow and oxygen metabolism in the cerebellum is observed in type 3 Gaucher disease [1306].

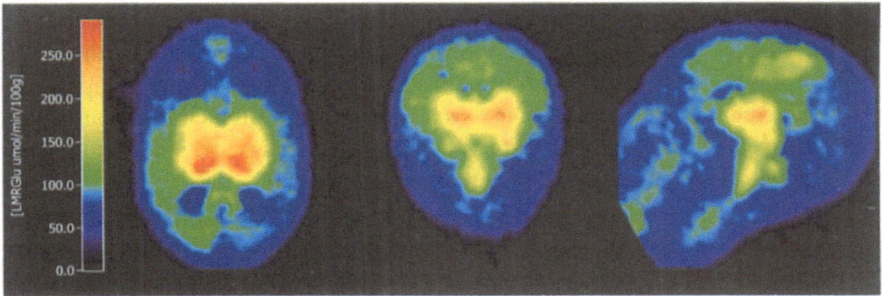

Fig. 2.46. FDG-PET (orthogonal slices through basal ganglia) in a newborn infant with acute severe mitochondrial encephalopathy (Leigh's disease) due to a complex-1 deficit. CMRglc is massively increased up to 250 µmol/100 g per min (see color scale, normal gray values are 40–60 µmol/100 g per min), in particular in the basal ganglia with subsequent tissue necrosis and death

Increased CMRglc is observed in Turner's syndrome (lack of one X chromosome or part of one X chromosome and of endogenous estrogen). However, normalized metabolism is significantly reduced in the insula and association neocortices bilaterally, which may be related to social-behavioral problems in these patients [1307].

Some metabolic disorders affect primarily the basal ganglia, with subsequent impairment also in cortex and cerebellum. Among these disorders is Leigh's disease, which leads to lactacidosis [1308], methylglutaconic aciduria [1309], and propionic acidemia [1310]. They may initially cause striatal hypermetabolism, (Fig. 2.46), indicating abnormally activated or nonoxidative metabolism, which then declines to abnormally low levels as striatal neurons degenerate. At later disease stages, metabolism of the whole brain is severely impaired.

2.7   Psychiatric Disorders

In the past, neurobiological research in psychiatry has been limited largely to the study of genetic traits, metabolic and endocrine alterations in blood and CSF, electrophysiological recordings, and the study of macroscopic and microscopic alterations of brain structure. There has been extensive discussion of the psychological abnormalities that accompany and can contribute to these diseases. In contrast to neurology, where postmortem and experimental studies laid a solid ground for classification and understanding of the essential pathophysiological alterations in most neurological diseases, this has been far more difficult for such psychiatric diseases as depression, schizophrenia, anxiety and personality disorders, and addiction.

Functional imaging, and in particular PET, opens up entirely new possibilities for the study of in vivo alterations of brain function that may underlie these disorders. Thus, we can now explore the neurobiological changes that accompany the mental and behavioral alterations in psychiatric diseases, their clinical course, and their changes during therapeutic interventions. They can be compared to molecular and biochemical indicators of genetic predisposition.

Those parts of the human brain that support neurological function underwent relatively little change during mammalian evolution. Thus, studies in experimental animals contribute a great deal to our understanding of diseases in these parts of the brain. However, prefrontal cortex and parts of the limbic system underwent major changes during human phylogenetic development, and the contribution of animal experiments to our understanding of their functions in humans is therefore limited. Thus, functional imaging studies of the prefrontal cortex [1311] and the limbic system are at the core of functional neuroimaging research in psychiatry.

Studies of brain function employing measurements of CBF or energy metabolism have provided considerable insights into functional abnormalities, often affecting frontal or limbic brain structures, but they are also plagued by wide variability. There are many sources for that variability, including inhomogeneity of clinical diagnostic groups, variability of current state (e.g., anxiety) [1312], influences of prior treatment [1313], different underlying genetic traits in different study populations, differences in ambient conditions, and the difficulty of controlling mental processes during examination. Thus, progress is expected to be slow, as all these factors need to be sorted out in careful experimental and clinical studies. In this volume, studies are only briefly summarized and it is not possible to discuss details of interpretation, for which the reader is referred to specialized review articles.

2.7.1   Depression

Depression is a very common disease with a lifetime prevalence of about 15 %, and it is very likely that some biological disorder contributes substantially to its etiology [1314]. Recent postmortem reports of cytoarchitectural alterations in prefrontal cortices in affective disorders, characterized by a decrease in the number or density of glia or size and density of some neuronal populations, putatively reflect aberrant neurodevelopment or cellular plasticity [1315]. The episodic nature of depression and the responsiveness to treatment permit functional imaging with PET in both early and later stages of depression and therefore could contribute to the clarification of biological alterations in depression. Besides studies of CBF and CMRglc at rest and during various activation paradigms, the function of the serotonergic system is of particular interest, because the action of the most effective antidepressant drugs is to increase serotonergic neurotransmission.

2.7.1.1   Cerebral Blood Flow and Cerebral Metabolic Rate (Glucose)

There have been many studies of resting CBF and CMRglc in depression, with varying results that are not easily understood or interpreted. Differences between studies may be ascribed to differing subject selection, symptom profiles, clinical states, medication status, and image acquisition (for reviews see [1316–1319]). A key focus of functional imaging with PET in major depressive disorders has been the differences from normal controls in brain function at baseline. Furthermore, studies have been aimed at elucidating changes before the initiation of and after response to specific pharmacotherapy and detecting differences in regional brain activity associated with particular depressive symptom factors [1320–1325]. A number of PET studies have identified primarily the prefrontal cortex, the temporal lobes, the cingulate cortex, and related parts of the basal ganglia and thalamus as an important neural correlate of depressive disorder. Specifically, the dorsolateral prefrontal cortex (DLPFC) has been found to show decreased activity in acute major depression (Fig. 2.47), whereas the ventrolateral prefrontal cortex (VLPFC) has most commonly been found to reveal increased activity. Hemispheric asymmetries have been noted, with predominantly left-sided impairment in some studies [1326, 1327] and right-sided impairment in others [1328].

Furthermore, blood flow and glucose metabolism in the prefrontal cortex have been found to be inversely correlated with disease severity [1326, 1329, 1330]. A correlation between left amygdala metabolism and stressed plasma cortisol levels is observed in depressive patients [1331]. Impairment in frontal metabolism appears to be common to different types of depression [1332], including drug-induced depression [1333] and secondary depression [1334].

There are many indications that alterations of CBF and CMRglc are more closely related to the affective and cognitive state during the examination than to the disease trait. Studies have revealed that variations in cerebral metabolic abnor-

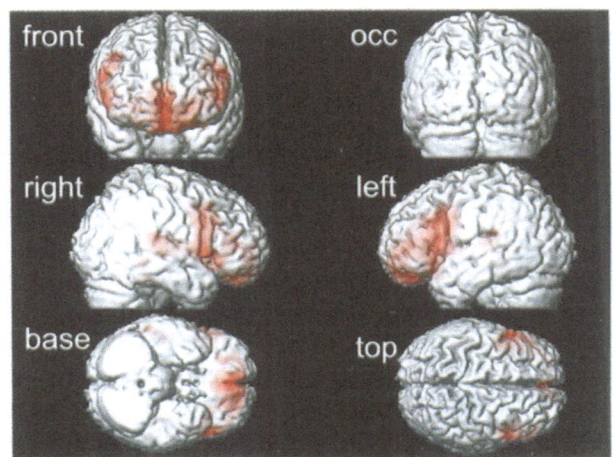

Fig. 2.47. Brain areas (marked by *red color*) with significant reduction of normalized CMRglc in a group of 84 patients with major depression, unipolar type, compared with an age-matched control group. (Courtesy of Dr. V. Holthoff, TU Dresden, Germany)

malities in prefrontal cortex activity may mediate different clusters of depressive symptom profiles [1319, 1335, 1336]. For example, hypoactivity of the dorsolateral prefrontal cortex has been firmly linked to psychomotor retardation and anhedonia [1337, 1338]. In normals, self-induced dysphoria leads to bilateral inferior and orbitofrontal activation in women, whereas men display predominantly left-sided activation in these areas [1339] A recent study in normals demonstrates a shift in limbic and neocortical activity upon self-induced dysphoria [1340]. In another study, left hippocampal dysfunction was associated with major depressive episodes not only in patients with depression, but also in those with obsessive-compulsive disorder [1327]. Thus, such studies provide insight into the neural networks involved in the regulation of affect and emotion that are dysfunctional in depression [1317, 1341–1344].

Pre- to posttreatment PET studies have indicated that these abnormalities in patients with ongoing affective symptoms resolve with successful treatment, but reports have been heterogeneous. Bench et al. followed patients from acute depression to the state of clinical remission and formulated the original concept of reversibility upon remission [1319]. Recovery from depression is associated with increases in regional blood flow in the left DLPFC and medial prefrontal cortex including anterior cingulate, the same areas in which focal decreases were described in the depressed state compared with normal controls. Results of later studies suggest an increase in DLPFC [1345, 1346] and a decrease in VLPFC [1325, 1330, 1343] with a variety of antidepressant pharmacotherapy or upon remission. In addition, studies have indicated that functional imaging data may predict clinical response to antidepressant treatment according to the distribution and also direction of aberrant brain activity [1320, 1325, 1347–1350]. High rostral anterior cingulate metabolism has been described as a predictor of good treatment response [1320, 1351], and this finding has been confirmed and extended in another study, which has shown that improvement of major depressive disorder symptoms is significantly correlated with lower pretreatment metabolism in the amygdala and thalamus and with higher pretreatment metabolism in the medial prefrontal cortex and rostral anterior cingulate gyrus [1352]. In vivo challenge designs for the serotonergic system have been proposed to predict treatment outcome [1353]. A recent study in patients suffering unipolar depression revealed significant regional decreases in brain metabolism upon remission with respect to their baseline scan in left prefrontal and anterior temporal regions, left anterior cingulate cortex, bilateral putamen, thalamus and cerebellum. There was significant asymmetry in prefrontal and anterior cingulate metabolism, with lower metabolism in the left hemisphere that persisted despite clinical remission [1354].

Cognitive performance is often impaired in depression [1355]. It is associated with reductions in frontomesial or frontolateral CMRglc [1328, 1356, 1357], at least in unipolar depression (Fig. 2.48). The issue is particularly relevant in late-life depression, because depression may be a risk factor for dementia [1358, 1359] and may be related to vascular lesions [1360]. Differentiation of the metabolic changes related to depression from those related to dementia is therefore an issue for ongoing studies.

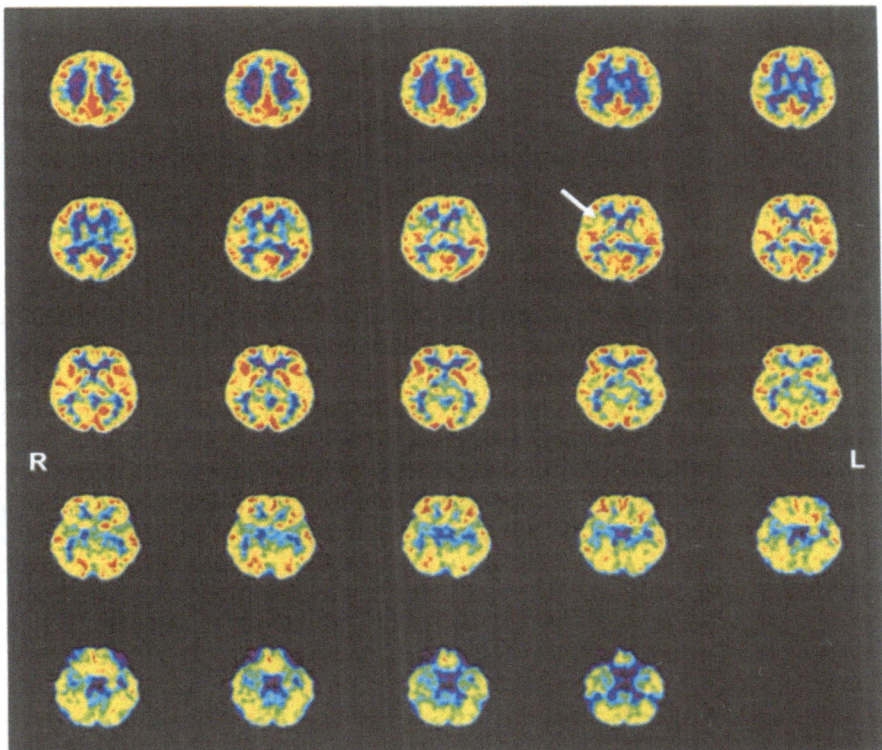

Fig. 2.48. Almost normal FDG-PET in a 60-year-old female patient with acute major depression (score 27 on 21 item Hamilton Depression Scale) associated with mild cognitive impairment (MMSE 24). There is a mild reduction of left frontal CMRglc (*arrow*) compared with the right side. This is clearly a different picture from that seen in early Alzheimer disease (which could lead to a similar MMSE score). (Courtesy of Clinic for Nuclear Medicine, PET Center Rossendorf, Germany, Dr. B. Beuthien-Baumann)

2.7.1.2   Serotonin Receptors

Abnormalities of 5-HT$_{1A}$ receptors have been described in depression. With PET ligands, a substantial reduction of the binding potential is found in the raphe and in the mesiotemporal cortex, and to a lesser extent also in other cortical areas [1361, 1362]. The magnitude of these abnormalities is most prominent in bipolar depressives and in unipolar depressives with bipolar relatives. The reduction does not depend on previous treatment with serotonin reuptake inhibitors [1363].

Some beta adrenergic receptor blockers (e.g., pindolol, penbutolol, and tertatolol, and also buspirone) that may be useful for augmentation of antidepressive therapy compete with 5-HT$_{1A}$ ligands at the receptors [1364–1367]. These effects are now also being studied in a more experimental setting with PET in small animals [1368]. Pindolol exhibits an in vivo selectivity for the 5-HT$_{1A}$ autoreceptors at the dorsal raphe nuclei, which appears advantageous for augmentation [1369]. High receptor binding has been observed with the novel full agonist DU125530 [1370].

No change of the 5-HT$_{2A}$ receptor binding potential (BP) has been found in depression of old age [162]. One study with ^{18}F-setoperone suggests that treatment by SSRIs could induce up-regulation of frontal 5-HT$_{2A}$ receptors [1371], whereas another study has found down-regulation after 6 weeks' paroxetine treatment in young depressive patients [1372].

2.7.1.3   Serotonin Transporters

Measurement of serotonin transporter (SERT) BP is of particular interest in depression, because it has recently been reported that a SERT promoter polymorphism may predispose to depression after stressful life events [1373]. Binding potential of SERT in the thalamus was significantly increased in patients with depression, whereas BP in the midbrain was normal [1374].

In a combined analysis of several long-term studies with ^{11}C-McN5652 PET, clinical doses of clomipramine and fluvoxamine occupied approximately 80 % of SERT. A daily dose of 10 mg of clomipramine hydrochloride is enough to achieve that level of occupancy in vivo, whereas 50 mg fluvoxamine is needed for the same effect [1375]. High SERT occupancy was also observed after 3–6 months of continuous treatment with paroxetine [1376].

2.7.1.4   Other Transmitter Systems

Left caudate dopamine function as measured with FDOPA differs between depressed patients with psychomotor retardation and those with impulsivity, suggesting a link between dopamine hypofunction and psychomotor retardation in depression [1377].

Bupropion is thought to treat major depression by blocking the dopamine transporter (DAT), because bupropion appears to have a selective affinity for the DAT. However, bupropion treatment occupies less than 22–26 % of DAT sites, suggesting that there is another mechanism involved in clinical treatment action [1378, 1379].

2.7.2   Fatigue

Fatigue can accompany depression and many other diseases, particularly cancer, but can also occur without identifiable cause. It is often difficult to classify clinically, because it can also be closely related to motivation and other psychological aspects. Reductions in CMRglc or CBF in anterior cingulate and orbitofrontal cortex, similar to but not as pronounced as those in depression, have been found in chronic fatigue syndrome [1380, 1381]. The data could indicate that the two diseases are closely related.

In cancer patients [1382], cortical hypometabolism has been observed as a consequence of chemotherapy or cerebral irradiation [1383, 1384], and it has also been linked to immunological factors, such as natural killer cell activity [1385, 1386]. Prefrontal cortical hypometabolism is also observed during low-dose interferon alpha treatment of hepatitis C, which can cause some degree of depression as a side effect [1387].

2.7.3   Schizophrenia

The search for an organic cause of this devastating and common psychotic disorder has motivated many PET studies. Since the pioneer isotope CBF studies by Ingvar and Franzen (1974, cited in [1401]), a dysfunction of the frontal lobe ("hypofrontality") has been a major research issue. From that starting point, a broad field of specific research issues has grown and has recently been reviewed [1388]. Since a large number of antipsychotic drugs are dopamine antagonists, and dopamine agonists may induce psychosis in normal individuals, studies of alterations of the dopaminergic system are at the core of many PET studies in schizophrenia [1389].

2.7.3.1   CBF and Energy Metabolism

Reduced frontal CBF or metabolism at rest or during a simple attention task has been observed in several PET studies [1390–1396], but not consistently in all studies [1397–1401]. Clinical patient characteristics found to be linked to hypofrontality are negative symptoms [1402], psychomotor poverty [1337], anxiety [1403], and treatment with neuroleptics [1404, 1405]. Further reduction of CMRglc is seen after the administration of amphetamines [1406, 1407], but the effect is smaller in patients with prominent negative symptoms [1408]. Abnormally low CMRglc in schizophrenia is also reported in the mediodorsal thalamic nucleus [1409].

Few studies have addressed the issue of the relationship between specific symptoms of psychosis and local brain function [1410]. A PET study performed during auditory verbal hallucinations demonstrated activation of thalamus, striatum, limbic structures, and temporoparietal association cortex [1411]. In an FDG-PET study comparing patients with and without auditory hallucinations, hallucination scores correlated with relative metabolism in the striatum and anterior cingulate regions [1412].

2.7.3.2   Activation Studies

Hypofrontality was not only found in single state subjects compared with controls, but also with functional activation of frontal cortex during specific tasks [1413–1415]. During high workload in a working memory task, patients show worse performance and less activation than normal controls [1416]. Left frontotemporal activation during episodic encoding and retrieval is disrupted in schizophrenia despite relatively intact recognition performance and right prefrontal function [1417]. Reduced activation is not only seen in frontal cortex, but also in other association areas, including temporal cortex and precuneus, and in cerebellum [1415, 1418]. Remission of hypofrontality is observed with recovery from acute schizophrenia [1414], suggesting that hypofrontality is more a state than a trait marker.

Schizophrenia is probably associated with a complex rearrangement of functional networks, rather than with simple localized dysfunction. For instance, better WCST performance is correlated with CBF increase in prefrontal regions for controls and in the parahippocampal gyrus for patients [1419]. During a working memory task, dorsolateral prefrontal activation is observed around the right superior frontal sulcus in healthy subjects, and ventrolateral prefrontal activation below the right inferior frontal sulcus is observed in schizophrenic patients [1420]. In a verbal fluency study, patients show the same pattern of activation as control subjects in most frontal brain regions, but they fail to show the normal decrease in blood flow when verbal fluency is compared with word repetition [1421].

In healthy volunteers, ketamine-induced psychosis is associated with focal activation (rather than deactivation) of the prefrontal cortex [1422]. Interregional correlation studies reveal strong frontostriatal connections in controls, but weak frontostriatal links in schizophrenic patients [1423].

2.7.3.3 Dopamine Receptors and Dopamine Release

A major controversy arose after a claim emerged, on the basis of studies with ^{11}C-methylspiperone with and without preloading by haloperidol, that the D2 receptor binding capacity was increased in schizophrenia [1424, 1425]. This claim was refuted using the more specific ligand RAC [1426]. Most subsequent studies have not found any significant abnormality in drug-naive patients [1427–1430]. The controversy has not yet been entirely resolved, but it appears that the ^{11}C-methylspiperone finding is not specific for schizophrenia [1431].

Findings are also partially discordant with regard to D1 receptors. Although it is agreed that there is no abnormality in striatum, one study employing ^{11}C-SCH23390 has found a decreased D1 receptor-binding capacity in prefrontal cortex, which was correlated with negative symptoms [1430]. In another study with ^{11}C-NCC112, increased binding in dorsolateral prefrontal cortex was a strong predictor of poor performance in a test of working memory [1432].

Dopamine release induced by amphetamine and measured with RAC appears to be increased during symptomatic schizophrenia [1433, 1434], but seems to normalize in remission [1435]. There is also evidence of increased baseline occupancy of D2 receptors by dopamine in schizophrenia, because RAC binding is increased in schizophrenics during pharmacologically induced acute dopamine depletion [1436]. This finding predicts a good clinical response to antipsychotic drugs. It corresponds well with observations of increased AAAD activity measured by FDOPA-

PET [1437–1441]. Increased dopamine synthesis and release do not seem to be due to an increase in dopaminergic neurons; dopamine transporter binding is normal in schizophrenics [1442].

2.7.3.4   Other Receptors

The serotonergic system may also be altered in psychosis, as indicated by serotonergic actions of hallucinogenic drugs [1443]. Most PET studies on serotonin receptor-binding capacity in schizophrenia have found no abnormality for 5-HT_{2A} receptors [1444–1446]. Only in one small sample has reduced 5-HT_{2A}-binding capacity been observed in the frontal cortex [1447]. Increased binding capacity has been observed for temporal 5-HT_{1A} receptors [1448].

2.7.3.5   Receptor Binding of Antipsychotic Drugs

As demonstrated with RAC-PET, neuroleptic drugs lead to 60-85% occupancy of striatal dopamine receptors at clinically effective doses [1449]. The atypical neuroleptics clozapine and quietapine tend to show somewhat lower occupancy, with considerable interindividual variability even at high doses [1450–1453]. Some other atypical neuroleptics, such as olanzapine[1454, 1455] and risperidone [1456, 1457], have similar properties but tend to show higher D2 receptor occupancy close to those seen after typical neuroleptics at standard doses. It has been estimated that antipsychotic action requires a receptor occupancy of 60%, whereas extrapyramidal side effects occur at 80% receptor occupancy or more [1458].

The time-course of receptor binding and the special properties of atypical neuroleptics are also of interest. Clozapine and risperidone have higher D1, 5-HT_1 and 5-HT_2 receptor binding than do typical neuroleptics [1459–1464]. Receptor occupancy of clozapine does not show a close relation to plasma levels [1461]. The 5-HT_{2A} receptor occupancy of quetiapine seems to be lower than that of clozapine [1465].

2.7.4   Anxiety Disorders

PET studies of anxiety disorders should be seen against the background of neuroanatomical models of fear and anxiety [1466]. Fear and anxiety in normal subjects have most frequently been associated with functional activations, mainly in amygdala, insula, anterior cingulate and orbitofrontal cortex [1467–1473]. Anxiety also is a personality trait that seems to be associated with low FDOPA uptake in the caudate [1474]. There is a significant negative correlation between 5-HT_{1A}-binding potential and anxiety in dorsolateral prefrontal, anterior cingulate, parietal, and occipital cortex [1475]. So far, no evidence has been found for a link between anxiety trait or state and benzodiazepine receptor-binding capacity in neocortex or cerebellum [1476].

Patients with generalized anxiety disorders show lower absolute metabolic rates in basal ganglia and white matter, but relative metabolism is increased in the left inferior area 17 in the occipital lobe, the right posterior temporal lobe, and the right precentral frontal gyrus [1477]. With benzodiazepine treatment, global CMRglc is reduced further, but the pattern is normalized.

2.7.4.1   Panic Disorder

In resting state studies of patients with panic disorder an abnormal hemispheric asymmetry of parahippocampal blood flow, blood volume, and oxygen metabolism and an abnormally high whole-brain metabolism has been described [1478]. Only hippocampal region asymmetry was confirmed in a subsequent study, in which decreases in the left inferior parietal lobule were also found [1479], even after treatment with imipramine [1480]. Left parietal hypofunction was also confirmed in another study, and a higher CBF increase than in normals was seen in this region after challenge with D-fenfluramine, which induces neuronal release of serotonin [1481]. Anticipatory anxiety in patients led to a higher CBF increase in the same regions as in normals (parahippocampal gyrus, superior temporal lobe, hypothalamus, anterior cingulate gyrus, and midbrain) [1482].

Unfortunately, some early studies of panic attacks (which can be elicited by lactate infusion in sensitive subjects) [1483, 1484] were invalidated by errors in effect location (innervation of temporal muscle was mistaken for temporal lobe activation) [1485]. In a subsequent study, a significant increase in glucose metabolism was found in the left hippocampus and parahippocampal area, associated with a significant decrease in the right inferior parietal and right superior temporal brain regions [1486]. An unexpected panic attack during a PET scan was associated with decreased rCBF in the right orbitofrontal (BA 11), prelimbic (area 25), anterior cingulate (area 32) and anterior temporal cortices (area 15) [1487].

Some abnormality of benzodiazepine receptors may contribute to panic disorder. In fact, a global reduction in benzodiazepine site binding throughout the brain has been seen with FMZ-PET [1488].

2.7.4.2   Phobic Disorders

Phobic fear seems to elicit different brain responses than unspecific anxiety. In one study phobic fear elevated the regional to whole-brain (relative) CBF in the secondary visual cortex but reduced relative CBF in the hippocampus, prefrontal, orbitofrontal, temporopolar, and posterior cingulate cortex [1489]. In a related study, phobic fear was associated with CBF changes in amygdala, thalamus, and striatum [1490].

Social phobia, such as anticipatory anxiety before public speaking, is accompanied by enhanced CBF in the right dorsolateral prefrontal cortex, left inferior temporal cortex, and left amygdaloid-hippocampal region [1491]. In responders to citalopram or cognitive behavioral therapy, regardless of the treatment approach, improvement is accompanied by a decreased rCBF response to public speaking bilaterally in the amygdala, hippocampus, the periamygdaloid, rhinal, and parahippocampal cortices [1492].

2.7.4.3   Obsessive-compulsive Disorder

In obsessive-compulsive disorder (OCD), most functional imaging studies support the concept that dysfunction of the orbital cortex and the striatum is responsible for this disorder [1493, 1494]. Resting glucose metabolic rates are significantly increased in OCD in orbital gyrus and caudate nuclei [1495–1497]. Some studies also suggest hypometabolism in various other regions, e.g., at the left parieto-occipital junction [1498]. In a symptomatic case caused by dysgerminoma, FDG-PET showed involvement of the caudate nuclei [1499]. Striatal dysfunction was also suggested by an activation study employing an implicit learning task, during which OCD patients did not activate right or left inferior striatum and instead showed bilateral medial temporal activation [1500]. There may be reduced lateral prefrontal CMRglc in OCD, which is found to be related to a selective attention deficit [1501].

In a study that related current symptoms to CBF, increases in the orbitofrontal cortex, neostriatum, global pallidus, and thalamus were related to urges to perform compulsive movements, while those in the hippocampus and posterior cingulate cortex corresponded to the anxiety that accompanied them [1502]. Obsessive stimulation (checking rituals) was associated with higher rCBF than neutral stimulation in orbitofrontal regions [1503]. Increased orbitofrontal activity is also seen during the compulsive state of craving in patients with drug addiction [1504] (see also section 2.7.5). In another study, symptom provocation was associated with CBF increases in right caudate nucleus, left anterior cingulate cortex, and bilateral orbitofrontal cortex [1505].

Treatment with trazodone or clomipramine may lead to some improvement in obsessive-compulsive symptoms, which is associated with a return of regional brain metabolism to a more normal level in regions of the orbital frontal cortex and the caudate nuclei [1506, 1507]. With paroxetine too, significant metabolic decreases are observed in the right caudate nucleus, right ventrolateral prefrontal cortex, bilateral orbitofrontal cortex, and thalamus [1336]. Other studies have also found significant treatment effects in multiple brain areas involving frontal-subcortical circuits and parietal-cerebellar networks [1508–1510]. Patients who respond to behavior therapy have bilateral decreases in caudate glucose metabolic rates that are greater than those seen in poor responders to treatment [1511].

With regard to the possible prediction of treatment effects, in one study a beneficial effect of paroxetine was associated with higher pretreatment glucose metabolism in the right caudate nucleus [1352]. In another, lower CBF in orbitofrontal cortex and higher CBF in posterior cingulate cortex predicted better treatment response to fluvoxamine [1512]. Conversely, higher metabolism in orbitofrontal cortex was associated with greater improvement after behavioral therapy [1513]. In one small series, response to therapy was correctly predicted in approx. 70 % of cases [1514].

OCD has proved very difficult to treat and can be refractory to pharmacological and psychological approaches. In these cases, stereotactic neurosurgical interventions have been advocated. Improvement of symptoms with stereotactic anterior cingulotomy is associated with higher preoperative CMRglc in right posterior cingulate cortex [1515]. In a patient treated with lesions of frontal white matter, this led to improvement of symptoms and long-term reduction of CMRglc in anterior cingulate gyrus, caudate, and thalamus [1516]. In a study with bilateral electrical stimulation of the internal capsule, improvement of symptoms was accompanied by a decrease of frontal metabolism during stimulation [1517].

2.7.4.4 Posttraumatic Stress Disorder

Attentional problems may underlie other symptomatology in posttraumatic stress disorder (PTSD). In a CBF study during a continuous performance task, PTSD patients with a history of substance abuse make more errors and show decreased parietal blood flow [1518]. A case report suggests that acute psychic trauma may cause widespread reduction of cerebral metabolism [1519].

Patients frequently have alterations in both declarative and nondeclarative memory function. In an activation study of emotionally valenced declarative memory, supportive evidence was found for a dysfunctional network that included hippocampus, medial prefrontal cortex, and cingulate [1520]. An analysis of functional connectivity during a working memory task suggests increased activation in the bilateral inferior parietal lobes and the left precentral gyrus, and decreased activation in the inferior medial frontal lobe, bilateral middle frontal gyri, and right inferior temporal gyrus [1521]. Chronic PTSD patients may show evidence of hippocampal damage (atrophy) associated with reduced hippocampal resting CMRglc (unpublished data) and failure of hippocampal activation [1522].

PTSD may be associated with vivid and fearful re-experience of traumatic situations, so-called flashbacks. This seems to be associated with particularly intensive activation of associative, limbic and vegetative brain regions. In one study, subjects with PTSD had increased rCBF in ventral anterior cingulate gyrus and right amygdala when generating mental images of traumatic combat-related pictures [1523]. In other PTSD patients, traumatic memories were associated with activation of right lingual gyrus, right thalamus, mamillary bodies, and right cerebellum [1524], with activation of primary and secondary visual cortex, posterior gyrus cinguli, and left orbitofrontal cortex [1525] or with increased activation of orbitofrontal cortex and anterior temporal pole [1526]. Perceptually induced symptom provocation promoted sensorimotor, amygdaloid and midbrain activation [1527]. A decrease of CBF in medial prefrontal cortex, an area postulated to play a role in emotion through inhibition of amygdala responsiveness, has also been reported [1528]. Flashback intensity was related to CBF increase in limbic regions and in the brain stem and insula, which probably represents involvement of vegetative functions [1529]. Memories of abuse in childhood elicit an increase in CBF in anterior prefrontal cortex, posterior cingulate, and motor cortex in sexually abused women with PTSD [1530].

2.7.5   Drug and Substance Abuse

Drug and substance abuse is a major health problem. Most addictive substances bind to specific receptors, which can be imaged by PET. Thus, functional imaging with PET is a very powerful tool for studying the biology of these disorders [1504, 1531].

The dopaminergic system is the main reward system in the brain, and thus it has been studied intensively in drug and substance abuse. This includes possible alterations of dopamine receptors, dopamine transporters, and dopamine release. Another transmitter system of obvious interest is the opiate system, because some opiates can cause severe addiction. In view of the strong genetic predisposition to addiction, many researchers have been looking for functional brain changes that may be related to this vulnerability. Lower CMRglc in many brain regions and reduced D2 receptor binding potential have been found in carriers of the D2 dopamine receptor A1 allele, which is associated with an increased risk of alcoholism and drug abuse and reduced central dopaminergic function [1532, 1533].

Another key region is the orbitofrontal cortex, which is essential for inhibitory control; it is related to personality traits and can be damaged by drug abuse [1534].

2.7.5.1   Alcoholism

Acute alcohol intoxication is associated with lower CBF and CMRglc than normal, the difference being more pronounced in cerebellum (and associated ataxia) than in thalamus and cerebral cortex [1535–1537]. In contrast to self-reports of the perception of intoxication, the reduction is more severe in men than in women [1538]. Alcohol ingestion also leads to a relative increase in the CBF in medial parts of the temporal lobes and in the anterobasal parts of the anterior cingulate cortex, including the septal region, which can be regarded as parts of the cerebral reward system [1539].

Chronic alcoholism leads to brain atrophy and neurological impairment. Cerebral glucose metabolism is globally reduced in subjects with severe chronic alcoholism [1540–1542]. If neurological function is still intact, global reduction may be mild and hypometabolism may be the main finding [1543, 1544]. Frontal hypometabolism may be associated with neuropsychological and behavioral alterations [1545, 1546]. Detoxification and abstinence usually lead to partial recovery of CMRglc [1547, 1548]. Chronic alcoholism may also lead to cerebellar degeneration with associated glucose hypometabolism [1549]. Dementia and the amnesic Korsakoff syndrome, which can be the result of chronic alcoholism, are associated with regional reductions in cortical CMRglc (see section 2.1.7.2).

In accordance with clinical reports, alcoholics have a blunted metabolic response to the administration of benzodiazepines [1550], which persists as a trend only in orbitofrontal cortex after detoxification [1551]. It is also observed in subjects with a positive family history for alcoholism and may thus be a genetic trait [1552]. In severe chronic alcoholism reduced ^{11}C-flumazenil binding was seen in frontal cortex and cerebellum in one study [1553], but another indicates that there are no major alterations of benzodiazepine receptor-binding capacity and affinity [1554]. Study reports should be interpreted with caution, because disulfiram, which has been used as an adjunctive agent in the treatment of patients with severe chronic alcoholism, can influence the results of PET studies of glucose metabolism and benzodiazepine receptor binding [1555].

Striatal D2 receptor binding potential appears to be reduced in chronic alcoholics [1556, 1557] and not to recover during abstinence [1558]. Dopamine transporters and FDOPA uptake are intact or even up-regulated [1557, 1559]. Nonetheless, a reduction of striatal monoaminergic terminals in severe chronic alcoholism is suggested by the observation of reduced binding of ^{11}C-dihydrotetrabenazine [1560].

Serotonergic mechanisms have been studied by application of *m*-chlorophenylpiperazine, which is a mixed serotonin agonist/antagonist. In normals it leads to an increase of CMRglc in many brain regions, which is blunted in alcoholics [1561].

2.7.5.2   Cocaine

Cocaine inhibits dopamine reuptake by dopamine transporters (DAT) into dopaminergic nerve terminals, which leads to higher synaptic dopamine concentrations [1562]. The potential for abuse of cocaine may be related to the increase in synaptic dopamine and, probably even more, to the time-course of transporter blockade [1563–1566]. The very fast uptake and clearance of cocaine from the brain is probably essential for its pleasurable and addictive effect [1567], which also depends on the route of administration [1568]. To block these effects, almost complete inhibition by a long-acting DAT inhibitor would probably be necessary [1569]. Oral methylphenidate in typical therapeutic doses occupies approx. 50 % of the DAT [1570]. The magnitude of the subjective euphoria produced by cocaine infusion can be reduced by selegiline. The effect is associated with an alteration of the CMRglc in hippocampus and amygdala, which suggests that these structures have an essential role in cocaine-induced euphoria [1571].

More detailed studies of the kinetics of cocaine and potential drugs used for treatment of dependence, and of their interaction with anesthetics and anticholinergics, have been performed with PET in nonhuman primates [1572–1576]. The potential of the high-affinity, selective and long-acting DAT inhibitor GBR12909 has been studied with PET in baboons, demonstrating up to 74 % blockade [1577] and attenuated synaptic dopamine release [1578]. An influence of social factors on D2 receptor capacity and vulnerability to cocaine dependence has also been demonstrated [1579]. Recently, protocols have been developed to conduct PET studies in conscious rhesus monkeys [1580, 1581].

Use of cocaine has been associated with the occurrence of cerebrovascular accidents. In a CBF study cocaine users showed areas of deranged CBF, as evidenced by patchy regions of defective isotope accumulation, throughout their brains. Chronic cocaine users showed decreased relative CBF in the prefrontal cortex compared with normal subjects [1582]. Another effect is prolonged reduction of norepinephrine uptake and storage capacity in the cardiac sympathetic nerve terminals, which has been measured by ^{11}C-hydroxyephedrine PET and may contribute to cardiac side effects [1583].

Behavioral changes can be explained by reduced rates of frontal CMRglc persisting for at least 3–4 months of detoxification even in neurologically intact cocaine abusers. Their severity is related to the severity of previous abuse [214] and to decreased D2 receptor availability [1584]. Cocaine abusers also show persistent functional activation in prefrontal neural networks (mainly orbitofrontal cortex) involved in decision-making [1585]. In another activation study, cue-induced anger in cocaine-dependent men was associated with decreased activity in frontal cortical areas involved in response monitoring and inhibition. The lack of this association in nicotine-dependent men suggests a possible deficit in anger regulation associated with cocaine dependence and a possible link between cocaine dependence, violence, and relapse [1586].

A study with FDOPA suggests that there is a delayed decrease in dopamine terminal activity in the striatum during abstinence from cocaine [1587]. There is also evidence of disturbance of the GABAergic system. Cocaine-abusing subjects have increased lorazepam-induced decrements in whole-brain CMRglc associated with increased sedation [1588]. In contrast, their response to alcohol appears blunted [1589].

During craving 1 week after cocaine withdrawal, CMRglc is generally increased. Intensity of cocaine craving was correlated with increased CMRglc in the prefrontal cortex and the orbitofrontal cortex [1590]. Similar effects are seen with methylphenidate-induced craving [1591]. Activation of the amygdala and of the temporal insula, a brain region involved with autonomic control, is also observed during craving [1592, 1593]. Craving may also be associated with increased μ-opioid binding, as demonstrated with ^{11}C-carfentanil in several brain regions of the cocaine addicts studied 1–4 days and up to 4 weeks after their last use of cocaine [1594]. Craving induced in detoxified subjects by a cocaine-related video led to CBF increases in limbic structures (amygdala and anterior cingulate) in cocaine-dependent but not in normal subjects [1595]. Activation by mental imagery of personalized drug use activated a network of limbic, paralimbic, and striatal brain regions, including structures involved in stimulus–reward association (amygdala), incentive motivation (subcallosal gyrus/nucleus accumbens), and anticipation (anterior cingulate cortex) [1596].

The urge to increase cocaine doses in chronic users may be related to the observation that detoxified cocaine abusers show decreased uptake of cocaine in brain but no changes in DA transporter availability [1597]. It may also be related to changes in the affinity of D1 and D2 receptors, as observed in an experimental study after binge-pattern cocaine administration [1598]. Recovery from dopamine receptor changes during sustained abstinence has been observed in rats [1599]. In a study of dopamine release (measured by competition with RAC) in response to intravenous methylphenidate, addicts showed reduced dopamine release in the striatum but an enhanced response in the thalamus, suggesting that thalamic dopamine pathways contribute to cocaine addiction [1600].

A study with ^{18}F-N-methylspiroperidol has demonstrated that postsynaptic dopamine receptor availability decreases with chronic cocaine abuse but may recover after a drug-free interval [1601]. However, cortical binding of the tracer remains unchanged, suggesting normal 5-HT$_2$ receptor availability in cocaine abusers [1602].

2.7.5.3   Amphetamine and Derivatives

Application of high doses of amphetamine or methamphetamine (2×2 mg/kg, i.m.) over 1–2 weeks in the vervet monkey produced extensive striatal dopamine system neurotoxicity with a decrease of 60–70 % in FDOPA uptake, corresponding to a similar reduction in striatal dopamine concentrations seen on biochemical analysis [1603]. There was partial recovery up to 32 weeks later. Similar findings were observed with ^{11}C-DOPA in rhesus monkeys [1604] and with ^{11}C-WIN-35428 in baboons [1605]. Damage can be prevented by pretreatment with GDNF [1606] but not with the glutamate antagonist MK-801, and not by hypothermia [1607]. There were associated behavioral changes, which persisted even after recovery of FDOPA uptake [1608]. Affiliative behavior was decreased for up to 6 months, while aggressive behavior was increased for 12 months. Similar findings have been observed in humans. Reduced striatal dopamine transporter density is found in abstinent methamphetamine and methcathinone users [1609, 1610]. It is related to psychomotor impairment [1611]. There is evidence of slow recovery with maintained abstinence [1612].

Repeated application of methamphetamine appears to sensitize the brain and is associated with an increase of 40 % in ^{11}C-methamphetamine uptake in dogs [1613]. This effect is prevented by pretreatment with haloperidol or cocaine [1614].

In Parkinson disease, dopaminergic lesions are associated with up-regulation of D2 receptors. In contrast, repeated administration of amphetamine in monkeys causes a long-lasting down-regulation of the D2-receptor density, which may be a neurochemical correlate to the abnormal movements, anhedonia, anxiety, and depression seen in psychostimulant abusers [1615]. In methamphetamine abusers an association between the level of dopamine D2 receptors and metabolism in the orbitofrontal cortex is found, which suggests that D2 receptor-mediated dysregulation of the orbitofrontal cortex could underlie a common mechanism for loss of control and compulsive drug intake in drug-addicted subjects [1616]. Whole-brain CMRglc is moderately increased in detoxified methamphetamine abusers [1617], and this effect is most pronounced in the parietal cortex (20 % higher than in controls), which is nearly devoid of dopaminergic innervation.

Excessive abuse of 3,4-methylene-dioxymethamphetamine (MDMA; "ecstasy"), which is often taken together with other substances, may lead to severe acute brain damage; this can include ataxia and amnesia, with associated cerebellar, thalamic, retrosplenial, and left medial temporal hypometabolism [1618]. Damage is not limited to the dopaminergic system. In a study of the effects of MDMA (5 mg/kg, s.c., twice daily for 4 consecutive days) in the baboon brain using PET and ^{11}C-McN-5652, a long-standing reduction of 5-HT transporter density with partial recovery was observed [1619, 1620]. Chronic abuse in humans results in decreased CMRglc in the bilateral frontal cortex [1621], but the CBF increase during functional activation by an attention task is not altered [1622]. Results must be viewed with caution, because there is a current controversy, which has led to the retraction of a paper on this topic, about whether MDMA alone can induce irreversible damage in primates, or whether contamination by more toxic compounds, such as methamphetamine, is required [1623, 1624].

During various activation tasks in healthy subjects, complex effects of low doses of dextroamphetamine (0.25 mg/kg) on activated CBF are seen, which could be interpreted as a tendency to "focus" neural activity for a particular cognitive task [1625, 1626]. During a continuous attention task, subcortical, limbic, frontal, and cerebellar CMRglc significantly increase after dextroamphetamine, whereas CMRglc of the temporal cortex significantly decreases [1627]. In a study of relatively high euphorigenic doses of D-amphetamine (0.9–1.0 mg/kg p.o.), a mania-like syndrome was produced concomitantly with a widespread increase in absolute cerebral metabolism, which was significant in the anterior cingulate cortex, caudate nucleus, putamen, and thalamus [1628]. These studies, taken together with studies in schizophrenia (see section 2.7.3), suggest that metabolic effects of amphetamines are heavily dependent on dose and possibly also on other variables.

2.7.5.4   Opiates

In a CBF activation study involving drug-related video cues in heroin addicts, self-reports of "urge to use" correlated strongly with activation of the inferior frontal and orbitofrontal cortex target regions of the mesolimbic dopaminergic system, which are implicated in conditioning and reward. Urge to use was also associated with activation in the right precuneus, an area associated with episodic memory retrieval, and in the left insula, which has been implicated in the processing of the emotional components of stimuli. Self-reports of feeling "high" correlated with activation in the hippocampus, an area relevant to the acquisition of stimulus-associated reinforcement [1629]. In another activation study opiate addicts responded differently than normals to nonmonetary, but not to monetary, rewards [1630].

The effect of methadone in former heroin addicts has been studied with ^{18}F-cyclofoxy. Specific binding was lower by 19–32 % in thalamus, amygdala, caudate, anterior cingulate cortex, and putamen. The effect was correlated with methadone plasma levels, suggesting that these lower levels of binding may be related to receptor occupancy with methadone [1631].

2.7.5.5   Nicotine

The effects of nicotine on CMRglc have been studied in tobacco smokers and normal volunteers. Nicotine causes a moderate reduction (by up to 10 % with i.v. infusion) of global CMRglc [1632, 1633], which appears to be in some contrast with its cognitive stimulating effects. Relative increases were seen in left inferior frontal gyrus, left posterior cingulate gyrus, and right thalamus [1633]. In related studies, relative CBF increases were observed in thalamus, pons, visual cortex, and cerebellum, whereas relative reductions were seen in the left anterior temporal cortex and in the right amygdala [1634–1636]. Regional brain activations and associations with cigarette craving are similar to findings with other addictive substances [1637].

Binding of nicotine to its receptors in brain has been studied in baboons: doses equivalent to those achieved by cigarette smokers led to an average 50 % receptor occupancy [1638].

In studies of the dopaminergic system no release of dopamine was induced by nicotine in monkeys [1639]. Although MAO-B activity is reduced in smokers [1640–1642], an acute dose of nicotine or smoking a single cigarette does not inhibit MAO-B in baboon brain in vivo [1643, 1644]. In another study a lower dopamine D1 receptor density in the ventral striatum was found in cigarette smokers than in nonsmokers [1645].

A CBF activation study suggests that the brains of smokers react to reward in a different way than those of nonsmokers, and in a regionally complex manner. In particular, this difference involves the regions of the dopaminergic system, including the striatum [1646]. This was also found in another activation study analyzing the effect of nicotine during performance of a working memory task [1647].

2.7.5.6   Hallucinogens

Psilocybin increased CMRglc in distinct right hemispheric frontotemporal cortical regions, particularly in the anterior cingulate and decreased rMRGlu in the thalamus [1648]. Studies of cannabinoid receptors are reported in section 3.17.

2.7.5.7   Phencyclidine

Phencyclidine is an anesthetic that is related to ketamine and has the potential for abuse and stimulating effects. An instability of metabolic rates has been observed in baboons under phencyclidine anesthesia [1649]. The effects of ketamine have been studied in human volunteers. It increases metabolic activity in the prefrontal cortex and produces psychotic symptoms, with conceptual disorganization in some subjects [1650]. The effects are thought to be related to the stimulation of

NMDA receptors. A PET study with RAC and ^{11}C-cocaine also suggests that PCP may be exerting some direct effects through the DAT and that GABA partially modulates NMDA-antagonist-induced increases in striatal dopamine [1651].

2.7.6   Personality and Behavioral Disorders

Personality disorders probably represent the extremes of the range of normal personality traits. Therefore, investigation into the neurobiological correlates of such traits may have considerable relevance for the understanding of these diseases. High 5-HT$_{2A}$ receptor density in cerebral cortex is associated with a strong tendency to avoid danger [1652].

2.7.6.1   Borderline Personality Disorder

As in several other psychiatric diseases, regional functional alterations have been found mainly in the frontal cortex. In a study of resting CMRglc, frontal and prefrontal hypermetabolism and hypometabolism in the hippocampus and cuneus was observed [1653], but frontal hypometabolism was described in other, earlier, studies [1654, 1655]. Decreased glucose uptake in medial orbital frontal cortex is associated with diminished regulation of impulsive behavior [1656]. Prefrontal cortex, cuneus, and anterior cingulate show alterations during activation study with memories of abandonment [1657]. Low 5-HT synthesis capacity in corticostriatal pathways was observed in a study with α-^{11}C-methyl-L-tryptophan [1658]. There is also a diminished response to serotonergic stimulation in areas of prefrontal cortex [1659].

2.7.6.2   Violence and Suicide

Impairment of frontal and temporal glucose metabolism is a frequent finding in violent offenders [1660–1663]. Reductions in prefrontal glucose metabolism have also been seen in murderers [1664, 1665].

Suicide is also associated with aggressive/impulsive traits, but also with hopelessness and depression, and often with comorbidity in terms of substance abuse and alcoholism [1666]. Postmortem findings indicate abnormalities of the serotonergic system in prefrontal cortex. PET demonstrates lower CMRglc in ventral, medial, and lateral prefrontal cortex in depressed high-lethality suicide attempters, which is even more pronounced after fenfluramine administration [1667].

2.7.7   Anorexia Nervosa and Bulimia

Anorexia nervosa is a psychiatric disorder with severe organic consequences. In the acute stage, when patients have lost a substantial proportion of their body mass, a variety of structural and functional alterations occur in the brain [1668, 1669]. Studies have highlighted a variety of findings, such as reduced CMRglc in the caudate nucleus [1670] and in frontal and parietal cortices [1671], with normalization of findings after weight gain [1672, 1673]. Milder focal changes and hemispheric asymmetries are observed in bulimia nervosa [1674–1676], which is not associated with major weight loss. Similar findings with normalization after recovery were observed in a CBF study [1677]. A reduction of global CMRglc is correlated with the amount of weight loss in these eating disorders and in depression [1678]. Reduction of CMRglc may be related to use of ketone bodies instead of glucose for energy metabolism in the brain during starvation [1679–1681].

In an activation study, patients with anorexia showed increased CBF in left amygdala, hippocampus, insula, and bilateral anterior cingulate when confronted with images of high-calorie drinks, suggesting psychological abnormalities in response to these stimuli [1682]. Increased limbic activation by demonstration of high-calorie food has been confirmed in another study, which also found evidence for greater activation in visual association cortex [1683].

Anorexia nervosa may be associated with an abnormality of the serotonergic system. A reduction in medial orbital frontal cortex $5\text{-}HT_{2A}$ binding is found in women who have recovered from anorexia or bulimia nervosa have [1684, 1685].

Imaging Brain Function

PET differs from other imaging techniques in requiring more expensive equipment and highly specialized personnel, not only for scanning but also for production of the radiotracers, which has to be done on site for the ultrashort-lived isotopes (mainly ^{11}C and ^{15}O) or requires fast delivery within few hours after production (with ^{18}F). The benefit of this complexity is that PET has the highest specificity and sensitivity currently available in imaging and measuring local physiological function. The mathematical models required for calculation of physiological parameters are described in section 4.9. Tables 3.1–3.3 provide an overview of tracers, their targets, and the references to sections in the text for clinical applications, physiology, and modeling.

3.1   Blood–Brain Barrier Transfer

Transfer of substances may be due to diffusion, which is determined by molecular size and lipophilicity, often expressed as the octanol/water partition coefficient [1686]. Substances that are essential for the brain but that cannot enter in sufficient quantity by diffusion are transported by carriers. The blood–brain barrier (BBB) carrier system with the largest capacity is GLUT1 for transport of glucose.

The inability of many substances to cross the BBB is apparently due to a system that protects the brain from potentially toxic substances that circulate in the blood. This system may be disturbed in disease, and it is therefore of interest to measure the permeability of substances that cannot usually enter the brain. A suitable tracer is Ga-68-EDTA, which is an inert complex too large to cross the intact BBB [1687]. Quantitation is usually based on a model with one tissue compartment (see section 4.9.1). Increased uptake is observed mainly in brain tumors and after pharmacological disruption of the BBB, but also after acute stroke and brain hemorrhage. It is still controversial whether uptake is also increased in neurodegenerative disorders, such as Alzheimer disease [1688].

Another tracer of potential interest is rubidium-82 ([82]Rb), which is injected as a cation and behaves as an analogue of potassium ions, which usually do not cross the BBB in large amounts [1689]. Measured transport is substantially larger than that of ^{68}Ga-EDTA. Cobalt-55 ([55]Co) has been introduced as a calcium analogue tracer for imaging traumatic, ischemic, and neurodegenerative brain damage [1690–1693], but it is not yet clear whether increased uptake actually reflects

Table 3.1. Tracers for transport, CBF, and metabolism (*AA* amino acids, *BBB* blood–brain barrier, *CBF* cerebral blood flow, *CMR* cerebral metabolic rate, glc glucose, *HSV-TK* herpes simplex virus thymidine kinase, *PSR* protein synthesis rate)

Tracer	Abbreviation	Target	Main clinical application[a]	Physiology[a]	Model[a]
[18]F-2-Fluoro-2-deoxy-D-glucose	FDG	CMRglc	All	3.4	4.9.2.2
[15]O-Water		CBF	All	3.2	4.9.1.2
^{68}Ga-EDTA		C	2.3	3.1	4.9.1.1
[82]Rb (ion)		BBB	2.3	3.1	4.9.1.1
[55]Co (ion)		BBB, Ca^{2+}		3.1	4.9.1.1
[11]C-O-Methyl-glucose	OMG	Glucose transport		3.1	4.9.1.5
[81]Kr (gas)		CBF		3.2	4.9.1.2
[15]O-Butanol / ^{11}C-Butanol		CBF	2.3.11	3.2	4.9.1.2
[18]F-Fluoromethane		CBF		3.2	4.9.1.2
[11]C-Albumin		Plasma volume		3.2	4.9.1.2
[62]Cu-Albumin		Plasma volume		3.2	4.9.1.2
[11]C-Carbon monoxide / ^{15}O-Carbon monoxide	CO	Erythrocyte volume	2.4	3.2	
[15]O-Carbon dioxide	CO_2	CBF	2.4	3.2	4.9.1.2.1
[15]O-Oxygen (gas)	O_2	$CMRO_2$	2.4	3.3	
β-1-^{11}C-Hydroxybutyrate		Ketone uptake		3.4	
[11]C-2-Deoxy-D-glucose		CMRglc		3.4	4.9.2.2
[11]C-D-Glucose		CMRglc		3.4	4.9.2.2
[11]C-Carbon dioxide		pH		3.6	
[11]C-5,5-Dimethyl-2,4-oxazolidinedione	DMO	pH		3.6	
O-(2-^{18}F-Fluoroethyl)-L-tyrosine	FET	AA uptake		3.7.1	4.9.1.5
[11]C-aminocyclohexane carboxylate	ACHC	AA uptake		3.7.1	4.9.1.5
[11]C-Methionine	MET	AA uptake, PSR	2.3	3.7.2	
[11]C-Leucine		PSR		3.7.2	4.9.2
L-1-^{11}C-Tyrosine		PSR	2.3	3.7.2	4.9.2
L-2-^{18}F-Fluorotyrosine		AA uptake, PSR		3.7.2	4.9.2
2-^{11}C-Thymidine		DNA synthesis		3.8	4.9.2
3'Deoxy-3'-^{18}F-fluorothymidine	FLT	DNA synthesis	2.3	3.8	4.9.2
[124]I-Iododeoxyuridine		DNA synthesis	2.3	3.8	4.9.2
[124]I-2-Fluoro-2'-deoxy-5-iodo-1-β-D-arabinofuranosyluracil	FIAU	HSV-TK	2.3	3.9	4.9.2.1

[a] Section numbers

Table 3.2. Tracers for the dopaminergic system

Tracer	Abbreviation	Target	Application[a]	Physiology[a]	Model[a]
L-6-^{18}F-Fluoro-3,4-dihydroxyphenylalanine	FDOPA	Dopamine synthesis	2.2	3.10.1	4.9.2.3
^{18}F-6-Fluoro-L-m-tyrosine,		Dopamine synthesis		3.10.1	4.9.2
^{18}F-4-fluoro-L-m-tyrosine					
^{11}C-DOPA		Dopamine synthesis	2.2	3.10.1	4.9.2
^{11}C-Cocaine		DAT	2.7	3.10.2	4.9.3
^{11}C-Methylphenidate		DAT	2.7	3.10.2	4.9.3
^{11}C-Nomifensine		DAT	2.7	3.10.2	4.9.3
^{11}C-WIN-35428, ^{11}C-PE2I	CFT	DAT	2.7	3.10.2	4.9.3
^{18}F-2-β-Carbomethoxy-3 β-(4-fluorophenyl)tropane		VMAT2	2.2	3.10.3	4.9.3
^{11}C-Dihydrotetrabenazine	DTBZ	D1	2.2	3.10.4	4.9.3
^{11}C-SCH-23390		D2, 5-HT2	2.7	3.10.5	4.9.3.1
^{11}C-N-Methylspiperone	NMSP				
^{18}F-N-Methylspiperone		D2, 5-HT2		3.10.5	4.9.3.1
^{18}F-Fluoroethylspiperone	FESP	D2, 5-HT2		3.10.5	4.9.3.1
^{11}C-Raclopride	RAC	D2	2.2	3.10.5	4.9.3.2
^{18}F-Fallypride		D2	2.7	3.10.5	4.9.3.2
N-1-^{11}C-Propylnorapomorphine		D2 (agonist)		3.10.5	4.9.3
^{11}C-L-Deprenyl		MAO-A	2.7	3.10.7	4.9.3

[a] Section numbers

Table 3.3. Neurotransmitter and receptor tracers (except dopaminergic system; *AChE* acetylcholine esterase, *GABA* gamma aminobutyric acid, *SERT* serotonin transporter)

Tracer	Abbreviation	Target	Application[a]	Physiology[a]	Model[a]
¹¹C-N-Methyl-4-piperidylbenzilate	NMPB	Muscarinic receptors	2.1	3.11.4	4.9.3
N-¹¹C-Methylpiperidin-4-yl propionate	MP4P, PMP	AChE	2.1	3.11.5	4.9.2
N-¹¹C-Methyl-4-piperidyl-acetate	MP4A, AMP	AChE	2.1	3.11.5	4.9.2
¹¹C-Alpha methyl-tryptophan	AMT	Serotonin synthesis	2.7	3.12.1	4.9.2
¹¹C(+)McN5652		SERT	2.7	3.12.2	4.9.3
¹¹C-3-Amino-4-(2-dimethylaminomethyl-phenylsulfanyl)-benzonitrile	DASB	SERT	2.7	3.12.2	4.9.3
¹¹C-WAY-100635	WAY	5-HT$_{1A}$ receptor	2.7	3.12.3	4.9.3
¹⁸F-Altanserin		5-HT$_{2A}$ receptor	2.1, 2.7	3.12.3	4.9.3
¹¹C-Flumazenil	FMZ	GABA-A/benzodiazepine receptors	2.4, 2.5, 2.6.2	3.13.1	4.9.1.4
¹¹C-PK-11195	PK	Peripheral benzodiazepine receptors	2.1, 2.2	3.13.2	4.9.3
¹¹C-Carfentanil		Opioid μ-receptor	2.5, 2.7	3.18	4.9.3
¹¹C-Diprenorphine	DPN	Opioid receptors	2.7	3.18	4.9.3
¹⁸F-Cyclofoxy		Opioid receptors	2.7	3.18	4.9.3

[a] Section numbers

pathologic cellular calcium influx or unspecific BBB damage. Bromide-76 is a relatively small ion similar to chloride that can pass the BBB, but does not enter cells in large amounts. It has been used to measure the extracellular space. Values for normal brain are between 18 % (cerebellum) and 28 % (frontal lobe), and are much higher in most brain tumors (up to 66 %) [1694].

The ability to enter the brain is essential for all centrally acting drugs. Although this can be estimated from experimental measurements in rodents, the situation may be different in primates. PET offers a noninvasive way of measuring the brain entry of labeled drugs in humans and other primates. Blood-to-brain transfer rates for compounds with complete extraction are in the order of $0.8\,min^{-1}$ in normal cortex, or $0.5\,min^{-1}$, corresponding to normal CBF (see section 3.2), which is the rate-limiting step for freely diffusible substances. Glucose is the substance with the highest capacity for carrier-mediated transport through the BBB, and corresponding transfer rates are in the order of $0.1–0.2\,min^{-1}$ in cortex. Most centrally acting drugs and related receptor-binding tracers, such as FMZ and RAC, have transfer rates between those of glucose and of freely diffusible tracers, often closer to the latter. For most compounds (except those with extremely high specific binding in brain), much lower transfer rates would imply that sufficient concentrations in brain cannot be reached because nearly all of the compound would have been excreted from the body before substantial transfer occurred. For PET tracers, slow blood–brain transfer also implies long measurement time (which requires at least the half-life time of ^{18}F), and often difficulties in reaching equilibrium (a requirement for most receptor ligands, see section 4.9.3.2) or in separating slow metabolic processes from similarly slow transfer rates by kinetic modeling (see section 4.9.2).

3.2   Cerebral Blood Flow

The most accurate and comprehensively validated approach to measurement of cerebral blood flow (CBF) relies on tracers (or indicators) that enter the brain freely by diffusion. Thus, their entry is not limited by BBB transfer but only by supply from blood flow (see section 4.9.1.2 for mathematical models used for quantitation). These properties are best fulfilled by inert gases, of which only krypton-77 (^{77}Kr) has a suitable positron emitting isotope with a short half-life of 73 min. It is difficult to produce and handle and has therefore rarely been used [1687]. Another tracer that comes close to this goal is ^{18}F-fluoromethane, which can be administered by inhalation or intravenously (dissolved in water). Butanol labeled with ^{11}C or ^{15}O is also freely diffusible and can be used, but undergoes significant and rapid metabolism in brain and body organs, so short scan times are used. For most practical purposes ^{15}O-water is used; it is almost freely diffusible and easier to use than the other tracers. Its extraction rate in normal gray matter ranges between 80 % and 90 %, which leads to moderate underestimation of CBF, particularly in high-flow areas [1695, 1696]. It is, however, sufficiently accurate to depict alterations in blood flow due to disease and most functional changes. In the early days of PET, ^{13}N-ammonia was also used to measure CBF [999], but this tracer was abandoned because of low first-pass extraction and the development of better alternatives. CBF measurements with intra-arterial injection of albumin microspheres, which can be labeled with ^{68}Ga [1697] or ^{11}C [1698] and are trapped in cerebral microvessels, are also of mainly historical interest.

Measurements of normal global CBF in absolute units converge on average values of approx. 50 ml/100 g brain tissue per min [1699], corresponding to a blood-to-brain transfer rate (k_1) for freely diffusible tracers of 0.5 min^{-1} (using an approximate specific mass for brain of 1 g/ml). Local CBF in gray matter is three- to fourfold that in white matter, but the difference is usually less pronounced on PET scans owing to limited resolution. Global CBF is held rather constant by autoregulation during physical exercise and changes of arterial blood pressure within the physiological range [1700]. It is, however, markedly altered by changes in arterial pCO_2, increasing during hypercapnia and decreasing during hypocapnia [1701]. This contributes to a relatively wide variance of absolute CBF values in normal human brain, which limits their sensitivity for functional alterations in diagnostic studies in individuals. The molecular mechanisms involved in CBF regulation are not yet entirely clear and may involve adenosine, extracellular pH, potassium ions, nitric oxide, and prostaglandins, as well as direct and indirect neural pathways.

Autoregulation involves constriction and dilatation of resistance vessels, which result in changes to local cerebral blood volume (CBV). The local intravascular plasma volume can be measured with PET by means of strictly intravascular tracers, such as albumin labeled with ^{11}C [1702] or ^{62}Cu [1703]. It can be compared with the local intravascular concentration of erythrocytes labeled by using ^{11}C carbon monoxide (CO) or ^{15}O-CO [1704, 1705], yielding a measure of local hematocrit [1702], which also changes during vasoconstriction. The ratio of CBV to CBF determines the transit time of intravascular indicators, which is often used as a surrogate indicator of CBF in dynamic CT and MRI techniques. A prolongation of transit time does not necessarily indicate a reduction in CBF, because it may remain constant because of local vasodilation [922].

There have been several studies on the effect of age on CBF. An age-related reduction of approx. 0.5 % per year in global CBF was observed with the ^{15}O-CO$_2$ steady-state technique [1706, 1707]. The decline tended to be most pronounced in frontal cortex. Similar findings, although differing in extent and anatomical distribution, have been obtained in subsequent studies [1708, 1709] and by other researchers [1710]. The effect was not seen in a study with partial volume correction [1711], suggesting that measurements may be influenced by age-related cortical atrophy. As a consequence, it is necessary to use age-adjusted reference values and consider the possible effects of atrophy in PET CBF studies of patients.

Local CBF depends on neuronal function. It is increased by increased neuronal activity and decreased by reduced or impaired neuronal function. Thus, measurement of local changes in blood flow during all kinds of physiological and cognitive tasks has become a major tool for the study of functional brain organization. These changes are usually associated with similar changes of glucose metabolism. Changes related to brain function are described separately in section 3.5.4.

3.3   Oxygen Consumption

The cerebral metabolic rate of oxygen ($CMRO_2$) is a fundamental parameter of brain function. It can be measured by the use of ^{15}O gas, which is given by either continuous or bolus inhalation [1712–1714]. Calculations are based on the fact that the oxygen concentration in tissue is very low and nearly all of the tissue signal stems from the metabolic product, ^{15}O-labeled water. Corrections have to be applied for intravascular oxygen (bound to hemoglobin) and for washout of ^{15}O-water which depends on CBF. For accurate corrections, separate measurements of CBV and CBF are required [1715, 1716], or estimates of these corrections can be obtained from dynamic measurements [1717–1719]. Further refinements to account for a nonmetabolic tissue oxygen pool have been suggested [1720, 1721].

Most studies indicate that $CMRO_2$ declines with age in a similar way to CBF [1706, 1707, 1709, 1710, 1722], but there has not yet been a study applying rigorous partial volume correction. CBF and $CMRO_2$ are lower in the neonatal period than in later childhood or in adult life, and increase significantly during early childhood. CBF and $CMRO_2$ falling to adult values during adolescence [1723]. In a small series of $CMRO_2$ measurements in newborns who had minimal or no detectable brain injury, $CMRO_2$ was considerably below the threshold for brain viability in adults, suggesting that energy requirements in fetal and newborn brain are minimal or can be met by nonoxidative metabolism [1196].

It has been noted that changes in $CMRO_2$ due to functional activation are much smaller than those in CBF and CMRglc [1724–1726], and this effect is the basis for the widely used brain oxygen level-dependent (BOLD) fMRI imaging of functional brain activation. The controversy about the physiological implications of these observations is still ongoing [1727–1730].

When a gaseous tracer is given by mask or mouthpiece, the methods require considerable cooperation from awake subjects. Signal-to-noise ratios are usually considerably worse than with FDG for measurement of glucose metabolism. Thus, nowadays, measurement of $CMRO_2$ is performed only if that metabolic parameter is specifically required. This is most frequently in studies of cerebrovascular disease (see section 2.4), in which impaired oxygen supply to tissue is the pivotal event, leading to metabolic failure within seconds in ischemic stroke.

3.4   Glucose Consumption

Glucose is the main substrate for energy supply of the brain by oxidation. It is transported into brain by the insulin-dependent carrier GLUT1, which is expressed in brain capillary endothelial cells and in cells of choroid plexus, ependyma, and glia. It is sufficient to transport two to three times as much glucose as is normally metabolized by the brain. Its expression is increased by hypoglycemia [1731]. Transport rates depend on plasma glucose levels in accordance with Michaelis-Menten kinetics for facilitated transport [1732]. Inherited GLUT1 deficiency leads to globally reduced cortical glucose metabolism, whereas basal ganglia appear to be less severely affected [1733]. The neuronal glucose transporter GLUT3 appears to be regulated in concert with metabolic demand and regional rates of cerebral glucose utilization [1734]. There appears to be another concentrative H^+-linked glucose transporter in the brain [1735]. Besides neurons, astroglia may also be involved in brain energy metabolism, with some exchange of lactate between neurons and glia [1736, 1737].

In circumstances of starvation only, ketone bodies can serve as an energy substrate instead of glucose to some extent. Their brain metabolism, which is limited by low-capacity carrier-mediated uptake, has been studied with R-β-1-^{11}C-hydroxybutyrate [1679, 1680]. Increased metabolism of ketone bodies is associated with reduced glucose consumption [1681, 1738, 1739].

The standard tracer for measurement of the cerebral metabolic rate of glucose (CMRglc) is ^{18}F-2-fluoro-2-deoxy-D-glucose (FDG) [1740]. Deoxyglucose has also been labeled with ^{11}C [1741], but the rapid isotope decay of ^{11}C is less suited for typical measurement times of up to 60 min. FDG is transported into tissue and phosphorylated to FDG phosphate, like glucose, but does not undergo significant further metabolism. Thus, it accumulates in brain in proportion to local CMRglc. After the first 10–20 min following i.v. bolus injection, during which transport effects dominate tracer distribution, the distribution of FDG in brain approximates local CMRglc. Methods for quantitation based on physiological modeling are described in section 4.9.2.2. Since FDG is an analogue tracer, conversion factors (the "lumped constant") are necessary to calculate CMRglc. To avoid this, PET studies using native glucose, which can be labeled with ^{11}C in various positions [1742–1745], have been performed. Because of the appearance and washout of labeled metabolites in tissue, the relation between tracer concentration in tissue and CMRglc is more complicated and time dependent than with FDG [1746] and therefore the method has not found widespread clinical use.

In normal brain, the spatial distribution of CBF and that of CMRglc are closely related and images look very similar. In contrast to CBF, CMRglc is not substantially influenced by pCO_2. This results in somewhat less physiological variability in absolute values, which are in the range of 23–30 µmol glucose 100 g per min as a normal whole-brain average [1747]. Typical resting state gray matter CMRglc values are in the range of 40–60 µmol glucose 100 g per min, and the corresponding level in white matter is about 15 µmol glucose/100 g per min. There are regional dif-

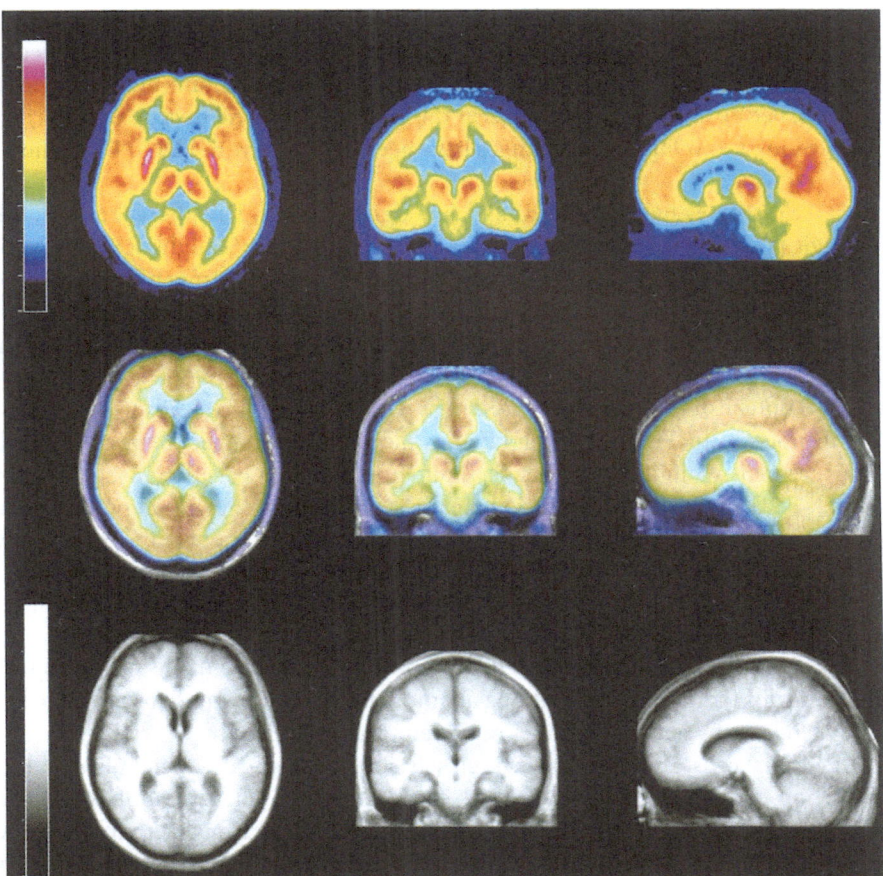

Fig. 3.1. Normal distribution of cerebral glucose metabolism in adults (average of 12 subjects, mean age 60 years). Individual MRIs were coregistered and all image data were spatially normalized and averaged. Orthogonal cuts through the left thalamus are shown (FGD PET, fusion, and T1-weighted MRI). The complete 3D data set is available on the provided CD-ROM, which also contains a sequence of transaxial PET-MRI fusion images from an individual normal volunteer as a video clip

ferences, with the highest values in striatum and parietal cortex close to the parieto-occipital sulcus (Fig. 3.1). Mesial temporal cortex and cerebellum have metabolic rates below the average rate in the gray matter.

In a similar way to CBF, local CMRglc measured with PET is influenced by age, with a predominantly frontal decline during adulthood [29–31, 33, 1748]. This has not been found in all studies [1749] and may also be influenced by partial volume effects due to age-related atrophy. An age-related decline in dopaminergic activity may also reduce frontal CMRglc [1750]. There are also prominent changes in glucose metabolism during maturation of the brain. After birth, glucose metabolism is low in association cortex and increases steadily to reach a more homogeneous adult pattern after 1 year of age. The absolute level of CMRglc in the cerebral cortex increases steadily from low values in infants to a maximum between 4 and 12 years, before declining to stabilize at the end of the second decade of life [1751, 1752]. Similar developmental changes have been observed with FDG-PET in monkeys [1753, 1754].

Several studies have reported mild but significant gender differences in local CMRglc [1755–1759]. One study suggests that the approx. 20% higher global CMRglc in women may be present during and related to the high estrogen levels on days 5–15 of the menstrual cycle [1760]. It has also been suggested that estrogen use may protect against regional cerebral metabolic decline in postmenopausal women [1761].

FDG uptake of the brain is reduced in hyperglycemia, which leads to a poor signal-to-noise ratio. FDG excretion kinetics may also be changed, requiring actual measurement of individual plasma input functions for calculation of CMRglc [1762]. FDG uptake may remain higher in brain tumors during hyperglycemia [649]. Images often display a reduced contrast between gray and white matter and therefore cannot be compared easily with images obtained in normoglycemia. Thus, it is generally recommended that clinical FDG-PET studies of the brain be perform in normoglycemia only.

Local CMRglc is also coupled to local brain function, but the physiological mechanisms that cause this coupling are more precisely known than for CBF. Glutamate release by neurons initiates rapid removal from the synaptic cleft and recycling by glial cells, and this process is closely coupled to glucose metabolism in the tricarbon cycle [1015, 1737, 1763, 1764]. In addition, the energy requirements of ATPases that maintain ion gradients are also related to neuronal activity [1765]. A close correlation between electrical activity and glucose consumption has been shown in several experimental models [1766]. There is a rise in the brain glucose-to-oxygen uptake ratio during functional activation that involves complex adaptations with altered fluxes into various oxidative and biosynthetic pathways [1767].

A major limitation of FDG-PET used to assess local brain function is the relatively long uptake period of at least 20 min that is required to record the metabolic increase related to functional activity by FDG accumulation. This excludes many functions that cannot be maintained over that time. On the other hand, the robustness of the method gives it a clear advantage over other more rapid methods for localization of motor function and language in a clinical context.

3.5   Influence of Brain Function on CBF and Metabolism

Because of the dependence of CBF and CMRglc on neuronal activity, examination conditions are crucial for interpretation of results. Major effects of the state of consciousness, sensory influences, anxiety, and sedative drugs on global levels and on spatial distribution need to be taken into account when clinical PET scans recorded in a "resting state" are interpreted. Local changes due to specific tasks or stimuli are used in activation studies to localize brain functions by comparison of two or more conditions.

3.5.1   Wakefulness and Sleep

There is a global decrease in CMRglc during slow-wave sleep [1768], and CBF is decreased particularly in the prefrontal cortex, dorsal pons and mesencephalon, thalami, basal ganglia, basal forebrain/hypothalamus, orbitofrontal cortex, anterior cingulate cortex, and precuneus [1769, 1770]. Even light stage-1 sleep is associated with decreased CBF in cerebellum, posterior parietal cortex, premotor cortex, and left thalamus [1771]. There is also a relative increase in occipital cortex, which may correspond to a wakeful dreaming state. A higher global CMRglc level is observed during non-REM dreaming in stage-1 sleep than in an awake resting state [1768].

During REM sleep, rCBF increases in pontine tegmentum, left thalamus, both amygdaloid complexes, anterior cingulate cortex, and right parietal operculum, but decreases in dorsolateral prefrontal cortex, parietal cortex, posterior cingulate cortex, and precuneus [1772]. Activation of extrastriate visual cortices with sparing of primary visual cortex has been noted [1773]. Increased CMRglc has also been observed in limbic and paralimbic regions, such as the hypothalamic area, amygdaloid complex, and orbitofrontal, cingulate, entorhinal, and insular cortices [1774]. The level of anxiety during REM dreams is reported to increase global CMRglc [1775]. The eye movements are associated with an increase in CMRglc in frontal eye fields and in dorsolateral prefrontal, cingulate, medial frontal, and parietal cortex [1776]. Deactivation of the heteromodal association areas (the orbital, dorsolateral prefrontal, and inferior parietal cortices) constitutes the single feature common to both non-REM and REM sleep states and may be a defining characteristic of sleep itself [1777].

There seems to be a paradoxical effect of visual stimulation during sleep: a relative decrease in CBF in the rostromedial occipital cortex is found during visual stimulation and slow-wave sleep. This decrease is more rostro-dorsal than the relative CBF increase along the calcarine sulcus found during visual stimulation in the awake state [1778]. A preferential reduction in resting CMRglc in the occipital areas, including visual and paravisual cortex, is observed during hypnosis [1779].

3.5.2   Effect of Drugs

There is a reduction in global CMRglc and CBF with most sedative and many anesthetic drugs, including propofol, sevoflurane, isoflurane, and halothane [1780–1787]. Only ketamine and related anesthetics may increase metabolic activity

[1650] (see also section 2.7.5.7). A reduction is seen with most antiepileptic drugs: when given at typical therapeutic doses carbamazepine is associated with a 12%, valproate with a 22%, phenytoin with a 13%, and phenobarbitone with a 37% reduction in CMRglc [1149–1152, 1788]. Sedative effects of benzodiazepines are often most pronounced in thalamus [1789]. A dose-dependent reduction in CMRglc brought about by zolpidem has been reported mainly in medial frontal cortex, cingulate gyrus, putamen, thalamus, and hippocampus [1790]. Lorazepam reduces whole-brain CMRglc by approx. 4%, with the largest effects in thalamus and occipital cortex [1791, 1792]. Midazolam significantly decreases rCBF in prefrontal, superior temporal, and parieto-occipital regions, but does not abolish rCBF activations induced by an explicit memory task [1793].

Reductions in CMRglc are also observed with several opioids, such as morphine [1794] and buprenorphine [1795]. With fentanyl, regional neuronal activation in cingulate and orbitofrontal and medial prefrontal cortices, as well as caudate nuclei is observed, whereas CBF decreases are noted in both frontal and temporal areas and the cerebellum [1796].

The effects of graded propofol anesthesia have been studied in association with vibrotactile stimulation. Propofol interferes with the processing of sensory information first at the level of the cortex before attenuating its transfer through the thalamus, where the response is abolished with loss of consciousness [1797].

Nonsedative drugs can also influence CBF and CMRglc. There is a marked CBF increase (without a change in CMRglc) caused by the carboanhydrase inhibitor acetazolamide [934, 940, 1798], which is related to an increase in arterial pCO_2, and by other vasodilating substances [1799]. Adenosine increases CBF in anesthetized, but not in awake humans [1800, 1801]. Scopolamine impairs memory functions but may increase resting CMRglc, mainly in parietal and occipital regions [1802, 1803], with a relative reduction in the attention-modulating structures thalamus and cingulate and basal ganglia [1804]. Conversely, cholinergic stimulation by nicotine reduces CMRglc by approx. 10% [1632]. Dexamethasone has no pronounced cognitive effects but may possibly reduce brain FDG uptake [647, 648].

Drugs may exert a neuromodulatory influence on changes in CBF and CMRglc during functional brain activation. This has been observed in some studies for anticholinergics [1805, 1806]. Dopaminergic drugs may also play a part. For instance, increases in rCBF in response to a memory challenge are attenuated by apomorphine in the dorsolateral prefrontal cortex and augmented in the retrosplenial region of the posterior cingulate. Conversely, buspirone attenuates blood flow increases in the retrosplenial region [1807].

3.5.3   The "Resting State"

Clinical CBF and FDG studies are usually performed with patients in a resting state, which means there is no specific sensory stimulation and patients are not engaged in any behavioral or physical task. This is an important point of difference from other radiological and nuclear medicine examinations and is the main reason why brain PET needs to be performed in specialized laboratories. Patients are typically lying comfortably in the scanner in the supine position, in a quiet examination room with dimmed lights. In many laboratories, patients are asked to close their eyes before the start of the examination and to keep them closed throughout. It is important to make subjects familiar with the surroundings prior to the examination to avoid any unnecessary anxiety or restlessness. With FDG, the initial tracer accumulation after i.v. injection does not need to be recorded by the scanner, and resting conditions can therefore be achieved in a preparation room before the patient is positioned in the scanner after 20–50 min. This temporary disruption of resting conditions is has little effect because of its short duration after most of the tracer accumulation has already been completed. However, even with the best care, it is obvious that the resting state will be subject to some variability, because some background noise is usually unavoidable (e.g., from fans of electronic equipment). Most importantly, there is no way of controlling the patient's mental activity during the examination [1808].

Other states have been tried as the standard reference condition. It was noted early on that sensory deprivation (achieved by use of earplugs and blindfolds) led to more asymmetric brain metabolism [1809] and could induce anxiety, and this is therefore not recommended. On the other hand, tasks demanding continuous attention have been suggested to keep mental processes controlled and to prevent subjects from falling asleep during the examination [1810, 1811]. A narrower variability of normal CMRglc was observed during a picture-viewing condition than in resting studies [1812], but these data have not yet generally convinced researchers to abandon the more easily applied "resting state".

Anticipatory or actual anxiety during the examination may influence CBF and CMRglc, mostly involving limbic regions [1466, 1813], but differences are usually small and often not significant [1814]. In one study, global CMRglc tended to be higher in the first than in the repeat study [1815], whereas the reverse was found in another study [1810], and the difference was not statistically significant in either study [1816, 1817]. Divergent results may be explained by an inverted U-shaped relation between anxiety and CBF [1818] (see section 2.7.4 on anxiety disorders for further data on regional effects).

3.5.4   Activation Studies

Specific brain functions can be localized by comparing CBF or CMRglc during performance of a specific task with a condition that does not include that task but is otherwise identical (or at least as similar as possible). This was made possible in the early 1980s by PET, initially using primarily FDG [1819, 1820]. With ^{15}O-water or other ultra-short-lived CBF tracers, multiple replications of conditions in the same subject became possible [1821, 1822], and these techniques were widely used, especially for the study of higher brain function in the rapidly growing field of cognitive neuroscience. This has become a research area in its own right, similar to neuroanatomy and neurophysiology, with only limited links to clinical neurology. In recent years fMRI has become the dominant imaging technique in this field because it does not involve ionizing radiation and is thus also easily used in normal controls, allows more rapid signal acquisition and more flexible experimental arrangement of stimuli.

There are advantages of PET, however, it provides a more physiologically specific signal, better quantitation, and a better signal-to-noise ratio and fewer artifacts in individual acquisitions. PET also provides not only the difference signal between activated and reference conditions, but also actual activated and reference regional values, which may show a much better correlation with task performance than the difference signal provided by fMRI [1823]. These advantages support its continued use, especially in pathophysiologically complex clinical situations such as brain tumors and cerebrovascular disease, where CBF responses to activation may be altered [1824]. These are reviewed in the appropriate clinical sections. For the large field of brain mapping by activation studies in normal controls, mainly in the context of cognitive neuroscience, the reader is referred to other recent publications [1825–1827].

3.6   Tissue Oxygen Pressure and pH

The measurement of tissue pH with ^{11}C-CO_2 employs a kinetic model that includes effects of tissue pH, blood flow, and fixation of CO_2 into compounds other than dissolved gas and bicarbonate ions [1828, 1829]. With known pH, tissue oxygen pressure (pO_2) can be calculated from the oxygen extraction fraction. The relation between oxygen saturation and pO_2 is obtained through the oxygen dissociation curve [1830].

An alternative that has been used to measure brain tumor pH is ^{11}C-5,5-dimethyl-2,4-oxazolidinedione (DMO) [715, 1831]. Its brain–blood partition coefficient depends on tissue pH and on arterial blood hematocrit and pH. With independent measurement of the latter, tissue pH can be imaged from the equilibrium distribution of the tracer. More accurate pH images were calculated from the data obtained after injection of DMO and measurement of extracellular water by injection of ^{76}Br [878].

3.7   Amino Acid Transport and Protein Synthesis

Amino acids are transported across the BBB by transporter enzymes, and to a lesser degree by passive diffusion. Generally, the L-isomers of the essential amino acids (leucine, valine, methionine, and for the brain also tyrosine) are of the greatest biological relevance and are therefore often used as tracers. Most of them are apparently substrates for the sodium-independent large neutral amino acid (LNAA) transport system, of which the transporter LAT1 is highly expressed at the BBB [1832–1834]. LNAA transport operates close to saturation under physiological conditions, and therefore tracer uptake depends on the amino acid plasma levels [1835]. LAT1 is up-regulated in glioma cells [1836].

Transport of amino acids across the BBB is slower than that of glucose. Blood–brain transfer rates for essential LNAA are in the order of $0.05–0.1 \, min^{-1}$, and even lower for the nonessential amino acids [1837]. Uptake of LNAA is higher in posterior parts of the brain and in cerebellum than in the anterior parts of the brain [1838].

Amino acids enter many metabolic pathways, which poses challenges for quantitative modeling. Traditionally, protein synthesis has attracted most interest, because of its obvious key function. Quantitative assessment of protein synthesis would be of interest in many brain disorders, such as degenerative diseases, stroke and trauma and subsequent regeneration, and in brain tumors. There have been pilot studies to address these issues, but it became clear that even if tracers could be found, that undergo little metabolism other than being used for protein synthesis the concentration of competing unlabeled amino acids at the level of transfer RNA would have to be taken into account [1839–1844]. Because these levels are usually not known with sufficient accuracy in vivo, quantitation of protein synthesis rate (PSR) has remained an elusive goal.

To date, the most conspicuous changes in amino acid uptake have been noted in brain tumors (see section 2.3). With kinetic studies, it became clear that the increase in amino acid uptake that is seen in most brain tumors compared with normal brain is present even at the start of the uptake curves. Modeling demonstrates that it is mainly a change in k_1, representing transport rather than metabolism (including protein synthesis) [1845]. This was confirmed further by use of amino acids that do not undergo major metabolism and do not enter protein synthesis [1846]. Thus, even for the dopamine precursors and metabolites ^{18}F-FDOPA [1847] and 3-O-methyl-6-^{18}F-fluoro-l-DOPA [1848] uptake is also increased in brain tumors, because it also uses the activated LNAA carrier. Thus, for current clinical applications, the use of most amino acid tracers is limited to the demonstration of altered amino acid transport.

3.7.1   Transport-only Tracers

Several chemically modified (unnatural) amino acids are transported to some extent by the carriers across the BBB (with increased uptake in brain tumors), although they are not incorporated into proteins [1849]. These include O-(2-^{18}F-fluoroethyl)-L-tyrosine (FET) [1850, 1851], which is transported mainly by the LNAA carrier. Absence of protein incorporation and other major metabolism simplifies quantitation. A one-tissue-compartment model with rate constants k_1 for transport from blood to tissue and k_2 for the reverse is usually used (see section 4.9.1). Peripheral extraction of amino acids is low. Thus, blood sampling from a heated dorsal hand vein such as is used for FDG studies (see section 4.9) is sufficient to provide an input function for absolute quantitation. In a comparison with ^{11}C-methionine (see section 3.7.2), uptake in normal brain tissue and brain tumors is slightly lower, but comparable with regard to tumor-brain contrast [1852]. The effect of aging has been studied with L-2-^{18}F-fluorophenylalanine. There was no change in blood-to-brain transfer, but there was an increase in efflux with age [1853].

There is also some interest in amino acids, such as α-aminoisobutyric acid and related compounds, that are preferentially transported by the A transport system [1854, 1855]. The A system has a lower capacity than the L system, but may be used to image brain tumors because there is no transport of these tracers across the normal BBB [1856]. A related compound is ^{11}C-labeled aminocyclohexane carboxylate (ACHC), which shows slow uptake in normal human gray matter [1857].

3.7.2   Tracers with Incorporation into Proteins

The most widely used tracer for incorporation into proteins is ^{11}C-methionine (MET), which is usually labeled in the methyl position. It is transported mainly by the LNAA system and subsequently undergoes complex metabolism that involves not only incorporation into proteins but also intermediary metabolism (Fig. 3.2) [1858]. There have been attempts at modeling the protein synthesis rate (PSR) [653, 1859, 1860], but they have not been validated convincingly. Intracellular metabolism of this tracer, however, has the practical advantage that it is largely trapped in tissue, which facilitates imaging. For clinical applications, lesion to contralateral brain uptake at intervals 10–60 min after i.v. bolus injection has been most widely used (see section 2.3.1.2). Normal uptake is higher in cortex than in white matter, and the occipital region and the cerebellum show higher uptake than frontal, temporal and parietal regions [1838]. There is a change of uptake associated with brain maturation [1861].

Uptake kinetics of MET into brain and tumor tissue have been studied by comparing the two stereoisomers, L-methionine (which is incorporated into proteins), and D-methionine (which is not, but is still a substrate for the carrier at the BBB). Higher trapping of the L- than the D-isoform was seen in pituitary adenomas [671] but not in gliomas [1862]. Specificity of transport was demonstrated by competition with branched amino acids [1863].

Other ^{11}C-labeled essential amino acids with somewhat less complex metabolism have been tested with regard to their potential for measuring the cerebral PSR. The include ^{11}C-leucine [1864] and L-1-^{11}C-tyrosine [1865–1868]. Reduced uptake and protein synthesis in phenylketonuria was demonstrated with L-1-^{11}C-tyrosine [1869].

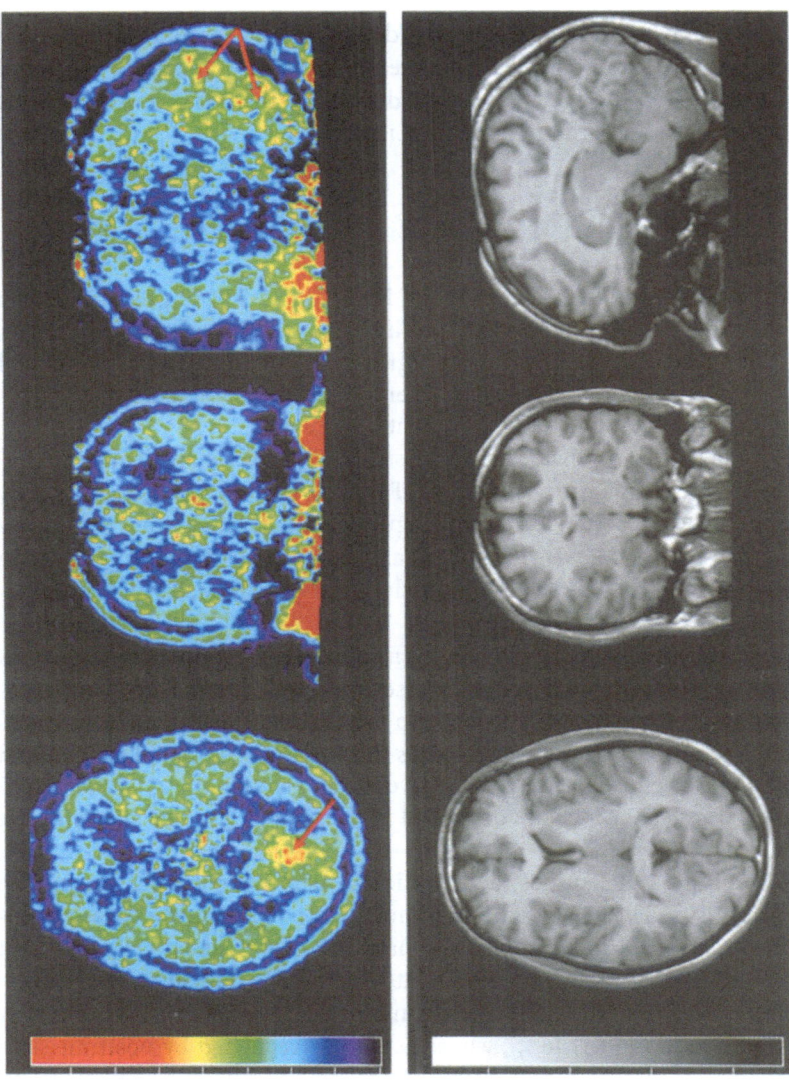

Fig. 3.2. Normal distribution of MET uptake in brain, shown as orthogonal cuts through the right thalamus with coregistered T1-weighted MRI. Uptake is higher in occipital cortex and in cerebellum (*arrows*) than in the rest of the gray matter

L-2-^{18}F-Fluorotyrosine is one of the few fluorinated amino acids that is incorporated into proteins [1849, 1870]. Owing to the longer half-life of ^{18}F compared to ^{11}C, it appears to be particularly well suited to the study of PSR, and it has been modeled successfully by a two-tissue-compartment model [1845]. By these means, the increased uptake in brain tumors has been shown to be largely due to increased transport rather than to increased PSR.

3.7.3   Precursors and Analogues for Transmitter Systems

The biogenic amine transmitter systems of dopamine and serotonin have amino acid precursors that can be labeled and used to estimate transmitter synthesis. They are described in the specific sections devoted to these transmitter systems (dopamine: section 3.10; serotonin: section 3.12).

3.8   Nucleosides and DNA Synthesis

Measurement of cell proliferation rates and DNA synthesis is of great interest in tumors, including brain tumors. Cell proliferation determines tumor growth, which is the ultimate cause of disease progression and death. Reduction of cell proliferation rates is therefore the goal of therapy, and measurement of these rates would provide a direct indicator of therapeutic success.

There are specific transporters for nucleosides at the BBB, in particular the concentrative Na^+ nucleoside cotransporter CNT2 [1871]. However, transport capacity and transport rates are several orders of magnitude smaller than those for glucose and large neutral amino acids [1872, 1873], and uptake may not be significant in normal brain within the 60- to 90-minute duration of a standard PET examination.

Most relevant tracers are analogues of nucleosides of the pyrimidine type, probably because purine nucleosides are also involved in second-messenger signal transduction and energy metabolism. In vivo the nucleosides undergo phosphorylation by kinases, forming nucleotides that are then combined to DNA and RNA. Methods are being developed to achieve specificity for nonhuman kinases to image gene expression, which are described in the section on molecular imaging (section 3.9).

The pyrimidine nucleotides deoxycytidine triphosphate and thymidine triphosphate are used in DNA synthesis. Although de novo synthesis of the pyrimidine nucleotides via uridine monophosphate is active in most cells, the high demand during DNA synthesis is accomodated mostly via the salvage pathway by phosphorylation of deoxycytidine and thymidine. A key enzyme in this pathway is thymidine kinase 1 (TK1) [674], whose activity is regulated by posttranscriptional mechanisms and is greater (about 10-fold) in S phase than in G1 phase [1874, 1875]. PET imaging has the goal of measuring the tracer accumulation due to this pivotal step and thus estimate proliferation rate.

This principle has been followed with 2-^{11}C-thymidine. Increased uptake of this tracer in gastrointestinal cancer is observed after inhibition of de novo nucleotide synthesis [1876]. Major metabolites must be considered in blood and tissue [1877], including degradation via thymine, dihydrothymine to ^{11}C-CO_2 which is exchanged between blood and tissue [1878]. Kinetic analysis was able to remove the confounding influence of ^{11}C-CO_2, the principal labeled metabolite of 2-^{11}C-dThd, and to estimate the flux of dThd incorporation into DNA [1879]. Because of its complexity, the quantitative model is not yet fully validated or suitable for clinical application. An alternative is labeling of thymidine by ^{11}C in the methyl group (methyl-^{11}C-thymidine), but modeling is even more complicated because the label goes mainly to organic acid metabolites and its uptake in brain tumors is not specifically related to DNA synthesis [1880].

Substantial progress was achieved with the synthesis of 3'deoxy-3'-^{18}F-fluorothymidine (FLT) [1881]. It is trapped in tissue after phosphorylation by TK1, resulting in high accumulation in tumors to give above-background tissue values [674]. There is little accumulation in DNA and little other metabolism. Close correlations were found between measures of FLT uptake and Ki-67 scores (a proliferation index) in lung cancer [675, 1882] and in lymphoma [1883]. There is also accumulation in human brain tumors [698]. The radiation dose with this tracer is comparable to that for other PET procedures [1884].

Other fluorinated thymidine analogues that are incorporated into DNA, in particular compounds with acronyms FMAU and FBAU, have been labeled with ^{11}C, ^{18}F, and ^{76}Br and are being tested for potential clinical use [1885–1888]. Another compound of interest is ^{124}I-iododeoxyuridine, which has been tested in brain tumors but has high background activity [676].

3.9   Molecular Imaging

Techniques for imaging of molecular mechanisms provide exciting possibilities for studying local tissue genomics and proteomics in vivo. In a wider sense this concept comprises all those many PET studies of the brain that measure the activity of specific enzymes (e.g., hexokinase by kinetic FDG studies, monoamine vesicular transporters by ^{11}C-dihydrotetrabenazine, and acetylcholine esterase by ^{11}C-MP4A) and the binding capacities of specific receptors, as described elsewhere in the text. Usually, PET does not measure the amount of protein expression directly, but assesses the functional capacity of the molecules, which is the ultimate result of proper expression. Standard in vitro antibody-based techniques for assessment of protein expression are limited to brain and tumor vessels but are not generally applicable with PET, because monoclonal antibodies do not cross the BBB and therefore do not enter the intact brain when injected into the bloodstream. The functional measures used in PET usually employ molecules of much lower molecular weight that are either small enough or lipophilic enough to cross the BBB passively or bind to specific transporters that carry them into the brain. Delivery and washout of substances with high blood-to-brain extraction typically depend on local blood flow (see section 3.2. for details). This needs to be taken into account for proper interpretation of their distribution and kinetics. Transport rates for substances with low extraction may also vary locally. Thus, quantitation of the activity of the target enzyme or receptor-binding capacity often requires kinetic analysis to separate this specific process from blood-to-brain transport (see section 4.9.2).

Application of these techniques to assess the expression of gene products that have been introduced artificially by gene transfer has found particular interest in recent years [1889]. Most frequently, expression of herpes simplex virus thymidine kinase (HSV-TK) has been studied, but studies have also been extended to the transfer and expression of other nonhomologue genes. In many instances, studies aim at the demonstration of expression in qualitative terms, which is appropriate in this situation with absent background expression, rather than quantitative measurement of gene expression. These studies are at the core of the modern concept of molecular imaging and are described in detail below. A related area of intense interest is specific binding of tracers to substances that occur in certain diseases, such as amyloid plaques in Alzheimer disease (see section 2.1). Fusion reporter genes can be engineered to be activated by specific protein-protein interactions, which can thus be imaged with PET [1890].

The modern concept of molecular imaging (see [1891] for review) also refers to local assessment of binding to specific RNA or DNA sequences. Since these techniques usually depend on complementary binding with labeled RNA or DNA and similar molecules, which in their native form enter normal brain slowly and in small amounts only, if at all, progress in this field is slow. A promising approach to overcoming the BBB is the coupling of these molecules to peptidomimetic monoclonal antibodies (mABs) that bind to endogenous BBB peptide receptor systems and are transported into the brain by receptor-mediated transcytosis [757, 1892]. By these means, antisense imaging of gene expression in rat brain tumors was achieved by a 16-mer antisense to the sequence around the methionine initiation codon of the luciferase mRNA [1893]. Monoclonal antibodies to the transferrin receptor have been used [1894], and the 83–14 murine mABs to the human insulin receptor appears promising because of its relatively high BBB permeability ($0.005 \, min^{-1}$) [1895–1897]. Coupling peptides to putrescine may also moderately increase BBB transfer [1898]. A dual-modality fusion reporter gene system consisting of *Escherichia coli* xanthine phosphoribosyltransferase has been developed for optical fluorescent imaging and for nuclear imaging with radiolabeled xanthine that easily crosses the BBB [1899]. Another possible approach involves binding to polybutyl cyanoacrylate nanoparticles coated with polysorbate 80. They adsorb apolipoproteins from the blood after injection and thus seem to mimic lipoprotein particles that could be taken up by the brain capillary endothelial cells via receptor-mediated endocytosis [1900].

With better understanding of those molecular processes that guide cell trafficking during development, disease, and regeneration, there is also interest in monitoring the paths of stably labeled cells. Compared with in vitro techniques, molecular imaging has the advantages of simultaneous assessment of molecular processes in different anatomical locations and of being noninvasive and repeatable, so as to provide multiple measurements over time.

3.9.1   Herpes Simplex Virus Thymidine Kinase Imaging

Herpes simplex viruses (HSV) carry the gene for their own variant of thymidine kinase (HSV-TK), which is expressed in transfected cells. HSV has attracted intense interest for gene therapy in the nervous system, because of its high capacity for infecting neurons and its potential use as a carrier of therapeutic genes. There are also drugs (acyclovir, gancyclovir) for treatment of HSV infections that are activated by HSV-TK. They provided a starting point for development of specific tracers to assess transfection by HSV.

The HSV-TK-specific tracer ^{124}I-2'-fluoro-2'-deoxy-5-iodo-1-β-D-arabinofuranosyluracil (FIAU) was developed on the basis of the drug fialuridine, which had been in clinical trials for treatment of hepatitis B but failed because of its toxicity at pharmacological doses [1901]. The long physical half-life of ^{124}I (4.2 days) is well matched to the relatively slow accumulation of the tracer due to trapping after phosphorylation by HSV-TK. The magnitude of FIAU accumulation in RG2TK+, W256TK+, and wild type tumors corresponds to the in vitro gancyclovir sensitivity of the cell lines used to produce these tumors, which indicates that the magnitude of FIAU accumulation reflects the level of *HSV1-tk* gene expression [1902]. The ability of this tracer to image the distribution and level of gene expression over time after *HSV-tk* gene transfer was demonstrated in implanted tumors [1903]. Accumulation rates in RG2TK+ xenografts and cells are much higher than for ^{18}F-FHPG and ^{18}F-FHBG (see below) [1904]. FIAU also successfully monitored transgene expression mediated by replication-conditional oncolytic herpes simplex virus type 1 mutant vectors [748]. Unfortunately, FIAU does not cross the intact BBB but it is taken up in malignant brain tumors and shows very slow washout with a half-life of several days [1905].

Analogues of the herpes virus virostatic drug acyclovir have also been developed for imaging HSV-TK activity. The most promising compound appears to be 9-(4-^{18}F-fluoro-3-hydroxymethylbutyl)guanine (FHBG) [1906–1909]. Apparently, however, there is no detectable transport across the intact BBB, and therefore further tracer developments are required for use in clinical neuroscience.

Tracers such as FIAU may also be useful to monitor other molecular processes. A herpes simplex virus type 1 thymidine kinase/green fluorescent protein (*TK-GFP*) dual reporter gene and a recombinant retrovirus bearing this TK-GFP reporter system have been constructed. With this system and FIAU it is possible to monitor the T-cell receptor-dependent nuclear factor of activated T cells [1910]. It has also been used to demonstrate elevated cellular levels of proteins fused to dihydrofolate reductase after exposure of cells to antifolates, which may be a very useful method of modulating gene expression in vivo [1911]. In a similar approach, a fusion gene was constructed from HSV1-sr39tk and renilla luciferase to combine PET and optical imaging for translation of techniques developed in cell culture into preclinical and clinical models [1912, 1913].

3.9.2   Reporter Gene Imaging

A general approach to noninvasive imaging of transgenes uses *cis*-linked herpes simplex virus thymidine kinase, which is controlled by the same promoter [1914]. Similarly, correlated hepatic D2 receptor and HSV1-sr39tk PET reporter gene expression was demonstrated in mice with microPET after transfection with an adenovirus containing a cytomegalovirus early promoter-driven transcription unit [1915].

Potential reporter genes other than *HSV-TK* code for the human type 2 somatostatin receptor [1916, 1917], which has been used for nonbrain tumor imaging with SPECT, and for the gene for the human dopamine D2 receptor, which is not expressed outside of the brain and can there serve as a reporter for successful gene transfer [1918]. It is not yet clear whether these approaches will also be applicable with PET in neurology. Somatostatin receptor expression in somatotrophic tumors of the anterior pituitary may be an area of clinical interest.

3.9.3   Oligonucleotides

Antisense oligonucleotides are promising pharmaceuticals because of their ability to selectively inhibit the expression of specific proteins. Their in vivo use is difficult because natural oligonucleotides undergo rapid metabolism and elimination, but chemical modifications are being developed to increase their stability. It has been demonstrated that they can be labeled with ^{18}F for PET imaging [1919]. Improved in vivo uptake into organs, including brain, has been shown by use of a synthetic anionic vector [1920].

3.9.4   Apoptosis and p53

Radiolabeled annexin-V was tested in control and camptothecin-treated (i.e., apoptotic) human leukemic HL60 cells [1921]. A key protein regulating apoptosis is p53, and therefore its expression is of interest in brain tumors, which may have lost p53 and therefore fail to enter apoptosis after cell damage. Endogenous expression of p53 has successfully been imaged in experimental tumors using a *cis*-p53-TK-GFP reporter system [1922].

3.9.5   Cell Trafficking

Cell trafficking in the nervous system is a process that has a time scale of days. Thus, tracers need to have appropriate half-lives. One of the longer lived positron-emitting isotopes is ^{64}Cu, with a 12.7-h half-life. ^{64}Cu-pyruvaldehyde-bis($N4$-methylthiosemicarbazone) has been used to effectively label C6 rat glioma cells ex vivo [1923]. TK1 imaging with FIAU has also been used to study the migration of ex vivo-transduced antigen-specific T cells [1924]. Mouse bone marrow-derived dendritic cells have been labeled with ^{18}F using N-succinimidyl-4-^{18}F-fluorobenzoate and their in vivo biodistribution and migration has been followed over 4 h [1925].

3.9.6   Angiogenesis

Expression of vascular endothelial growth factor (VEGF) was assessed with an immunoglobulin G1 monoclonal antibody known as VG76e, labeled with ^{124}I [1926]. HuMV833, a humanized version of a mouse monoclonal anti-VEGF antibody (MV833) that has antitumor activity against a number of human tumor xenografts, was labeled with ^{124}I and the distribution and biological effects of HuMV833 was studied with PET in patients in a phase I trial [751]. Preliminary results showed a heterogeneous tracer distribution between and within patients and between and within individual various solid tumors. Angiogenesis can be stimulated by hypoxia via activation of the hypoxia-inducible transcription factor 1α. This activation and subsequent up-regulation of VEGF were also imaged after transduction of C6 and RG2 glioma cell by dual-reporter vector and expression of the TK-GFP reporter gene [750].

3.10   Dopamine System

Dopaminergic neurotransmission has a central role in many brain functions. It is necessary for proper movement coordination, and degeneration of the nigrostriatal dopamine system causes Parkinson disease and related syndromes. Pulsatile dopamine secretions elicit pleasant feelings and form a strong reward system that appears to play a central role in drug abuse. In striatum dopamine receptors of type 2 (D2) are dominant, whereas in the mesolimbic and mesocortical projections D1 and D3 receptors also play an important part in cognition and emotion. Thus, the dopamine system has attracted much interest and several PET tracers have been used clinically to study the presynaptic (dopamine synthesis, transport and storage) and the postsynaptic (receptor) side (Fig. 3.3).

3.10.1   Precursors and Analogues of Dopamine

The most widely used analogue for the dopaminergic system is L-6-^{18}F-fluoro-3,4-dihydroxyphenylalanine (FDOPA) [275]. The effective dose per unit of administered FDOPA activity is 0.0199 mSv/MBq, with the highest organ dose to the bladder wall surface (0.150 mGy/MBq) [1927]. The tracer is transported across the BBB, as are the other large neutral amino acids (see above), and is then decarboxylated by aromatic amino acid decarboxylase (AAAD) to ^{18}F-dopamine, which is stored

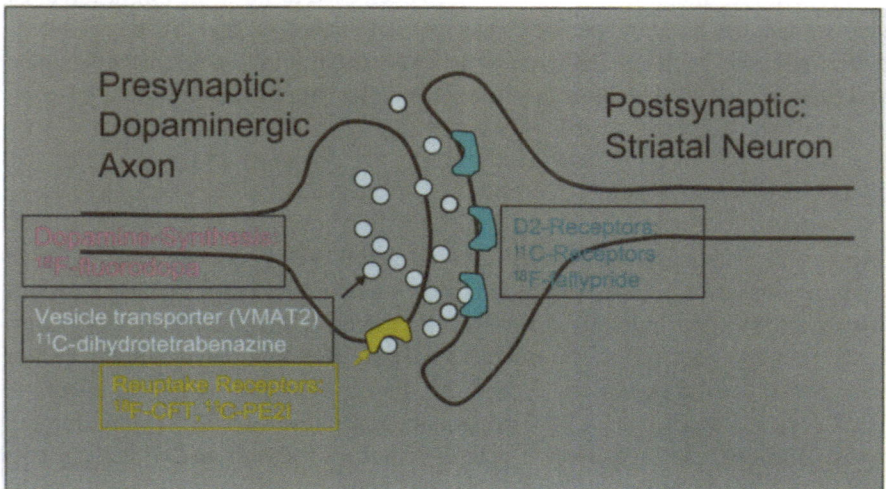

Fig. 3.3. Schematic illustrating a dopaminergic synapse and related radiotracers

in dopamine vesicles. In a similar way to dopamine, it is metabolized further by catechol-O-methyltransferase (COMT) and by monoamine oxidase (MAO) [1928]. The metabolites leave the brain only slowly [1929], and thus activity accumulates during measurement times of typically up to 90 min. Accumulation is highest in brain regions rich in aromatic amino acid decarboxylase, these being the midbrain, caudate, and putamen. There is a slow loss of metabolites from striatum, which has been studied in monkeys with extended measurement times of up to 240 min [1929]. Metabolite loss is increased in MPTP-lesioned monkey striatum.

Decarboxylase activity is reduced in diseases with degeneration of dopaminergic neurons, mainly Parkinson disease and related disorders (see section 2.2). It has been demonstrated that striatal uptake of FDOPA is strongly correlated with in vitro measurements of AAAD activity [1930]. In vivo estimates of AAAD are generally as little as one-tenth those measured in vitro, which may be due to transport restrictions limiting substrate availability to AAAD within the neuron [1930]. FDOPA uptake is not strictly specific for dopaminergic cells but also includes many regions with relatively high concentrations of norepinephrine and serotonin [1931].

Accurate quantitation of AAAD activity is difficult owing to interference from the metabolite 3-O-methyl-6-^{18}F-fluoro-L-dopa (OMFD), which is generated in body and brain and can cross the BBB, probably also via a carrier mechanism [1932–1934]. In principle, it would require arterial (or hot venous) blood sampling with separate determination of metabolite activity. Such studies demonstrate that forward transport rates for FDOPA and OMFD from plasma to striatum are very similar [1935]. Thus, for detection of reduced uptake in Parkinson disease and related disorders, a simplified kinetic procedure has proved sufficient [295]. It relies on kinetic data uptake over 90 min after i.v. bolus injection of ^{18}F-fluorodopa. The influx constant is determined according to the Patlak approach [1936] modified so that uptake in occipital cortex, which is relatively free of dopaminergic innervation, is used as the reference input function (see section 4.9.2). In order to reduce peripheral decarboxylation of ^{18}F-FDOPA, a decarboxylase inhibitor (e.g., carbidopa 100 mg) is given about 1 h before tracer injection. As Parkinson disease progresses, the bias due to OMFD becomes more important, and the influx constant is more closely related to DA storage capacity and less closely to the DA synthesis rate [1937].

Several other substrates for AAAD have been tried as alternatives to FDOPA. ^{18}F-6-Fluoro-L-m-tyrosine has been suggested because it is not a substrate for COMT [1938–1940]. It has been used to demonstrate that AAAD activity does not decline with age [1941], and reductions have been demonstrated in MPTP-lesioned monkeys [1942, 1943]. It has low affinity for vesicular transporter enzymes and is therefore not stored in transmitter vesicles, which makes it inferior to FDOPA for imaging of presynaptic dopaminergic system integrity [1944, 1945]. ^{18}F-4-Fluoro-L-m-tyrosine has also been studied, with similar results [1946]. Another compound of potential interest is ^{18}F-fluoro-β-fluoromethylene-m-tyrosine [1947, 1948].

An obvious approach to study DOPA decarboxylation is labeling of DOPA with ^{11}C. Whereas labeling in the carboxyl position does not lead to specific uptake due to early loss of the metabolite ^{11}C-CO_2, ^{11}C-DOPA labeled in the beta position accumulates in striatum in a similar way to ^{18}F-FDOPA, and influx rates can also be measured by the Patlak approach with occipital reference tissue [1949]. In comparison with FDOPA, accumulation can be measured in practice only over shorter times up to 60 min, owing to loss of signal from decay of ^{11}C and loss of labeled metabolites. There may be less disturbance of the signal from labeled COMT, which is found in lesser amounts in plasma than with FDOPA [1950].

3.10.2   Dopamine Transporter

Dopamine reuptake from the synaptic cleft is mediated by the dopamine transporter (DAT) [1951]. DAT is essential for recycling of dopamine back into the presynaptic neuron. Its inhibition, e.g., by psychostimulants cocaine and methylphenidate, leads to a massive rise in synaptic dopamine levels. Clinical interest in measuring DAT levels is therefore mainly centered on studies related to drug abuse (see section 2.7.5) or to Parkinson disease and related disorders with degeneration of dopaminergic neurons (see section 2.2).

Cocaine has been labeled with ^{11}C and ^{18}F [1952–1955], but it has relatively low specificity [1956]. Methylphenidate has similar affinity for DAT to cocaine [1564, 1957, 1958] and increases the synaptic dopamine concentration to a similar degree [1566]. Owing to its slower kinetics, the addictive potential is not as high as that of cocaine [1959, 1960]. Specific binding of methylphenidate has been demonstrated by comparing the active with the inactive enantiomer, both labeled with ¹¹C [1961]. Nomifensine was the first drug for DAT imaging that was labeled with ¹¹C [310], with subsequent clinical use mainly in Parkinson disease, but interest in it waned due to its low affinity and specificity.

Other analogues of cocaine, such as RTI-55 (β-CIT), RTI-121, RTI-364, RTI-330, RTI-357, WIN-35428, GBR-12935, and PE2 have been labeled with iodine-123 for SPECT and with ¹¹C for PET [1962–1971]. Subsequently, some have also been labeled with ¹⁸F [1972–1975] and with ⁷⁶Br [1976]. Kinetic properties of these tracers and DAT occupancy by drugs have been studied using a small-animal PET scanner [1977]. In vivo measurements of DAT occupancy by such rapidly clearing drugs as cocaine requires the use of radiotracers with similar kinetics to the drug itself [1978]. Absence of measurable lipophilic labeled metabolites and the occurrence of transient equilibrium within the time of the PET measurement indicate that ¹⁸F-β-CIT-FP is superior to ¹¹C-β-CIT-FP as a PET radioligand [1979]. With a related compound, ¹¹C-β-CIT-FE, the antiaddictive drug bupropion has been shown to act as a DAT blocker [1379]. Other recently developed compounds with high selectivity for the dopamine transporter are ¹¹C-β-CPPIT (derived from RTI-177) [1980] and ¹¹C-PE2I [1981]. In a human subject, the latter reached a striatum-to-cerebellum ratio of 10 at peak equilibrium 40–50 min after i.v. injection.

Increased selectivity for dopamine transporters has been demonstrated for ¹⁸F-labeled 2-β-carbomethoxy-3-β-(4-chlorophenyl)-8-(2-fluoroethyl)nortropane (FECNT) [1982]. It was used to study the interaction of anesthetic agents with DAT, which had also been demonstrated with ¹¹C-WIN-35428 [1983]. Experiments indicate that DAT is trafficked into the cell by isoflurane without changing the total amount of DAT in the striatum. The PET data are consistent with this finding, provided that intracellular DAT acquires a conformation that has low affinity for FECNT [1984]. There is also an interaction between DAT and ketamine [1575]. There seems to be little effect of drugs that cause dopamine release on apparent DAT-binding capacity [1985].

An age-related decline of DAT in basal ganglia has been demonstrated with ¹¹C-cocaine [1986] and with ¹⁸F-2-β-carbomethoxy-3-β-(4-fluorophenyl)tropane (CFT). The annual decline with the latter compound in normals was 2.1 % for the putamen and 2.9 % for the caudate nucleus [372]. A more detailed study of aging effects in the dopaminergic system including pharmacological challenge conditions has been performed in monkeys [1987].

3.10.3   Vesicular Monoamine Transporter

Function of synaptic vesicles requires action of vesicular transporters which mediate vesicular uptake of dopamine and other monoamines. Dihydrotetrabenazine (DTBZ), which is a weak dopamine receptor antagonist and a potent inhibitor of the vesicular monoamine transporter type 2 (VMAT2) [1988], has been labeled with ¹¹C.

3.10.4   D1 Receptors

The benzazepine ^{11}C-SCH 23390 is a potent and selective D1 antagonist that was developed as the first clinically useful tracer for these receptors [1989]. A major drawback for practical applications is that it takes more than 4 h to reach equilibrium in vivo [1990], so that a kinetic approach is necessary to quantify receptor binding. It has been used mainly in competition studies to examine binding of neuroleptic drugs to D1 receptors, usually in comparison with binding to D2 receptors [1461, 1991–1993], and for studies of Parkinson disease [1994] (see also section 2.2.1.6). An age-dependent decrease in D1 receptor binding is observed in humans in the caudate (6.9 % per decade) and putamen (7.4 % per decade) [1995], but no age effect is found in monkeys [1987].

A more recently developed tracer is ^{11}C-NNC112, for which both kinetic and graphic analyses provide receptor measures in good agreement with the known distribution of D1 receptors (striatum > limbic regions = neocortical regions > thalamus) in humans [1996]. Related compounds are ^{11}C-NNC 756, which has been used for studies of D1-dopamine receptor occupancy and pharmacodynamics in man [1997, 1998], and ^{11}C-NNC 687 [1999]. ^{11}C-SCH 39166 has also been evaluated as a PET ligand for central D1 dopamine receptor binding and occupancy [1998, 2000, 2001].

3.10.5   D2 Receptors

The first tracers to come into clinical use to study D2 receptors were spiperone derivatives N-methylspiperone (NMSP) labeled with ^{11}C or [18]F [2002, 2003] and ^{18}F-ethylspiperone (FESP) [2004, 2005]. They bind to D2 receptors with high affinity (K_D 0.1 nM), and also show affinity to serotonin 5-HT$_2$ receptors [2006–2008], which are more abundant in cortex than D2 receptors. Owing to their tight binding, displacement from D2 receptors is not achieved by other D2 ligands [2009]. Binding is competitively reduced by pretreatment with butyrophenones but not by substances with lower affinity, such as the atypical neuroleptics [2010]. Another consequence of this essentially irreversible binding is that tracer accumulation (at small doses without saturation effects) not only depends on binding capacity but also on delivery, which depends mainly on local blood flow (see information on modeling in section 4.9.3.1). Kinetic studies using two levels of nonradioactive competing compound have been used to measure maximum binding capacity (B_{max}) and affinity (dissociation constant K_D) separately [2011, 2012]. Spiperone has also been labeled with Br-76, allowing extended measurement times of up to 25 h [2013].

When the influence of blood flow on binding of spiperone tracers became clear to most researchers, interest in the benzamide ^{11}C-raclopride (RAC) rose in spite of its lower affinity. It has medium affinity to D2 receptors (K_D 1.2 nM), is displaced more easily and does not bind to serotonin receptors [2014]. It was demonstrated that D2-dopamine receptors are present in the human neocortex, although their density is very low [2015]. Apparent binding capacity is measured after the tracer that is bound to receptors has approached equilibrium with plasma levels (see section 4.9.3.2) [2016–2018], minimizing the dependence on blood flow [2019]. RAC has been widely used to study receptor occupancy by neuroleptics [1452, 2020–2023].

For routine analysis of clinical RAC studies, no arterial cannulation is required [2024]. Compared with spiperones, there is also a difference with regard to B_{max} of D2 receptors, which is not readily explained by different tracer kinetics but is also seen in in vitro studies [2025]. The striatal B_{max} is about 30–40 nmol/l for raclopride [1426, 1429], but only 15–25 nmol/l for NMSP [2026]. This difference is also seen in vitro with cloned D2 receptors [2025].

A decrease of D2 receptors with age has been observed in humans [2027, 2028]. The D2 receptor binding potential appears to be significantly lower in women than in men [2029]. The variability of D2 receptor binding capacity and reproducibility of the method has been studied in various conditions [2030–2033]. Polymorphisms of the D2 receptor gene may contribute to differences in binding capacity [1532, 2034].

Nemonapride (YM-09151-2) is another benzamide with similar kinetic properties to raclopride that has been labeled with ^{11}C and has been used for human D2 receptor studies [397]. It is not entirely specific for the D2 receptor but also has apparent affinity to sigma receptors [2035].

The short half-life of ^{11}C poses some limitations on D2 receptor studies (time too short for longer displacement studies, dependence on local cyclotron), which could be overcome by suitable benzamides labeled with [18]F. A promising candidate is ^{18}F-fallypride [2036, 2037] which has excellent specific-to-unspecific binding contrast. Its derivative ^{18}F-desmethoxyfallypride (DMFP) [2038] has a striatum-to-cerebellum ratio of 3:1 after 60 min in humans [2039]. D2 receptor occupancies of commonly used neuroleptics have been measured in the monkey brain with ^{18}F-fallypride [2040].

3.10.6   Studies of Synaptic Dopamine Release

Binding of RAC to striatum is sensitive to endogenous levels of dopamine. A reduction in the RAC-binding potential (BP) can be caused by direct competition of the tracer with dopamine at the D2 receptor. This effect can be analyzed quantitatively by the occupancy model using two separate RAC injections or during a continuous tracer infusion [2041]. Alternatively, a kinetic analysis of a single RAC scan employing a linear extension of the simplified reference region model that accounts for changes in ligand binding during the study has been proposed [2042]. The occupancy model is well founded, but is not without problems related to more complex processes, such as agonist-induced change of receptor affinity and receptor internalization [2043–2045].

Pharmacological challenges with amphetamine, methylphenidate, cocaine, dopamine reuptake inhibitor, or tetrabenazine, which lead to increased synaptic levels of dopamine, result in decreased RAC binding [1562, 2041, 2046–2048]. The effect is most pronounced in the ventral striatum, and its magnitude correlates positively with the hedonic response to dextroamphetamine [2049, 2050]. It is not accompanied by major blood flow changes that could disturb measurement of the D2 receptor binding potential [2051]. In normal humans, an i.v. bolus of 0.3 mg/kg of methamphetamine results on average in a 24% fall in striatal binding of RAC, while in dopamine-deficient Parkinson disease (PD) patients only a 10% fall is seen [420]. It has been estimated from animal microdialysis studies that a 1% reduction in striatal ^{11}C-raclopride binding is equivalent to an increase of at least 8% in synaptic dopamine levels [1433].

A reduction of D2 receptor binding capacity is also seen with NMSP in the baboon after pretreatment with amphetamine [2052]. Amphetamine-stimulated reduction of specific binding is also observed with ^{18}F-fallypride [2053].

Dopamine release can also be modulated by drugs that do not influence dopamine reuptake directly. Ketamine may increase dopamine release and cause hallucinations, but results with regard to the correlation between the two are still conflicting [2055–2057]. Changes in endogenous dopamine concentrations resulting from drug-induced potentiation of gamma aminobutyric acid (GABA)ergic transmission by administration of gamma-vinyl-GABA or lorazepam have been measured with PET and RAC [2058]. A decrease in RAC binding suggesting an increase in dopamine concentration is also observed after administration of the serotonin-releasing agent and reuptake inhibitor fenfluramine [2059]. Nicotine in high tobacco-smoking-related doses did not release sufficient brain dopamine to displace RAC in the striatum [1639]. Alfentanil (a μ-opioid receptor agonist) reduced subjective pain intensity and associated striatal dopamine release that was associated with a mechanical pain stimulus [1273].

Physiological dopamine release associated with reduced RAC binding was observed during functional stimulation tasks such as playing a video game [2060]. Even placebo saline infusion in expectation of a reward may elicit dopamine release in the ventral striatum [2061]. Dopamine release in the caudate nucleus was also elicited by repetitive transcranial magnetic stimulation of prefrontal cortex, probably due to the close functional connections and reciprocal projections between these brain structures [2062].

Altered receptor binding after pharmacological challenge has also been observed with NMSP, but differences from the alterations after RAC have been noted. For instance, MK-801 increases NMSP binding but not RAC binding, pretreatment with reserpine increases RAC binding but decreases NMSP binding, and the two ligands yield different values for B_{max} [2063]. Possible explanations include differential influence of affinities to dimeric versus monomeric forms of the receptors, access to external and internal receptors, tracer specificity, and interference from CBF effects [2044]. Competition between dopamine release and RAC binding is not detected in cats under halothane anesthesia. It might reflect a halothane-promoted conversion of D2 receptors to a state of lower affinity for DA [2064].

Labeled agonists, which are usually not used for imaging purposes because of their generally lower affinity, may be more sensitive to competition from endogenous dopamine. Agonists, like dopamine, bind only to receptors in a high-affinity state and may therefore reflect endogenous dopamine release more accurately [2065]. No PET studies definitively confirming this concept are available as yet. Promising agonist tracers are apomorphine derivatives [2066] such as N-1-^{11}C-propylnorapomorphine, which demonstrates a striatum/cerebellum ratio of 2.8 and a binding potential similar to that of the antagonist RAC in mammals [2067, 2068].

In contrast to the findings for striatal D2 receptors, there are no measurable effects of amphetamine and reserpine on D1 dopamine receptor binding, which may be due to low receptor occupancy in physiological conditions [2069, 2070].

3.10.7   Monoamine Oxidase

Monoamine oxidases (MAO) degrade biogenic amines and thus terminate their synaptic action. The enzymes are expressed in aminergic neurons and in glia. MAO activity can be imaged by tracers derived from the drugs clorgyline and L-deprenyl, which as so-called suicide inhibitors bind covalently to the enzyme [2071]. Covalent binding involves cleavage of a C–H bond, and the kinetics of this process can be modified selectively by substitution of the proton by deuterium. Thus, comparison of the kinetics of tracers according to whether or not they are deuterium-substituted can be used to analyze binding specifically, even if binding kinetics are otherwise difficult to separate from transport processes [2072].

MAO-B is specific for dopamine. Its inhibitors are being used in Parkinson disease to enhance the action of L-DOPA. It has been studied with ^{11}C-labeled L-deprenyl (also known as selegiline), an irreversible MAO-B inhibitor. Specificity of imaging can be enhanced by deuterium substitution [2073], and thus deuterium-substituted ^{11}C-L-deprenyl (DED) is now preferred and provides reproducible quantification of MAO-B by kinetic measurement [2074]. It is being used to measure MAO-B inhibition by drugs [2075] and in the brains of smokers (due to an unidentified component of smoke) [1640]. Probably because of its preferential expression in glia, increased focal binding is seen in patients with epilepsy [1103], and there is an age-dependent increase in binding in normal brain [2076].

Up to now, there have been few studies with labeled clorgyline, which is an irreversible inhibitor of MAO-A [2077]. It has been labeled with ^{11}C and with ^{18}F [2078, 2079]. Comparison of ^{11}C-clorgyline and deuterium-substituted ^{11}C-clorgyline yielded evidence of non-MAO A binding of clorgyline in the white matter in human brain [2080].

Another approach to imaging MAO-A involves befloxatone, which is a competitive and reversible inhibitor of MAO-A that has been labeled with [11]C. PET studies in baboons showed rapid uptake into the brain, with high tracer concentrations in thalamus, striatum, pons, and cortical structures that could be blocked by pretreatment with moclobemide. Thus, the tracer appears promising for PET studies of MAO-A in humans [2081].

3.11   Cholinergic System

3.11.1   Acetylcholine Synthesis

The rate-limiting step of acetylcholine synthesis is probably choline uptake in presynaptic cholinergic nerve terminals by the high-affinity choline uptake system. PET tracers are being developed for this transporter system [2082] but have not yet been validated for clinical use.

3.11.2   Vesicular Acetylcholine Transporter

Benzovesamicol is a substance that binds specifically to the vesicular acetylcholine transporter [2083]. Owing to its potential toxicity, the tracer doses that can be used are strictly limited. It has been successfully labeled with ^{123}I and used with SPECT in humans [145]. ^{18}F-Labeled derivatives were also developed [144, 2084], but have not been successfully implemented because it takes several hours to achieve a low nonspecific background in vivo.

3.11.3   Nicotinic Receptors

A natural candidate for the study of nicotinic receptors is ^{11}C-nicotine. Its initial distribution matches that of blood tracers [2085]. Attempts to measure specific binding have involved comparison of the uptake of stereoisomers (+)-(R)- and (–)-(S)-N-methyl-^{11}C-nicotine [2086–2088] and kinetic analyses [2089], without clear success. Problems with high nonspecific binding have also plagued investigators using several other potential tracers [2090].

Progress may come from more specific ligands, such as epibatidine and derivatives, which have been labeled with ^{11}C and [18]F [149], but they are difficult to handle because of their high toxicity at small doses. (+/–)-Exo-2-(2-[18]F- fluoro-5-pyridyl)-7-azabicycloheptane, a high-affinity nAChR agonist and epibatidine analogue, was evaluated in a baboon. Kinetics were compatible with PET imaging; high uptake was observed in thalamus and hypothalamus/midbrain, intermediate uptake in the neocortex and hippocampus, and lowest uptake in the cerebellum, which corresponds with the known densities of nAChR [152]. High specific binding was also demonstrated for ^{18}F-norchlorofluoroepibatidine [150].

Another potential tracer with lower toxicity is 2-^{18}F-fluoro-A-85380 [147, 2091], but kinetics are too slow to be practical with ^{18}F in man [2092, 2093]. To overcome this limitation, an analogue of A-85380 has been labeled with ^{76}Br [2094]. Long-lasting occupancy of central nicotinic acetylcholine receptors after inhalation of tobacco smoke has been demonstrated with 2-^{18}F-fluoro-A-85380 in monkeys [154].

3.11.4   Muscarinic Receptors

The goal of imaging muscarinic receptors, which are the dominant postsynaptic type of cholinergic receptors in the brain, is one that has already been pursued for many years [2095, 2096]. Tracers that have been tested in primates include N-^{11}C-methyl-benztropine [2097, 2098], ^{11}C-quinuclidinyl benzilate [2099], ^{11}C-labeled alpha tropanyl benzilate [2100], ^{11}C-scopolamine [2101], ^{11}C-N-methyl-4-piperidyl-benzilate (NMPB) [2102–2104], and ^{11}C-tropanyl benzilate [2105, 2106]. N-(2-^{18}F-Fluoroethyl)-4-piperidyl benzilate may have advantages owing to longer possible measurement times [159]. The agonist ^{11}C-milameline (CI-979) showed transient binding, and revealed displacement in cortex and a paradoxical increase in striatum on the addition of a cold compound [2107].

So far, NMPB has been regarded as the most promising compound, although full kinetic modeling is required to differentiate transport from receptor binding [2108]. It may bind preferentially to the M4 subtype of muscarinic acetylcholine receptors [2109]. Normal aging is associated with a reduction in muscarinic receptor binding in neocortical regions and thalamus, but there are no consistent changes in AD [158].

More recently 3-(3-(3-^{18}F-fluoropropyl)thio)-1,2,5,thiadiazol-4-yl)-1,2,5,6-tetra-hydro-1-methylpyridine (FP-TZTP) has been introduced as a M2 subtype-selective muscarinic cholinergic ligand with potential suitability for the study of AD [2110]. Spatial distribution and blocking studies support its suitability for human studies (Fig. 3.4) [2111]. Compared with younger subjects, older subjects have significantly larger volumes of distribution of FP-TZTP (means and standard deviations) throughout much of the cerebellum, cortex, and subcortex [2112], possibly due to lower synaptic concentration of acetylcholine.

3.11.5   Acetylcholine Esterase

Anatomical studies indicate that AChE activity in cerebral cortex is mainly due to expression of this enzyme in cholinergic neurons and their axons [2113]. Thus, it seems that tracers that measure AChE activity should be useful to detect loss of AChE activity due to degeneration of cholinergic neurons. ^{11}C-Physostigmine has some potential for quantitation of cerebral AChE activity [141, 2114]. A tracer with

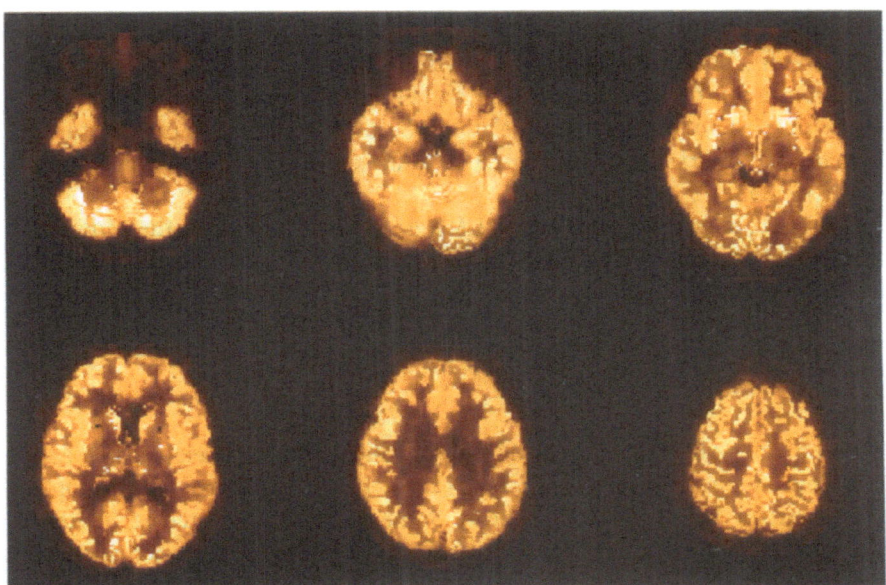

Fig. 3.4. Parametric images of the distribution volume (DV) of ^{18}F-FP-TZTP, an M2 muscarinic agonist, in a young normal volunteer. Images are partial-volume corrected. The cerebellum and basal ganglia have the highest DV. Courtesy of Dr. Robert M. Cohen, National Institute of Mental Health, NIH, Bethesda, MD

high cerebral extraction that is hydrolyzed by AChE and thus permits measurement and imaging of local AChE activity in humans is N-^{11}C-methylpiperdin-4-yl propionate (MP4P, or PMP)[119,134,2115]. This is hydrolyzed by AChE, and to some degree also by butyrylcholinesterase (BChE), and additional compounds are being developed with specificity for BChE [2116]. A related compound that is highly specific for AChE is ^{11}C-labeled N-methyl-4-piperidyl-acetate (MP4A, or AMP) [133, 2117, 2118]. Cerebral AChE activity appears to remain stable with age or even to increase slightly [119, 2119].

Trapping of MP4A and MP4P in tissue after hydrolysis by AChE is quantified by a two-tissue-compartment model (see section 4.9.2.1) [2120]. Very high activity of AChE in basal ganglia and cerebellum has the consequence that virtually all tracer that is transported by blood flow is trapped in these structures. This prevents accurate quantitation of AChE in these regions, but also opens up the possibility of using these structures as a reference to derive the plasma activity kinetics and avoid arterial sampling [2121–2123].

Instead of tracers hydrolyzed by AChE, tracers that bind to AChE have also been introduced for PET imaging. They include N-benzylpiperidinebenzisoxazole ^{11}C-CP-126998 [142, 2124] and the reversible inhibitor ^{11}C-donepezil [2125, 2126].

3.12   Serotonin System

In the central nervous system, serotonin is involved in eating, sleep, sexual behavior, impulse control, circadian rhythm, and neuroendocrine function. Serotonin is also a precursor of the hormone melatonin. Serotonin function is probably altered in many psychiatric disorders, in particular in depression and obsessive compulsive disorders.

The effect of serotonin on cortical CMRglc has been studied by the administration of fenfluramine in normal volunteers [2127]. It leads to increased CMRglc in left prefrontal and left temporoparietal cortex.

3.12.1   Serotonin Precursors

^{11}C-Alpha methyl-tryptophan (AMT, MTrp) was originally developed to measure serotonin synthesis in vivo with PET [2128]. It is an analogue of tryptophan, the precursor for serotonin synthesis. Some AMT is converted to alpha ^{11}C-methylserotonin, which is trapped in serotonergic neurons because it is not degraded by monoamine oxidase [2129]. The Patlak graphic approach used on a pixel-by-pixel basis in normal human subjects reveals high uptake values (suggesting high serotonin synthesis capacity) in putamen, caudate, thalamus, and hippocampus. Among cortical regions, the highest values are measured in the gyrus rectus of the inferior frontal lobe, followed in declining order by transverse temporal gyrus; anterior and posterior cingulate gyrus; middle, superior, and inferior temporal gyri; parietal cortex, and occipital cortex. Values are 10–20 % higher in women than in men [2130], and there is a gradual increase in uptake from ages 2 to 15 years in childhood [2131]. Studies in monkeys, however, indicate that the AMP model for the computation of serotonin synthesis rates is heavily dependent on free TP concentration in plasma and cells and that AMT uptake rates may not accurately reflect actual serotonin synthesis rates [2132]. In pathologic conditions, its uptake may also depend on the synthesis of quinolinic or kynurenic acid via the kynurenine pathway. Increased levels of serotonin and quinolinic acid have been observed in epileptogenic lesions, raising the possibility that MTrp-PET may localize the epileptogenic area [1098, 2133].

5-Hydroxy-L-tryptophan labeled with ^{11}C in the β-position (HTP) is a substrate for the second enzymatic step in serotonin synthesis [2134]. Kinetic tracer uptake studies in normal human subjects indicate a component of irreversible tracer accumulation, suggesting the use of HTP-PET measurements to investigate serotonin synthesis [2135].

3.12.2   Serotonin Transporter

The serotonin transporter (SERT) belongs to the same family as the transporters for dopamine (DAT; see section 3.10.2) and norepinephrine (NET), all of which provide reuptake of neurotransmitters into presynaptic neurons. Interest in SERT is based on the essential role of the serotonin system in depression and on the fact that many antidepressants enhance extracellular serotonergic levels by blocking reuptake.

The first tracer for imaging SERT was obtained by labeling the serotonin reuptake inhibitor McN5652 with ^{11}C. ^{11}C-(+)McN5652 was found to be a promising candidate as a PET radiotracer for studying 5-HT uptake sites in vivo [2136]. It has been evaluated in baboon and human brain [2137–2139]. Massive reductions in specific binding of ^{11}C-(+)McN5652 are seen in the pons and occipital cortex after MDMA (3,4- methylenedioxymethamphetamine)-induced lesions of SERT in baboons [1619]. It is suitable for quantification in areas with high receptor density (midbrain, thalamus, and striatum) only [2140].

The ^{18}F-fluoromethyl analogue of (+)McN5652 has been validated as specific for imaging of SERT in rats by competition with various other inhibitors [2141]. Because of faster kinetics than the ^{11}C-labeled compound, it can reach binding equilibrium during a study length of 120 min [2142].

3-Amino-4-(2-dimethylaminomethyl-phenylsulfanyl)-benzonitrile (DASB) labeled with carbon-11 is a recently introduced radiotracer, which binds specifically to SERT with nanomolar affinity and high selectivity [2143]. In a comparison of five different SERT ligands, DASB displayed significantly lower affinity, but had clear advantages over McN5652 because of its faster uptake and clearance and higher free fraction, which allow shorter scan duration to derive time-independent estimates of regional distribution volumes [2144, 2145]. Studies indicate that DASB is a suitable PET radioligand for measuring drug occupancy of SERT and has potential for monitoring in vivo changes in serotonin levels [2146]. During treatment with clinical doses of paroxetine or citalopram, approximately 80 % of SERT binding sites are occupied [2147]. Procedures for parametric imaging of the SERT BP using a reference tissue model have been developed for this tracer [2148].

3.12.3   Serotonin Receptors

3.12.3.1   Serotonin 1A Receptors

Serotonin 1A (5-HT$_{1A}$) receptors are present in high density in the hippocampus, septum, amygdala, hypothalamus, and neocortex of human brain. On stimulation, they open K$^+$ channels, and in some areas also inhibit adenylate cyclase. A comprehensive review of PET receptor ligands has been given elsewhere [2149].

Carbonyl-^{11}C-N-(2-[4-(2-methoxyphenyl)-1-piperazinyl]ethyl)-N-(2-pyridinyl) cyclohexane carboxamide (^{11}C-WAY-100635, WAY) has high affinity for the receptor and provides good delineation of 5-HT$_{1A}$ receptors in human brain with PET

[2150]. It has fewer metabolites than WAY-100635 labeled in the methyl position. A limitation of WAY is its relatively low blood-brain extraction, and therefore tracer development by modification of the molecule continues [2151]. In healthy volunteer subjects, measurement of BP by kinetic analysis using the arterial plasma input function has emerged as the method of choice because of its higher test–retest reproducibility, lower vulnerability to experimental noise, and absence of bias. Graphical analysis slightly underestimated BP and a reference tissue method (with cerebellum as reference) underestimated k3/k4 and the underestimation was apparent primarily in regions with high receptor density [2152]. In a sample (N = 61) collected from multiple centers [2153], BP values varied greatly across subjects (range 2.9–6.8), probably due to radiochemical, demographic, physiological, and personality variables that had been described in smaller series [1475, 2154–2156]. Age may have an indirect influence on BP: the degree of tracer displacement by a serotonin agonist was more marked in young than in aged monkeys [2157]. Anxiety also seems to be associated with changes in 5-HT$_{1A}$-binding potential [1475].

The receptor occupancies by robalzotan (NAD-299), DU 125530, clozapine, and tandospirone have been studied with ^{11}C-WAY [1370, 1460, 2158, 2159]. NAD-299 has also been labeled with ^{11}C and used for visualization of the 5-HT$_{1A}$ receptor [2160, 2161].

^{18}F-Labeled analogues of WAY-100635 have been developed to increase clinical applicability. The selectivity of fluorine-18-labeled 4-(2-methoxyphenyl)-1-(2-(N-2"-pirydynyl)-p-fluorobenzamido)ethylpiperazine (^{18}F-MPPF) has been established in animals and humans [2162–2164]. Affinity is lower than that of ^{11}C-WAY, and there seems to be potential for measurement of competition effects with endogenous serotonin [2165]. In human volunteers, B$_{max}$ of 2.9 pmol/ml and a K$_d$ of 2.8 nmol/l were found in hippocampal regions, K$_d$ and distribution volume in the free compartment were regionally stable, and the Logan binding potential was linearly correlated to B$_{max}$ [2166]. It has been shown that ^{18}F-MPPF binding can be modulated by modifications of extracellular serotonin in the rat hippocampus, suggesting that ^{18}F-MPPF binding might constitute an interesting radiotracer for PET in evaluating the serotonin endogenous levels in limbic areas of the human brain in health and disease [2165, 2167]. Another potentially useful tracer is ^{18}F-FCWAY, in which the cyclohexanecarbonyl group acid is replaced by a *trans*-4-fluorocyclohexanecarbonyl group (FC) [2168]. There are metabolites that may induce significant biases in estimates in regions with low specific binding and need to be considered in quantitative models (Fig. 3.5) [2169].

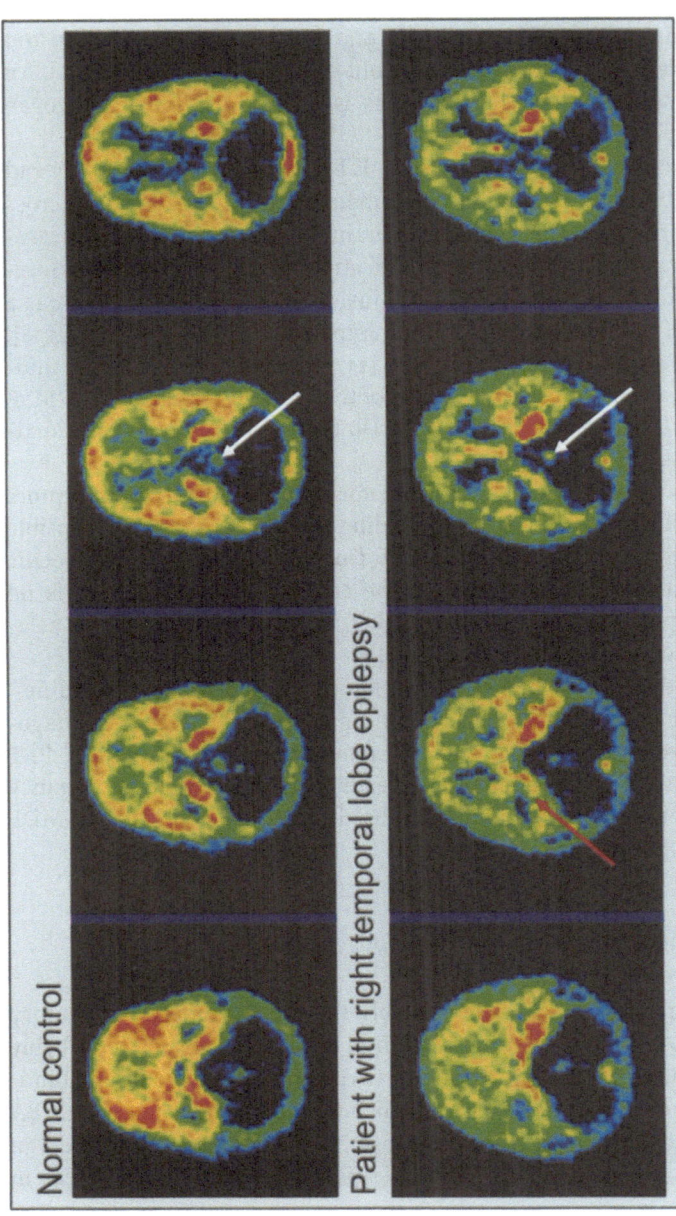

Fig. 3.5. Summed 40- to 60-min images obtained with ^{18}F–FCWAY on a GE Advance tomograph (3D mode, resolution 6–7 mm) in a patient with right temporal lobe epilepsy (TLE) and a normal control. Asymmetric binding, with a decrease in the right temporal lobe (red arrow) is seen the a patient with right TLE. Binding in the brain stem raphe (white arrow) is visualized in both subjects. Some ^{18}F activity is also present in the skull owing to tracer defluorination. Images courtesy of Dr. William H. Theodore, NINDS, NIH, Bethesda MD

3.12.3.2   Serotonin 2A Receptors

Serotonin 2A (5-HT$_{2A}$) receptors are present in all neocortical regions, the density being lower in the hippocampus, basal ganglia, and thalamus. The cerebellum and structures of the brain stem are virtually devoid of 5-HT$_{2A}$ receptors [2170]. An increase in prefrontal cortex 5-HT$_{2A}$ receptors is observed following estrogen treatment in postmenopausal women [2171].

^{18}F-Altanserin is one of the most selective 5-HT$_{2A}$ antagonist radiotracers and has good reproducibility [2172]. Radiolabeled metabolites of ^{18}F-altanserin cross the BBB and are distributed nonspecifically in brain [2173] but do not bind to serotonin receptors [2174]. Steady state for quantification of specific binding in forebrain can be achieved within 2h by a bolus-infusion technique [2175]. There is a decline in binding of ^{18}F-altanserin to serotonin type 2A receptors with age [2176], which persists after partial volume correction [2177] and is steepest during middle age [2178]. There is also significant binding of the D2 receptor ligand ^{18}F-NMSP to 5-HT$_{2A}$ receptors that can be displaced by MDL 100,907 [2179] and by the anxiolytic drug deramciclane [2180].

^{18}F-Setoperone is another PET radioligand for neocortical 5-HT$_{2A}$ receptors [2181–2183]. Determination of the cortical binding potential may not be possible with a reference region approach, because in normal human subjects nonspecific binding is smaller than that estimated for the cerebellum [2184]. There is no change of BP after pretreatment with paroxetine, whereas it is reduced by nefazodone probably because of competition [2185].

$R(+)$-Alpha-(2,3-dimethoxyphenyl)-1-(2-(4-fluorophenylethyl))-4-piperidine-methanol (MDL 100,907), a highly selective and potent 5-HT$_{2A}$ receptor antagonist, has been labeled with ^{11}C [2186] and has suitable kinetic properties for PET studies [2187]. A dose of 20mg of cold drug leads to >90 % receptor occupancy [2188]. The age effect seen with ^{18}F-altanserin has been confirmed with ^{11}C -MDL 100,907 in monkeys [2189].

3.13   Gamma-aminobutyric acid (GABA)

Gamma-aminobutyric acid (GABA) is probably the most important inhibitory neurotransmitter. GABA transmission is impaired in focal epilepsy, and probably also in anxiety and other psychiatric disorders. GABA receptors have several binding sites, including one for benzodiazepines which is used for PET imaging. GABA receptors are also sensitive to ischemic damage, which has aroused interest in studying them as a way of determining viable tissue in hypoperfused areas in acute stroke [860].

3.13.1   Central Benzodiazepine-binding Sites

The tracer most widely used for central benzodiazepine binding sites is the antag-onist flumazenil (RO-15–1788). Early studies with ^{11}C-flumazenil (FMZ) demon-strated that different areas of the healthy human brain showed approximately 10-fold variation in tracer binding, which corresponded to the previously known distribution of benzodiazepine receptors in these regions. The highest degree of binding was obtained in the medial occipital cerebral cortex, followed by other cortical areas, cerebellum, thalamus, striatum, and pons, whereas there was little binding in the striatum [2190]. It appears to reflect mainly binding to the alpha$_1$ subtype [2191]. Displacement by other receptor ligands suggested low nonspecific binding that was achieved as soon as 20 min after tracer injection [2192, 2193]. Thus, quantitation of the binding potential can be achieved in equilibrium by a rather simple one-tissue-compartment model [2194]. Owing to a high extraction rate, initial tracer uptake represents CBF with subsequent rapid approximation of the distribution volume that represents receptor binding [2195]. The metabolites in plasma are mainly polar and do not cross the BBB [2196, 2197]. For clinical use to detect focal reductions in FMZ in stroke and epilepsy (see sections 2.4, 2.5) activity images obtained from the time frame of 10–20 min are essentially equiva-lent to parametric images of the distribution volume (DV; see section 4.9.1.4) [2198]. Analysis of tracer kinetics in combination with various amounts of unlabeled tracer allow differentiation between receptor density and affinity (see section 4.9.3) [2199–2203].

FMZ binding is age dependent. The highest values are reached at approx. 2 years of age, with a subsequent decline by 25–50 % until adult values are reached at age 14–22 years [2204]. In contrast to findings in rats, alcohol does not directly affect central benzodiazepine receptor binding in man [2205].

^{11}C-RO-15–4513, a partial inverse agonist at the benzodiazepine receptor site [2206] that binds preferentially to the α5 receptor subtype, showed uptake that was relatively greater in limbic areas, in particular in the anterior cingulate cortex, hip-pocampus, and insular cortex, but lower in the occipital cortex and cerebellum, in comparison to FMZ [2191, 2207]. It may be of particular interest for studies of memory function and memory-enhancing drugs. An ^{18}F-labeled analogue of flumazenil, 5-(2'-^{18}F-fluoroethyl)-flumazenil could be of interest for clinical use, but its metabolism is very rapid, with some lipophilic metabolites, and it has lower affinity [2208, 2209].

3.13.2   Peripheral Benzodiazepine Receptors

Peripheral benzodiazepine receptors are present in the brain in microglia and in brain tumors [2210]. The tracer most frequently used is ^{11}C-PK-11195 (PK) [2211], mostly for imaging of microglial activation [168, 2212]. PK also binds with high affinity to the acute phase reactant alpha$_1$-acid glycoprotein, which is found in large amounts in the plasma of patients with acute inflammatory diseases. This could significantly alter the free plasma concentrations of the ligand and contribute to its variable kinetic behavior [2213].

3.14   Glutamate and NMDA Receptors

Glutamate is the main excitatory neurotransmitter in cortex, and alterations of glutamatergic neurotransmission are associated with many neurological diseases. There is also a neurotoxic potential of glutamate at higher concentrations. Unfortunately, tracers to study this system directly are not yet available, and most developments so far (^{11}C-MK801 and derivatives, ^{18}F-fluoroethyl-TCP, ^{11}C-ketamine, ^{18}F-memantine) have been stalled by low specificity [2214–2217]. Preliminary studies with the glycine(B) site antagonist ^{11}C-L-703,717 demonstrated preferential binding to a cerebellar NMDA receptor subtype consisting of GluR ε3 subunit in vivo, but not in vitro [2218].

Indirect assessment of glutamatergic neurotransmission may be possible by studies of CMRglc, because cerebral glucose consumption appears to be closely coupled to glutamate reuptake and recycling (see section 3.4). There have also been studies on the modulation of the dopaminergic system by glutamate in striatum with RAC [2057] (see also section 3.10.6).

3.15   Adenosine Receptors

Currently, the A(1) adenosine receptor (A(1)AR) and the A(2A)adenosine receptor (A(2A)AR) are regarded as clinically significant targets for tracer development, because of their neuromodulatory role and possible alterations in epilepsy, stroke, movement disorders, and schizophrenia [2219]. Xanthine analogues ^{18}F-CPFPX and ^{11}C-MPDX have been developed as a tracers for the A(1)AR [2220, 2221]. Uptake in humans is high in putamen and mediodorsal thalamus, intermediate in most cortical regions, and low in midbrain, brain stem, and cerebellum. A(2A)AR imaging has been performed with ^{11}C-KF18446 and PET in the rat brain [2222].

3.16   Histamine Receptors

It has been known for many years that there are cerebral histamine receptors, but their physiological and pathophysiological role is not clear. ^{11}C-Doxepine has been used as a tracer to study these receptors in humans [2223], and reduced BP is seen in AD [2224].

3.17   Cannabinoid Receptors

The CB1 cannabinoid receptor is expressed in the brain at levels sufficient for it to serve as a potential target for in vivo imaging using PET or SPECT. To date, tetrahydrocannabinol (THC) labeled with ^{18}F [2225] and tracers based on the cannabinoid antagonist, SR141716A, have been tried. Rodent data obtained with these radiotracers in vivo have demonstrated that both the behavioral and neurochemical effects of cannabinoids occur at very low levels of receptor occupancy. More recently, an agonist radiotracer based on the structure of aminoalkylindole cannabinoids has been examined for in vivo labeling of CB1 receptors. Although rodent studies have indicated that in vivo imaging of CB1 receptors is feasible, at present this receptor has yet to be successfully imaged in a human PET study [2226]. Work continues on the development of central cannabinoid receptor ligands with low lipophilicity [2227].

3.18   Opioid Receptors and Sigma Receptor

PET imaging of opioid receptors began with labeling of morphine, codeine, heroin, and pethidine with ^{11}C. The kinetics of brain uptake were analyzed, and it was shown that pethidine had the most rapid and extensive uptake, followed by heroin, codeine, and morphine in order of declining lipophilicity [2228]. Owing to their complex metabolism and some nonspecific binding, these tracers were not optimally suited to receptor imaging, and more specific tracers have since been developed.

^{11}C-Carfentanil [2229, 2230] binds preferentially to μ-receptors, which are probably the primary site for the pleasurable reward feeling caused by opiates. Highest concentrations after i.v. injection are seen in the basal ganglia and thalamus [2231]. Quantification has been based on the standard equilibrium model with two tissue compartments (see section 4.9.3.2), and a region/occipital cortex ratio can be used when tracer kinetic modeling is not feasible [2232]. The μ-opioid receptor BP has been found to increase with age in neocortical areas and the putamen. Sex differences, with higher μ-opioid binding in women, have been observed in a number of cortical and subcortical areas [2233].

^{11}C-Diprenorphine (DPN) binds to μ- and to non-μ-receptor sites and does not reach equilibrium in the time required for a typical PET study. A full kinetic analysis has been suggested for quantification [2234–2236]. Compared with carfentanil it shows greater binding in the striatum and in the cingulate and frontal cortex [2237, 2238], including the cortical projections of the medial pain system [2239]. To allow longer measurement times, N-(3-^{18}F-fluoropropyl)-N-nordiprenorphine [2240] and 6-O-(2-^{18}F-fluoroethyl)-6-desmethyldiprenorphine (^{18}F-DPN) were developed. High affinity to opioid receptors was demonstrated for ^{18}F-DPN, and parametric images showed a binding pattern equivalent to that of ^{11}C-DPN.

[11]C-Buprenorphine has similar properties but shows prolonged nonspecific cerebellar uptake [2241, 2242].

Cyclofoxy is an antagonist that binds to μ- and κ-receptor subtypes with a distribution very similar to that of naloxone and high uptake in caudate, amygdala, thalamus and brain stem. It has been labeled with [18]F [2243]. Kinetic studies with compartmental modeling are being used for quantification of receptor binding [2244, 2245]. There appears to be a pronounced age-related decline of thalamic binding in women [2246].

The regional distribution of $N1'$-(^{11}C-methyl)naltrindole in vivo in mouse brain correlates with established α-opioid receptor densities in vitro [2247]. In human brain the highest binding is observed in neocortex (insular, parietal, frontal, cingulate, and occipital), caudate nucleus, and putamen [2248]. The tracer shows irreversible binding characteristics, which makes separation of binding effects from CBF changes difficult [2249].

Although several synthetic opiates bind to the sigma receptor, it is not an opioid receptor, and other drugs and addictive substances, in particular phencyclidine (see section 2.7.5.7), also bind to it. A tracer that binds to central sigma receptors in cortex and cerebellum and to striatal D2 receptors is ^{11}C-nemonapride (NEM, YM-09151-2) [2035]. Another candidate is ^{11}C-labeled 1-(3,4-dimethoxy-phenethyl)-4-(3-phenylpropyl)piperazine (^{11}C-SA4503), which has much higher affinity to sigma-1 than to sigma-2 receptors [2250, 2251].

3.19   Steroid Receptors

Steroid receptors in the brain are involved in many basic regulatory processes, including gene expression. Therefore, quantitative imaging of these receptors has been an important goal. Several fluorinated ligands for corticoid receptors have been tested, but so far none has been validated for human studies [2252–2254]. High uptake of an estrogen receptor ligand, 16α-^{18}F-fluoro-17β-estradiol, is observed in the pituitary and hypothalamus, but quantitation has not been possible in other brain regions [2255]. High uptake is also seen in some meningiomas [2256].

3.20   Substance P

The SP (NK1 receptor) antagonist (SPA) aprepitant (also known as MK-0869) has been labeled with [18]F. PET imaging studies in rhesus monkeys and humans demonstrate high affinity for the NK1 receptor, low nonspecific binding, and good BBB penetration. PET studies in humans have led to the prediction that very high levels of central NK1 receptor occupancy (>90%) by the unlabeled drug will be associated with therapeutically significant antidepressant and antiemetic effects [2257].

3.21   Secondary Neurotransmitters

Brain incorporation of labeled arachidonic acid has been shown to be increased in awake rats by pharmacological activation associated with phospholipase A2 signaling. PET and ^{11}C-arachidonic acid may be useful to measure signal transduction in the human brain. Brain incorporation of ^{11}C-arachidonic acid has been quantified and shown to be independent of flow in monkeys [2258] and in young healthy humans [2259]. Increased uptake of ^{11}C-diacylglycerol is seen after muscarinic cholinergic receptor (mAChR)-stimulation in rats and monkeys and is interpreted as the phosphoinositide response, which is an effector in the production of second messengers [2260].

Data Acquisition, Reconstruction, Modeling, Statistics

4.1   Positron Emitters and Tracers

PET is based on the detection of coincident 511 keV gamma rays that originate from positron electron pair annihilation. Positron emitting isotopes typically have nuclear masses that are smaller than those of stable isotopes, and most of them are very short lived (Table 4.1). Their short half-lives have the advantage that relatively large physical activity doses (with ^{11}C and ^{18}F typically 370–740 MBq) lead to rather low effective biological radiation doses [2261]. Thus, images can be based on high total count numbers with accordingly low stochastic noise, whose variance is essentially proportional to the square root of total counts according to the Poisson distribution [2262]. Longer half-lives in the order of days may be required to study such slow processes as cell trafficking or DNA synthesis. ^{64}Cu and ^{124}I are the most frequently used isotopes that fulfill this requirement.

Table 4.1. Isotopes used in brain PET

Isotope	Physical half-life	Remarks
^{11}C (carbon)	20.4 min	Very versatile, usually replacing natural carbon in molecules
^{13}N (nitrogen)	10.0 min	Rarely used in brain studies
^{15}O (oxygen)	2.05 min	Mainly for studies of blood flow (activation studies) and oxygen metabolism
^{18}F (fluorine)	109 min	Convenient half-life for most metabolic and receptor studies
^{64}Cu (copper)	12.7 h	For processes with slow kinetics, molecular imaging; rarely used
^{76}Br (bromine)	16.1 h	Also emitting unpaired high-energy γ-rays
^{68}Ga (gallium)	68.3 min	Produced from ^{68}Ge generator
^{82}Rb (rubidium)	1.3 min	Produced from ^{82}Sr generator
^{124}I (iodine)	4.2 days	For processes with slow kinetics, molecular imaging, also emitting unpaired γ-rays

The preparation of PET tracers is an expensive and demanding endeavor that requires highly specialized equipment and personnel. Many aspects of it exceed the scope of this book by far, and the reader is referred to the specialist literature [2263–2269]. Radiotracers are radiopharmaceuticals, and their production is subject to legislation by national and international regulatory agencies, such as the U.S. Food and Drug Administration (FDA). Use of radiotracers is associated with exposure to ionizing radiation for the subject being studied, and potentially also for the personnel. There is an associated risk of inducing cellular DNA damage. Guidelines have been published by the International Commission for Radiation Protection (ICRP), and international and national laws and regulations apply. Most frequently, radioactivity doses of PET tracers for human use are in the order of 370–740 MBq, and these typically result in an equivalent dose of 5–20 mSv. This is a range at which the risk for adverse health effects is so small that it is regarded as safe even for use in normal volunteers if there is appropriate clinical or scientific justification, consent is obtained from the subject of study, and (in the case of scientific studies) permission is granted by the appropriate ethics committee and (in most countries) national regulatory agencies.

Quality requirements for PET tracers include absence of contamination by other radionuclides (radionuclidic purity), chemical purity (in particular, absence of toxic organic solvents), and radiochemical purity. The pharmaceutical quality must ensure sterility, apyrogenicity, and stability, and control pH and osmolality. With fulfillment of these basic requirements, clinically significant side effects related to tracer injection are extremely rare.

A critical issue for accurate measurement of saturable biochemical processes is the specific activity, which indicates the amount of the labeled pharmaceutical relative to the total amount of pharmacologically active (labeled and unlabeled) pharmaceutical [2270]. For most radiotracers that measure metabolism, a specific radioactivity of at least 10 GBq/µmol is considered standard. However, this level is not sufficient to measure the binding potential of neuroreceptors with high-affinity ligands, because even small amounts of the binding drug can occupy a significant proportion of receptors and thus reduce the proportion of receptors that is accessible for the labeled drug. For ligands with moderately high affinity with a dissociation constant (K_D) in the order of 1 nM, such as ¹¹C-raclopride (RAC), injection of 50 nmol (e.g., 500 MBq at 10 GBq/µmol) will probably occupy approx. 5% of striatal receptors in a normal man. Margins are even narrower for tracers with very high affinity (with K_D in the order of 0.1 nM), such as ¹¹C-WAY-100635. Thus, for high-affinity receptor ligands, specific activity should be in the order of 100 GBq/µmol or higher. In small animals, receptor occupancy by the ligand may be especially critical [2270, 2271]. Thus, tracers need to be prepared with extremely high specific activity for rats and mice, and sensitive PET scanners are necessary to allow a low injected dose that will keep pharmacological mass effects to a minimum.

4.2   Scanners and Detector Systems

At the core of PET scanners are scintillation detectors associated with coincidence electronics to detect the two 511-keV gamma rays that are emitted at an angle of 180° by a positron-electron annihilation event. Such a coincidence event defines the line of response (LOR), eliminating the need to use collimators that would limit the angle of incoming rays, which are necessary in the case of conventional isotopes emitting single gamma rays. These lead to an enormous increase of the solid detection angle, resulting in a net sensitivity advantage in the order of 10 for PET over SPECT [2262], even if detection is limited to coincidences within the transaxial planes.

Detectors are typically arranged as a hexagonal, octagonal, or circular "ring." The width of the detector ring is larger in whole-body scanners than in dedicated brain scanners, and both can be used for brain studies. Some older scanners had axial fields of view (FOV) of less than 12 cm, which is required to scan all clinically essential parts of the brain in a single session. Modern scanners, such as the ECAT HR and ECAT HR+ (Siemens-CTI) and the ADVANCE (General Electrics) have an axial FOV of 12–15 cm at a near-isotropic spatial resolution of 4–6 mm in the center of the FOV [2272–2276].

Usually inorganic scintillators are used as detector materials, with bismuth germanate (BGO) being the most widely used in current PET scanners. Other scintillation crystals, such as sodium iodide [2277], cesium fluoride [2278], barium fluoride [2279, 2280], and gadolinium oxyorthosilicate (GSO) are also used [2281]. In recent years, lutetium oxyorthosilicate (LSO) has become available for commercial PET scanners [2282, 2283], and owing to a higher light output and a shorter decay time than BGO this achieves 2- to 3-fold increase in noise equivalent count rate [2284]. GSO is used as an alternative; it has less favorable physical characteristics than LSO but its performance may be similar in some practical applications [2281]. Combinations of crystals with different decay times are being used to maximize sensitivity and resolution by including measurements of depth of interaction in the crystals [2285].

Crystals are usually arranged as block detectors, with groups of detectors sharing a common base and a common array of photomultipliers [2272, 2286, 2287]. This arrangement has brought about a substantial improvement in spatial resolution, which depends mainly on detector size. With very fast detector crystals, time-of-flight information can be used to improve the signal-to-noise ratio, but the effect is more relevant to whole-body applications than to brain scanning [2280, 2288, 2289].

Former scanner constructions had septa to separate detector rings and block scattered γ-rays. Coincidences were recorded only within the transaxial planes defined by these rings, a technique called 2D acquisition. Progress in detector and electronics technology has allowed optional retraction or complete abandonment of these septa [2290–2292], so that it is now possible to record coincidences that occur in opposite detectors not belonging to the same ring and transaxial plane;

this is called volume acquisition or 3D acquisition [2293–2295]. It leads not only to a very substantial gain in detection efficacy, especially in scanners with a large axial FOV [2296], but also to a large increase of the contribution of scatter to recorded events. Quantitative image reconstruction (see section 4.3) is much more complex than for 2D acquisition.

Alternative low-cost designs for clinical imaging usually employ sodium iodide crystals [2297], which are commonly used in SPECT cameras. The clinical utility of such scanners operating in 3D mode (e.g., PENN PET 300H) has been demonstrated [1043]. Rotating SPECT cameras have also been used successfully for positron coincidence detection in oncology [2298, 2299], but their clearly inferior sensitivity (at least one order of magnitude lower than with dedicated BGO-PET scanners) and spatial resolution (7–7.6 mm FWHM) [2300] severely compromise their application in neurology. Reduced sensitivity at lower cost than with full-ring scanners is also provided by rotating devices with fewer BGO crystals than full-ring scanners [2301, 2302].

An alternative to scintillation crystals is the multiwire proportional chamber technology, which is inferior in sensitivity but can reach very high spatial resolution [2303–2305]. It is now mainly being used in small-animal scanners and in nonmedical industrial scanners.

The highest spatial resolution that is currently possible with PET is in the order of 1–10 µl, i.e., about 1–2 mm in each dimension [2306–2308]. It has been reached by several small-volume scanners, most of which have been built as research prototypes for studies of small rodents [2309–2311]. A few are also commercially available, including the LSO-based micro-PET [12, 2312–2314] (Concorde Microsystems Inc., USA), which also offers high sensitivity, and the wire-chamber system quad-HIDAC PET (Oxford Positron Systems Ltd., UK), with which ultrahigh resolution is possible although sensitivity is lower [2315]. With the increase in the volume of experimental PET work in brain research, molecular imaging, and drug development, these systems are becoming more commonly used. Experimental research in larger animals, especially mammals such as cats, dogs, and pigs, and nonhuman primates is often performed in the high-resolution multipurpose clinical and experimental scanners described above.

High resolution is also an important clinical goal in brain imaging, because most gray matter brain structures have an extent of only a few millimeters in at least one spatial dimension. With current clinical PET scanners this results in significant spillover of signal from and to adjacent white matter, which often has a quite different physiological signal (metabolic rates, receptor densities, etc.). This results in a bias of quantitative values (usually signal reduction in the order of 20–30% in gray matter) that is known as incomplete recovery or partial volume effect [125, 2316]. The effect is accentuated in the presence of the atrophy that accompanies many neurological disorders. There have been attempts at correction [2317–2319], which are useful in scientific studies but depend on segmentation of coregistered, high-quality MR images and are not yet robust enough for clinical studies (see section 4.10.2 for correction algorithms). The problem is particularly notorious in the study of hippocampal [2320] and brain stem structures, because they are so small and located so deep in the skull (requiring large corrections for attenuation). Functional changes in these structures have pivotal roles in many diseases, such as epilepsy, most neurodegenerative diseases including AD and PD, and probably also psychiatric diseases. Therefore, improvement of the techniques for quantitative functional imaging of these structures is urgently needed. With high-resolution acquisition, high sensitivity is also crucial because at $1\,\mu l$ resolution, at least 10^6 voxels within a human brain need to contain sufficient counts, requiring total count numbers in the order of 100×10^6. Thus, long measurement times of up to 60 min and appropriate tomography designs with high-sensitivity crystals [2321] are required to achieve that goal.

Functional PET images often require coregistered CT or MRI images for anatomical localization. There have been developments directed at integrating CT and PET in one scanner, and such scanners are now becoming commercially available [2322]. Since CT of the brain does not demonstrate much anatomical detail, a combination of PET and MRI, which is still at a preliminary stage [2323], would be of more interest for neuro-PET. The current standard solution is to record brain PET and MRI independently and coregister 3D data sets afterwards (see section 4.7).

Separate detector systems are necessary to measure activity in blood samples. This is usually done by manual sampling, separation of plasma from blood cells by centrifugation, and measurement of tracer concentration in plasma samples in a cross-calibrated well counter. For good representation of arterial tracer, sampling has to be done as rapidly as possible during the first 2–3 min after a bolus injection, after which sampling intervals can be gradually increased up to several minutes with data interpolation between samples. As an alternative to manual sampling, devices for continuous blood sampling have been developed [2324–2328], and noninvasive methods of measuring arterial activity have been suggested [2329].

4.3   Data Acquisition

In standard static data acquisition, coincidence events are counted for each LOR over the scanning interval. The LOR counts are usually represented and ordered as sinograms, with angle Φ and position x_r along the radial axis as coordinates. With 2D acquisition a sinogram is recorded for each slice. With 3D data acquisition, a second angle and associated radial position are required to store the data [2330]. Image reconstruction is based on these sinograms.

Cumulative counting of events for each LOR over fixed scanning intervals does not allow sorting of individual events. Thus, if something has changed during the scanning interval (e.g., the functional status of the patient or the position of the head in the scanner) it cannot be distinguished. This is only possible with data acquisition in list mode, in which for each event, the LOR and time are recorded in a single long list for the entire study. Other characteristics of the event, e.g., measured energy, can also be recorded. In 2D mode, list mode typically requires much more data transfer and storage than does standard acquisition, and has therefore rarely been performed. In 3D mode, the data volume in list mode may be quite comparable to that in cumulative mode, because there are very many LORs and a substantial proportion of them are associated with very small numbers (0 or 1) of events during typical acquisition times. With list mode data, categorization into time frames is done after acquisition but prior to reconstruction, and thus can be changed more flexibly to optimize time frames in dynamic studies. Advanced reconstruction algorithms may even process list-mode data without prior rearrangement [2331] and might accommodate linear models of tracer physiology for direct reconstruction of parametric images.

4.4   Image Reconstruction

Quantitative image reconstruction needs to take into account
- Differences in scanner sensitivity among LORs (correction by normalization).
- Random coincidences, which occur when two γ rays from two *different* positron annihilations are sensed by a detector pair, so that a false or random coincidence count is collected (randoms correction).
- Scattered events, which occur when an annihilation photon traveling in tissue is deflected by a collision with an electron, so that its direction changes. This results in incorrect positioning of the coincidence line (scatter correction).
- Attenuation of γ rays, due to their interaction with tissue, resulting in their loss and in a decrease in the number of detected decay events (attenuation correction).

Corrections for these effects are typically integrated into reconstruction procedures, and can be achieved using a large number of possible methods of implementation and combinations of procedures. The corrections can be done for PET with much better accuracy than for SPECT. Thus, reconstructed images allow *absolute* radioactivity measurements. It is not possible in this context to give even a reasonably comprehensive account of this field. Procedures for image reconstruction are usually supplied by scanner manufacturers, and we will therefore only briefly address the main issues. More comprehensive and detailed reviews can be found elsewhere [2295].

The basis for data normalization is usually provided by acquisition of a very large number of events (typically more than 100 for each LOR) from a cylindrical homogeneous phantom filled with radioactivity or a rotating rod source. It is more complex for 3D than for 2D owing to the much larger numbers of LORs and possible combinations of detectors.

In PET, attenuation is constant for each LOR (which is not the case with SPECT and contributes to the robustness and accuracy of PET). Attenuation can be estimated directly from PET images by thresholding to determine the contour of the head or by fitting transaxial ellipses to the head, which yields reasonable imaging results in brain scanning, with the largest errors in the basal parts of the brain. More accurate, and therefore preferable, methods rely on individual measurements of attenuation by rotating rod or point sources of ^{68}Ge/^{68}Ga or ^{137}Cs (see [2332] for review). With combined CT-PET scanners, attenuation correction can be based on X-ray transmission scans with appropriate adjustments for the energy difference [2333].

Correction for scatter is the most demanding and complex part of the correction procedures [2334]. Estimation of the amount of scatter that contributes to observed coincidences can be derived from energy window manipulations (scattered events tend to have lower energy than true events) or registration of non-coincident events, or by measurements of scatter from line or point sources within a scatter medium. Such measurements are usually used to adjust mathematical models that represent the physics and geometry of scanner and object, and are then incorporated into reconstruction algorithms [2335–2338].

The large increase of scattered events in 3D acquisition relative to 2D is a major concern. Proper implementation of correction algorithms is even more important for 3D than for 2D acquisition and is required for quantitatively correct results in human and experimental brain studies [2271, 2339]. Scatter from outside the FOV, e.g., from body organs with high activity, also requires attention and should be reduced as much as possible by physical measures [2340]. Scatter reduces the gain in efficiency to some degree. A PET scanner optimized for and operating exclusively in 3D mode with a large axial FOV (23.4 cm) achieved 5.8 % absolute efficiency for a line source in air, and 10 % for a central point source (with thresholds of 350–650 keV). For a uniform 20-cm-diameter cylinder the efficiency was 69 kcps/kBq per ml after subtraction of a scatter fraction of 42 % [2273].

Image reconstruction can be achieved by two essentially different approaches. The first is an analytic approach, according to which reconstruction is achieved by back-projection that essentially inverts the physical process of projection of original events into LORs. The second approach consists of iterative procedures, in which estimates of the original tracer distribution are refined by iterative approximation.

The standard approach is known as filtered back-projection, which is commonly used in all types of computer-assisted tomography [2341]. It is commonly implemented by the use of Fourier transforms and filtering in frequency space. Current implementations of filtered back-projection are very fast and well developed for 2D PET imaging. They are not directly applicable to 3D imaging, mainly because 3D projections are not complete (owing to missing LORs at the cranial and caudal opening of the ring). Rearrangement of 3D LORs into 2D LORs has been achieved by single-slice rebinning [2342], which assigns oblique LORs to the mid-transaxial plane. A more advanced and accurate technique is Fourier rebinning (FORE) with its related algorithms [2343, 2344].

Iterative reconstruction should provide a higher accuracy of quantitation (see [2345] for review). This has been demonstrated for images containing large target-to-background ratios, which are common in whole-body oncological studies [2346, 2347]. The advantage is less clear for standard brain studies with less sharp contrasts, but its potential for improving spatial resolution and increasing the statistical power in ^{15}O-water-PET activation studies over those obtained with FBP reconstruction has been demonstrated [2348, 2349]. A commonly used algorithm is called "ordered subsets expectation maximization" (OSEM) [2350–2352]. If performed in 3D it usually results in extremely long reconstruction times, which can be reduced by rebinning and by parallel processing in a computer cluster [2353]. It is also possible, at least in principle, to use prior anatomical information derived from MRI to sharpen contrasts between different anatomical compartments [2354] and to incorporate spatially variant spatial resolution of the scanner [2355]. Corrections for head movements can also be implemented [2356] (see also section 4.5). Image noise typically increases during iterations together with improvement in image contrast, and can lead to considerable degradation of reconstructed image quality. Analysis of statistical image properties may help to find the number of iterations that optimized the signal-to-noise ratio in reconstructed images [2357–2359]. Iterative reconstruction can be used to optimize spatial resolution by implementing a maximum a posteriori probability algorithm (MAP) that accounts for all factors that degrade quantitation and resolution on a volume basis and thus recovers the full intrinsic resolution [2360]. By these means, 1-μl resolution has been achieved in a small animal scanner [12].

4.5   Motion Detection and Correction

Head motion, which can occur during long acquisition times even in very cooperative normal subjects and even more so in patients, degrades image resolution and can severely distort local and regional kinetic data. Thus, correction for head movement is essential for high-resolution studies, whereas standard static PET scans aiming at large brain regions, e.g., with FDG for detection of metabolic impairment in association areas due to AD, are more robust. Several image processing software packages provide routines for coregistration of multiple frames in a kinetic study that can reduce these problems at least with regard to kinetic and functional activation studies. Head movement that has occurred within a frame still leads to reduced spatial resolution. Therefore, devices for online detection of head movement and registration of these data in list mode are being introduced, particularly for use with high-resolution scanners [2361–2363].

4.6   Data Visualization

Effective means of data visualization are important to get a good overview of the reconstructed images, to identify interesting and relevant brain locations, and to communicate results. Images that provide high contrast at anatomical borders, such as most standard MRI sequences, are usually best visualized by using an intensity display in which brightness (e.g., on a gray scale) corresponds to pixel intensity. Representation of quantitative PET values (which may not reflect anatomical structures) is also possible with brightness scales to detect "hot" and "cold" areas. For a more accurate assignment of quantitative values to image pixels, color scales are more useful. By covering a large part of the available color space (with its three different color channels), e.g., by the popular spectrum color scale (used in many of the figures in this volume), the viewer can assign values to colors more accurately than values can be assigned to a single-intensity channel. The assignment of the numbers to the colors is provided by a color bar. One should always be aware that the color scales tend to produce arbitrary borders at certain color changes. For instance, the red color in the spectrum color scale is usually perceived as clearly distinct from the neighboring yellow, whereas the green-blue transition is much less conspicuous. Thus, an image can be made to look very different just by shifting the window of the color scale. For visual interpretation of functional images, it is always a good idea to visualize it with several window settings of the color scale and also by using an intensity scale.

PET images have three basic spatial dimensions, x, y, and z. Although data are usually still reconstructed as multiple transaxial slices along the z-axis, with pixels as image elements, there are usually no gaps between slices, and the data should therefore be regarded as volume data consisting of volume elements (voxels). Imaging software should provide a convenient means of representing the data volume in an intuitive and representative way, typically as multiple planes that are either all oriented in parallel, or as three selected planes that are orthogonal to each other (transaxial, coronal, and sagittal). There may be additional dimensions along the time axis (from dynamic data acquisition or multiple follow-up studies in the same subject) or representing multiple modalities (different PET or SPECT tracers, different MRI acquisition protocols). There should be tools to extract and highlight the important features of these time series or modality comparisons. The principles of data processing to achieve this are described below.

Image fusion is a basic visualization technique that is needed to combine functional PET data (in color scale) with structural MRI data (in gray scale) for accurate anatomical localization. It is particularly important for studies of patients with brain lesions, such as tumors or vascular lesions.

4.7   Image Coregistration

There are many situations in which coregistration of functional PET images and other images from the same subject is crucial [2364–2373]. There are two basic types of intrasubject coregistration:
- All images from same image modality (intramodality): this is typically the case when a dynamic image series has been acquired and we have to adjust for some (usually slight) head movement within this series. The situation also occurs

when multiple follow-up measurements of the same subject and same imaging modality are to be compared.

- Images from different modalities (intermodality): in this case multiple images obtained with different techniques (e.g., different PET tracers, or PET and MRI) have been obtained in the same subject. Coregistration may be required for accurate spatial correlation of functional changes or for attribution of function to macroscopic anatomy or specific pathological brain lesions.

Intrasubject coregistration typically only allows for translation and rotation defined by six independent transformation parameters (e.g., translations along main axes and rotations around main axes). Results can conveniently be visualized by image fusion.

Coregistration methods have also been integrated into procedures for stereotactic fractionated radiation treatment planning [2374]. Some software packages provide additional surface rendering of anatomical details from high-resolution segmented MRI with color coding of cortical functional activity beneath that surface by ray tracing [2375–2377].

Intersubject registration is another type of registration that refers to the adaptation of different brains (with different shape) that have been scanned with the same modality, most often by matching each subject to a template for anatomical standardization as a basis for regional data extraction or voxel-based statistical calculations.

4.7.1   Fiducial Markers

Traditionally, coregistration has relied on fiducial markers that are in a fixed geometric configuration relative to each other and to the subject's brain and can be identified in both data sets. In principle, a minimum of three markers (not all lying on a straight line) should suffice to define a unique geometric transformation for coregistration. For instance, small pellets filled with contrast agents and positron emitting substances have been used. In practice, this approach has proved difficult to implement, because fiducial markers fixed to human head skin can shift several millimeters together with the skin. The outer ear canal has also frequently been used, but accurately modeled plugs are required for high precision (Fig. 4.1). Bite plates are another possibility, but are uncomfortable and cannot be used in subjects who have lost their teeth (as many older patients have).

The gold standard for spatial coregistration is provided by stereotactic systems that are rigidly fixed to the skull. In principle, functional images at fixed geometry relative to the stereotactic frame can be obtained with proper adaptation of PET headholder systems. However, the technical complexity and discomfort associated with a fixed stereotactic system and the fact that very accurate coregistration with MRI can be achieved without that, as described below, have prevented wider use.

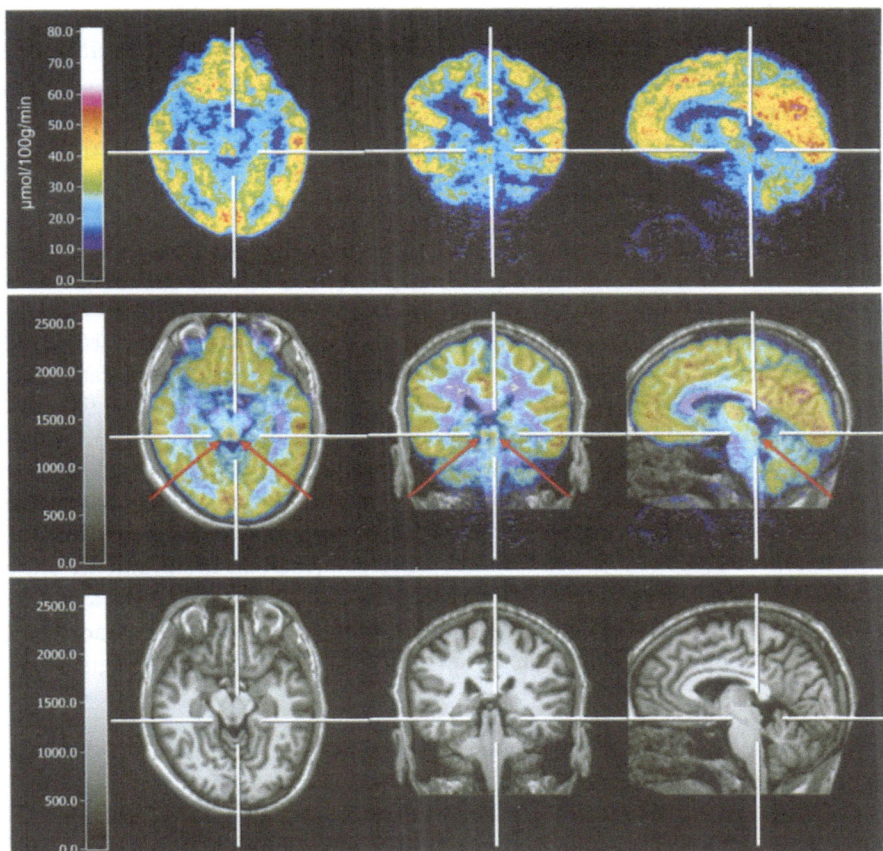

Fig. 4.1. FDG-PET (with quantitation of CMRglc) obtained in a subject in a normal resting state (eyes closed, ears not plugged) with a high-resolution scanner (ECAT HRRT): high metabolic activity of left and right inferior colliculi (marked by *arrows* on fusion image) is clearly separable from the other side and from superior colliculi, which are inactive owing to lack of visual stimulation. Orthogonal 1-mm cuts [positron emission tomography (PET), fusion, and T1-weighted magnetic resonance imaging (MRI)] through left inferior colliculi are shown

4.7.2   Matching of Brain Structures

Many coregistration procedures rely on the use of external fiducial markers [2378], and their use may in fact be necessary in body parts that lack reliable intrinsic features for coregistration. For neurological applications, the complex anatomy of the brain itself probably provides the most natural and reliable marker for coregistration. If images are assumed to be virtually identical then a correlation analysis over all image points provides a measure of matching accuracy

[2379]. Alternatives are the sum of absolute or squared differences between corresponding image points or stochastic sign changes [2380, 2381].

If similarity of images in terms of voxel values cannot be assumed, which is the case in coregistration of images obtained by different modalities, more complex measures of image similarity must be used. Some approaches are based on extraction of anatomical features, such as the outer contour of the head [2382]. Multivariate cost functions have also been suggested to assess multimodal brain coregistration [2383]. A very robust method to match the same structures that are presented in different ways by different imaging modalities, without making specific assumptions about anatomy, is maximization of mutual information. This measure is derived from information theory and provides a very general and powerful criterion in this situation, because no limiting constraints are imposed on the image content of the modalities involved [2384]. It has been demonstrated that subvoxel accuracy for PET-MRI coregistration with respect to a stereotactic reference solution can be achieved completely automatically and without any prior segmentation, feature extraction, or other preprocessing steps [2385].

Extracerebral structures (skull, scalp, and nasopharynx) are typically within the FOV of brain scans. Scalp and nasopharynx display much more variability, even in repeated examination in the same individual, than the brain, and therefore some coregistration procedures can be misled by an attempt to achieve a perfect match for these structures. It is therefore recommended that the extracerebral structures be removed by segmentation procedures prior using coregistration algorithms [2367, 2386].

4.7.3   Algorithms for Maximization of Similarity

Early programs relied on multiple orthogonal displays of images and visual judgment of correspondence [799, 800]. Estimation of correspondence was facilitated by fusion display, transfer of threshold contours from one image to the other, and other comparative displays. This approach permits rapid and fairly accurate coregistration of data sets by experienced users, and it is still useful to ensure the plausibility of the results of automated procedures [2387, 2388]. However, it does not provide reproducible criteria for the quality of coregistration.

Maximization of similarity can be done without user intervention by computer algorithms. To increase processing speed and convergence, similarity is typically first checked at reduced spatial resolution and proceeds in multiresolution steps to maximum resolution at the end of the iterative process [2384]. It is also appropriate to smooth images for reduction of high-amplitude high-frequency noise that would obscure matching of gross anatomical structures. Various optimization strategies, such as Powell, simplex, steepest descent, conjugate-gradient, quasi-Newton and Levenberg-Marquardt methods, can be used to determine the best coregistration. Algorithms have been tested and optimized for speed and convergence by various authors [2389]. With current PC processor speed, accurate coregistration of 1-mm³-voxel whole-brain data sets can be achieved by such algorithms within a few minutes.

Some computer algorithms may fail if starting conditions are far removed from coregistration. They also cannot guarantee in general that the globally best fit, rather than a local maximum of similarity, will actually be identified within a reasonable time. A final visual check of the coregistration result is therefore always advisable.

4.7.4   PET/CT and PET/MRI Scanners

A special case of intermodality coregistration is present with PET/CT systems that use X-Ray CT scans for attenuation correction, which may require some correction for (slight) movements between scans. In a more general vein, the question of whether CT scans obtained separately can then be coregistered with emission scans and be used for attenuation correction has also been considered [2332]. For brain PET, interest in such approaches is limited, because CT does not provide good gray/white matter contrast, and PET/MRI would therefore be a more valuable (but technically more challenging) combination for neurology [2333]. New challenges arise from multimodal molecular imaging in small animals Fig. 4.2) because of the constraints of spatial resolution and the corresponding scarcity of identifiable anatomical markers [2390, 2391].

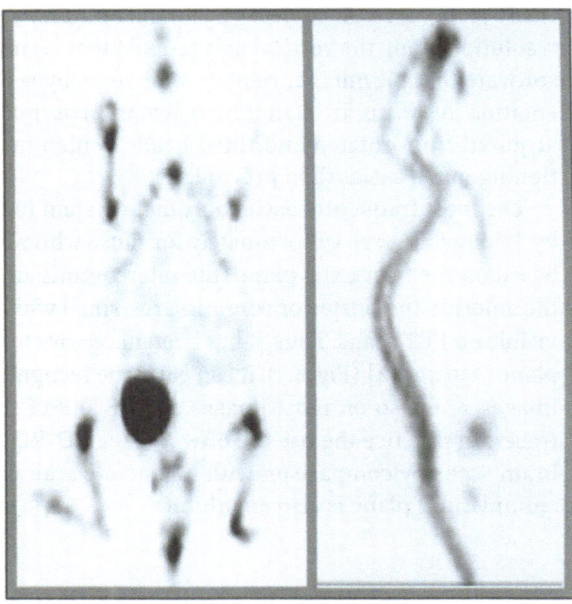

Fig. 4.2. High-resolution microPET ^{18}F bone scan of a mouse. Individual vertebral bodies and ribs can be distinguished

4.8   Anatomical Standardization

Most PET tracers provide images with a good to excellent contrast between gray and white matter, permitting easy orientation with regard to the major gray matter structures. Thus, localization of the cerebral lobes, putamen and caudate nucleus, thalamus, brain stem, cerebellar vermis and hemispheres is usually easily possible on PET images from subjects with no alterations in gross brain anatomy. Most experienced readers of PET images will be able to recognize these structures on transaxial, coronal, and sagittal slices. Thus, straightforward display of reconstructed images is often sufficient for qualitative visualization of local blood flow, metabolism, or receptor binding.

Limitations of this simple approach become evident if slight deviations from the normal distribution of local values are of interest. This is the case in many clinical situations in which we wish to detect mild abnormalities of function before major changes occur that would be associated with irreversible structural changes. Anatomical standardization may provide a means of permitting local quantitative comparisons that will overcome this limitation.

4.8.1   Orientation of Transaxial Slices

A traditional way of presenting PET images in a standard way is to use a defined orientation of transaxial slices. With early PET scanners and thick slices this was most commonly done by acquiring scans in parallel to the orbitomeatal or canthomeatal plane, because the markers for these traditional radiological planes are easily recognized on subject's head. Current PET scanners often provide a spatial resolution along the vertical axis (z-axis) that is similar to in-plane resolution and software that permits reorientation of slices by reslicing at any angle. Thus, presentation of scans in standard orientation is possible even if data have been acquired from rotated and tilted heads (which makes comfortable patient positioning much easier than previously).

The most frequently used coordinate system for brain orientation was defined by Talairach [2392]. Unfortunately for those who would like to use it with PET, it is based on the transaxial plane (the intercommissural or AC–PC plane) defined by the anterior and posterior commissures, small white matter structures that are not visible on PET scans. Thus, it has been necessary to define a PET surrogate for that plane [2393, 2394] (Fig. 4.3). It can easily be recognized on normal FDG-PET brain images, and also on most images of CBF and of the initial distribution of other traces. In practice the use of software (see CD-ROM) that allows reorientation of brain scans by comparison with a template scan oriented according to the intercommissural plane is also an option.

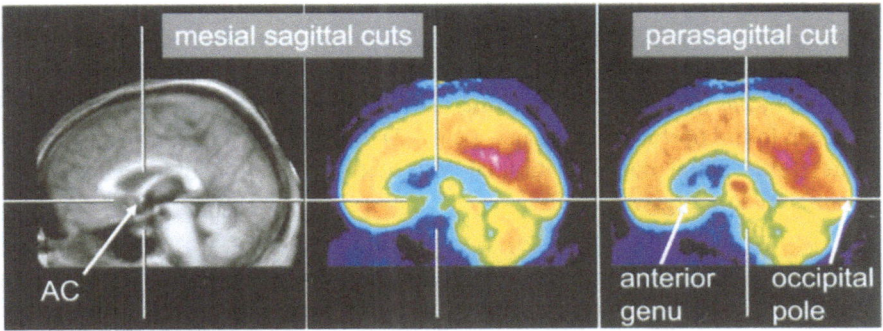

Fig. 4.3. Approximation of AC-PC line in PET images by the horizontal part of the anterior genu of the cingulate gyrus and the occipital pole, as suggested by Minoshima et al. [2393]

4.8.2   Matching of Individual Brains to Image Templates ("Spatial Normalization")

Spatial normalization, which means the deformation of individual brain anatomy to match a template, is essential for data analysis across different individuals on a voxel basis (statistical parametric mapping, see section 4.10.5), and also facilitates the use of standard volumes of interest [2395] (see also section 4.10.1). Several software packages for 3D image processing allow matching of volume data sets to a template. Separate templates should be created for each tracer [2396, 2397], in order to avoid confounding of anatomical differences by major functional differences. Frequently, normal brain scans that have been oriented and scaled to match a standard position and size are used as templates. More modern versions are derived from anatomical atlases that are based on multiple subjects rather than on a single subject, as described below.

Automatic matching of individual brains to atlas templates is usually achieved by coregistration procedures (see section 4.8.3) that allow not only for translation and rotation (defined by six independent transformation parameters), but also for scaling along the three main axes (adding three more independent parameters) and, in most instances, also for linear distortion (affine transformation with 12 independent parameters). Additional nonlinear transformations can be applied to adjust for local variations [2398–2402]. Nonlinear transformations substantially improve coregistration of cortical landmarks [2403]. It should be noted that image smoothing that is usually applied in the context of parametric imaging (see sections 4.10.4, 4.10.5) also serves the purpose of mitigating residual anatomical inaccuracies that could otherwise cause large artifactual variability at gray matter borders. More advanced techniques to relate individual images and brain atlases use Bayesian schemes to incorporate prior knowledge of the variability in the shape and size of heads [2404], cortical warping with constraints on the position of the major invariant cortical sulci, and correct topological representation of the cerebral cortical mantle [2405–2410].

Disease-related or age-related changes (such as enlargement of CSF spaces, vascular lesions, or brain tumors) may cause distortions of normal anatomy, and this issue has not generally been solved. A procedure based on thin-plate spline transformations has been shown to be rather insensitive to cortical atrophy [2400], and has been found useful for the diagnosis of AD by PET [123]. Further progress may come from spatial normalization of MRI with application of transformation parameters to coregistered PET images.

4.8.3   Atlases for Identification of Anatomical Structures

Identification of anatomical structures is relevant if we wish to determine the anatomical location of a particular point or area on a PET scan. On spatially normalized scans, this is often done by reference to the standard coordinates, which are then labeled by reference to a standard atlas. Alternatively, for more systematic quantitative analyses in anatomical structures, we may wish to use a set of volumes of interest that delineate those structures.

Brain atlases based on CT, MRI or PET scans obtained from normal subjects and the use of interactive procedures and automatic algorithms to register them in a common coordinate space have been developed by several groups [2398, 2411–2415]. There is considerable interindividual variation in the position of major cortical anatomical landmarks with respect to a standard coordinate system [2416, 2417], so it is preferable to have atlases based on multiple rather than single subjects. A standard brain that is close to the Talairach atlas brain and coordinate system was generated from 305 normal MRI brain scans at the Montreal Neurological Institute (MNI) [2418]. It was adopted by the International Consortium of Brain Mapping (ICBM) [2419] as an international standard and is now known as the MNI standard brain. It has been parcelated manually into multiple volumes that represent anatomical structures and can be used as an atlas template on spatially normalized brains [2395]. The ultimate goal is a probabilistic atlas that is both visualizable (i.e., contains recognizable brain images) and quantitative (e.g., linked to tabular databases that refer to micro- and macroscopic structure and brain function) [2420, 2421].

Traditionally, the Talairach atlas is used to identify anatomical structures for coordinates obtained on images that have been matched to a standard template. The anchor point of the coordinate system is the anterior commissure (with coordinates 0, 0, 0). The coordinates in the system originally proposed by Talairach are not identical to the x, y, z coordinates provided by reference to digital templates. Talairach's system divided the brain into six sectors that were scaled independently, whereas most image processing systems use only global scaling factors for each axis [2394, 2422]. Moreover, the Talairach brain is not in exactly the same position and does not have the same shape as modern digital templates, in particular the MNI standard brain, which is particularly evident and relevant in the posterior parts of the brain. Approximate coordinate transformation functions have been suggested [2423].

As noninvasive PET and MRI studies have become possible in nonhuman primates, atlases are now also being developed for these species [2424].

4.9   Physiological Modeling

Quantitative measurement of local tracer activity is an essential precondition for, but does not by itself provide, quantitation of local physiological function. To achieve this, we must relate local tracer activity to physiological and biochemical processes. In quantitative terms, this is done by mathematical modeling of tracer transfer between compartments. These represent either distinct anatomical spaces that cannot spatially be resolved at the instrument level (e.g., capillaries, extracellular and intracellular tissue), or different chemical states of the tracer that reside at the same place (e.g., deoxyglucose and deoxyglucose phosphate).

Quantitative models usually require data on the arterial tracer concentration that is available to the brain for uptake ("input function"). The most accurate means of achieving this is arterial cannulation and blood sampling, which is, however, invasive. Short-term cannulation of the radial artery, after testing for proper collateralization by the Allen test, is generally considered safe for use in human volunteers, with a small risk of reversible impairment and a very small risk of permanent damage [2425]. Arterial cannulation may not always be possible (e.g., in the case of effective anticoagulation or coagulation disorders) or technically successful, and it certainly involves substantial inconvenience to the subjects. Thus, there has been a constant quest for less invasive alternatives. Blood samples of tracers with low peripheral extraction, such as FDG, can be obtained by sampling from a heated dorsal hand vein ("hot venous sampling"), providing tracer concentrations that are very close to arterial values. Reference tissue techniques (mainly used in receptor studies; see sections 4.9.2.3, 4.9.3) or restriction of analysis to the relative distribution of CBF [by use of the ^{15}O-water bolus technique (section 4.9.1.2) and CMRglc (section 4.9.2.2)] may obviate the need for blood samples in clinical studies entirely. Other approaches try to extract input curve parameters (that are common to all voxels) from the kinetic tissue parameters measured with PET in brain tissue [2426]. Recent progress in scanner resolution may open up new possibilities for the measurement of arterial concentrations directly from brain PET images.

Tracer concentrations within each (anatomical or chemical) compartment are denoted by C_x, where index x denotes the compartment (commonly used indices are a: arterial, v: venous, t: tissue, f: free tracer in tissue, m: metabolized tracer in tissue, b: tracer bound to receptors). The speed of transfer between compartments is described by transfer rate constants k_i. General assumptions are that concentrations are spatially homogenous within each compartment for each image resolution element but will typically change over time and are therefore often written as a function of time, $C_x(t)$, whereas rate constants are constant during the duration of measurement. It is obvious that these assumptions and the segmentation of

Table 4.2. Parameters in modeling equations

Parameter	Description
C_a	Tracer concentration in arterial blood plasma (whole blood for ^{15}O oxygen gas)
C_v	Tracer concentration in venous blood plasma (whole blood for ^{15}O oxygen gas)
C_t	Total tracer concentration in tissue
C_f	Concentration of free tracer in tissue
C_m	Concentration of metabolized tracer in tissue
C_b	Concentration of tracer in tissue that is bound to specific receptors
C_r	Concentration of tracer in reference tissue
k_1	Kinetic constant for tracer transfer from blood to tissue
k_2	Kinetic constant for tracer transfer from tissue to blood
k_3	Kinetic constant for tracer transfer from free to metabolized or bound state
k_4	Kinetic constant for tracer transfer from metabolized or bound to free state
F	Cerebral blood flow (CBF)
λ	Tissue/blood partition coefficient
E	Tracer extraction fraction (during capillary passage)
PS	Permeability surface product
V_{max}	Maximum initial reaction velocity
K_M	Michaelis-Menten constant (substrate concentration at half-maximum velocity)
LC	Lumped constant (for converting deoxyglucose to glucose metabolic rates)
B_{max}	Maximum receptor-binding capacity
k_{on}	Receptor-ligand association rate constant
k_{off}	Receptor-ligand dissociation rate constant
f_2	Fraction of free ligand that is available for specific binding
SA	Specific tracer activity
BP	Binding potential
K_D	Equilibrium dissociation constant
DV	Distribution volume

physiological processes into compartments are approximations and that real life is much more complicated. Therefore, mathematical models need to be validated for each tracer before being used in clinical studies. For ease of reference a list of abbreviations used in this chapter is given in Table 4.2.

4.9.1   Blood and Homogeneous Tissue (One Tissue Compartment)

This situation is the simplest and most common dealt with in tracer kinetic modeling (Fig. 4.4). K_1 describes the speed of transfer from blood to brain, which depends on the (constant) properties of the transport process (e.g., the activity of a carrier enzyme) and the concentration of the tracer in arterial blood, $C_a(t)$, which is also called "input function" (see also Chapter 3). For most tracers, it is not actually the concentration of the tracer in whole blood that is relevant, but the concentration in plasma, because the tracer is not taken up by blood cells or, if it enters blood cells, does not pass from them to tissue during capillary passage. Thus, in most cases, $C_a(t)$, is actually the tracer concentration in plasma. There are no saturation effects caused by the minute amounts of tracer substance that could change transport rates, and therefore actual tracer transfer from blood to tissue is simply given by k_1 multiplied by $C_a(t)$. The same is true generally for the reverse transport, which is the product of k_2 and tracer concentration in tissue, $C_t(t)$. In actual measurements, we will usually be able to measure arterial blood concentration by blood sampling, and tissue concentration (together with a contribution from blood within each voxel, typically approximately 5 % of total volume) by PET. In mathematical terms, the change of tissue concentration over time is described by the differential equation:

$$\frac{dC_t}{dt} = k_1 \cdot C_a - k_2 \cdot C_t \tag{4.1}$$

Rate constants k_1 and k_2 are given per time unit, typically per minute. We assume throughout that all concentration measurements in blood and tissue are made per volume (usually per milliliter). Alternatively, the transfer rate constant from blood to brain may be found in the literature in units of milliliters of blood per gram of tissue and minute, and it is then denoted by a capital letter as K_1. The conversion is made by use of the specific weights for blood and tissue [2427]. The standard solution of Eq. 4.1 is the convolution of the time course of arterial concentration by the transport process:

$$C_t(t) = k_1 \int_0^t C_a(\tau) e^{-k_2(t-\tau)} d\tau \tag{4.2}$$

Fig. 4.4. Schematic of model with one tissue compartment

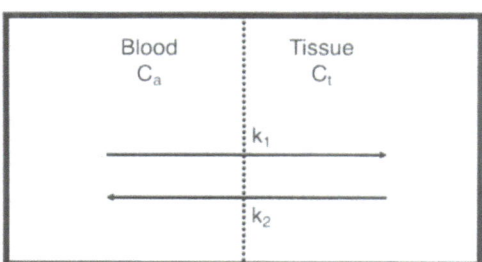

In the classic and most general situation, we know values of $C_a(t)$ and $C_t(t)$ at multiple time points from measurement, and wish to determine k_1 and k_2. This can be achieved by nonlinear regression analysis, also called curve fitting. This is an iterative optimization procedure for estimating k_1 and k_2, in which the sum of squared differences between measured C_t values and C_t values calculated from C_a, k_1 and k_2 is minimized [2428]. If equation parameters k_1 and k_2 (or similar terms from other models) are determined for each voxel, the resulting images are called parametric images.

In practical implementations, it is necessary to consider that tissue concentrations are measured with PET not at discrete points in time, but over certain time intervals determined by frame duration. Best accuracy is achieved if both sides of the equation are integrated over these measurement time intervals before fitting. Measurements of tissue activity are associated with random error due to stochastic isotope decay. According to the Poisson distribution, the absolute variance is proportional to total counts, and therefore the relative error is smaller at high activity and long measurement intervals. To account for this, data points should be weighted by the inverse of relative variance in the fitting procedure.

With manual blood sampling, it is necessary to interpolate $C_a(t)$ between measured data time points. Linear interpolation is frequently used, but the possibility that with bolus injections peak arterial activity might be missed and be underestimated by linear interpolation of adjacent data points must be borne in mind. Therefore, automated continuous arterial blood sampling is preferable for such measurements [2324, 2429, 2430]. With rapid bolus tracer administration, differences in arrival time of the tracer in the brain and in the peripheral blood sampling device (typically a few seconds later) and bolus dispersion in sampling tubing must be taken into account, as both of these can have a significant impact on results if not properly corrected for [2431, 2432]. There have also been attempts to derive arterial blood activity from PET measurements of heart or cerebral vessels, but this is not yet generally possible in practice. In certain situations, sufficiently accurate values can be obtained by the use of average input functions to represent C_a [1798, 2433].

The contribution of intravascular tracer activity to tissue activity, as measured with PET, needs to be considered, particularly in cases in which C_a is high relative to C_t. Thus, measured activity, C_{PET}, is then modeled as a weighted sum of C_a and C_t, where the relative weight is obtained either from independent measurement of local blood volume (e.g., by ^{15}O-CO, see section 3.3).

4.9.1.1   Transport across the BBB

A standard application of this simple model is the measurement of transport across the BBB (see section 3.1). It is rarely used in its most general form in clinical practice, but is encountered with slight modifications in many measurement situations explained below.

4.9.1.2   Cerebral Blood Flow

Freely diffusible tracers are the most suitable for CBF measurement (see section 3.2). Their kinetics can be modeled by modification of the standard one-tissue-compartment model (Fig. 4.5). Rate constants k_1 and k_2 are now no longer related to transport at the BBB, which is complete due to free diffusion of the tracer into tissue during capillary passage. Free diffusion also leads to equilibration of tracer concentration in tissue, C_t, with that of outflowing venous blood, C_v. Thus, the concentration in tissue is the same as in venous blood multiplied by the tissue/blood partition coefficient, λ, of the tracer. Consequently, tracer delivery to tissue (flow times arterial concentration) and loss from tissue (flow times venous concentration) depend entirely on blood flow, F. With replacement of the venous concentration by tissue concentration divided by the distribution coefficient, this results in the following differential equation for the change of tissue concentration:

$$\frac{dC_t}{dt} = F \cdot C_a - \frac{F}{\lambda} \cdot C_t \qquad (4.3)$$

It has exactly the same structure as Eq. 4.1, with F replacing k_1 and F/λ replacing k_2. With this replacement, Eq. 4.2 also describes the dependency of tissue activity on CBF. Thus, in principle, the units of CBF are the same as of the rate constants (e.g.,

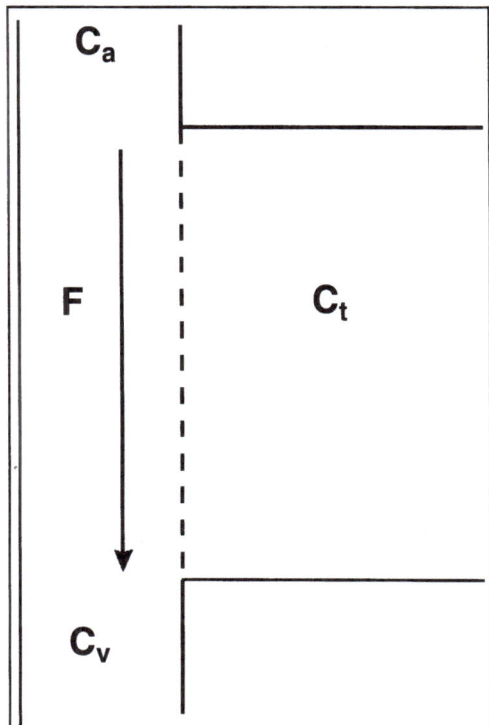

Fig. 4.5. Schematic of model for cerebral blood flow (CBF) measurement with freely diffusible tracers

min⁻¹). For historical reasons, CBF is often given in units of milliliters (blood) per milligram (tissue) per minute (ml/g per min), and λ in grams per milliliter, whereas it is given as absolute numbers in our notation. Fortunately, brain tissue has a specific weight of close to 1 g/ml and, thus, CBF values per minute and as milliliters per gram and minute are numerically very similar.

Typical (schematic) time–activity curves after rapid venous bolus injection of a tracer with distribution coefficient of 1 (^{15}O-water with a distribution coefficient of approx. 0.8–0.9 [1696, 2427], is close to that) and normal average brain blood flow (0.5 min⁻¹) [2434] compared with reduced blood flow (0.2 min⁻¹) are shown in Fig. 4.6. They illustrate that initial tracer activity in brain is directly related to blood flow. Subsequent washout is more rapid at high blood flow, resulting in cross-over of curves and higher residual tracer activity in tissue with lower blood flow after a few minutes.

Since the first application of the basic principles of local CBF measurement in man [2435], the assumptions of equilibration of venous blood and tissue and of homogeneous tissue tracer concentration have been checked critically in many experiments and error calculations [2436]. As always with modeling, they are only approximately valid, but it has been shown that errors are not large for freely diffusible tracers (underestimation by less than 10 % in most cases) [2437–2439].

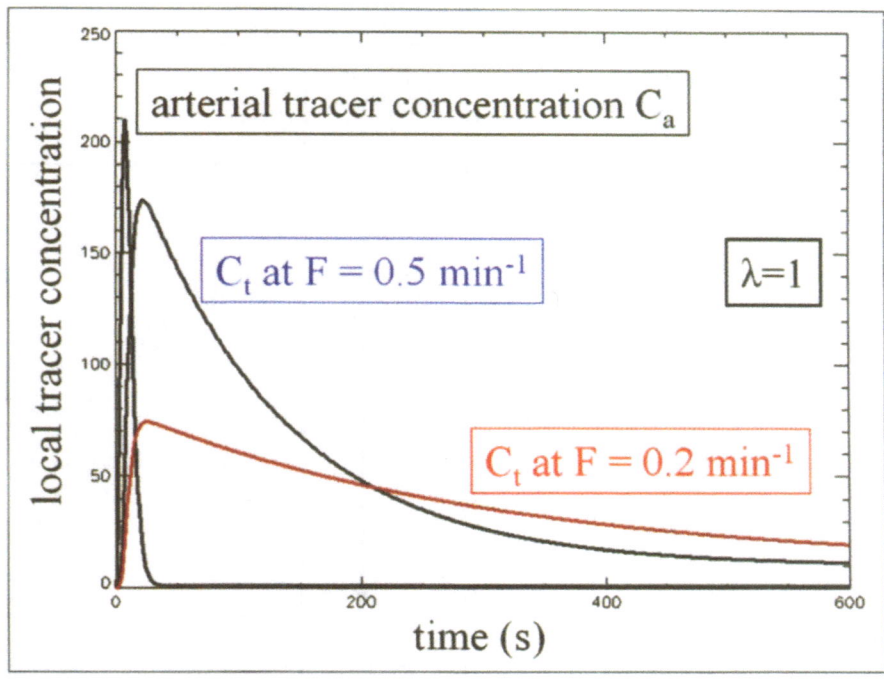

Fig. 4.6. Schematic illustrating time–activity curves of cerebral ^{15}O-water activity after a rapid i.v. bolus at normal and reduced CBF

Several different practical implementations of CBF measurements with PET based on this model have been suggested (see ref. [2440] for a detailed methodological discussion). They can be ordered with respect to the part of the dynamic curve they mostly rely upon.

4.9.1.2.1 The Influx Approach

The influx approach relies on the very first part of the curve, when washout effects are still small (Fig. 4.7). When we neglect washout (the second term in Eq. 4.1), the exponential term in Eq. 4.2 disappears and tissue activity is proportional to blood flow (and the integral of arterial activity). This implies that the initial tracer distribution approximates the CBF distribution. If we wish to quantify CBF in absolute units, we need to measure C_t and C_p in absolute units. This approach has been validated for clinical use with single-frame measurement times of 40–60 s duration, and because of its structural similarity with the autoradiographic microsphere technique has been dubbed the "autoradiographic" technique [2441]. In its original version, it involves a schematic correction for the reduction of tissue activity due to early washout, but it is also often used without that correction, especially for brain activation studies. It has become a widely used technique

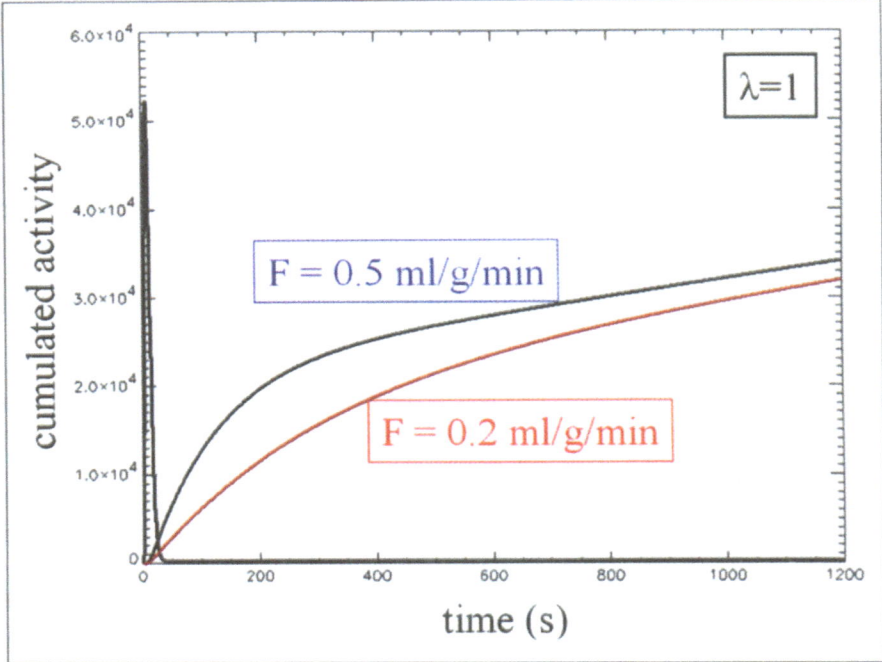

Fig. 4.7. Schematic showing cumulated activity curves of cerebral ^{15}O-water activity, as used with influx "autoradiographic" technique, after a rapid i.v. bolus at normal and reduced CBF. The distinction between high and low CBF signal is best at the initial parts of the curve

because it is easy to implement, it has a short measurement time, and there is a straightforward relation between measured activity and CBF. Limitations are caused by the number of counts that can be collected during the short measurement interval (often resulting in relatively noisy images), and it requires a scanner that can process high count rates (which is now standard but was not the case in some earlier scanner types). Proper correction for bolus delay and dispersion in peripheral blood samples is crucial for accurate quantitation of absolute values [2442, 2443]. A simplified version without blood sampling that provides relative values (mainly for activation studies) has also been developed [2444]. Optimum sensitivity for detection of local CBF changes in functional activation studies is achieved by the use of 3D data acquisition and fractionation of the total injected dose into multiple injections of approx. 370–500 MBq each, with replication of activation conditions [2445]. A short phase of data acquisition lasting approx. 45 s starting when the activity bolus reaches the brain minimizes signal loss through rapid tracer washout in high-flow areas [2446].

4.9.1.2.2 Fitting the Kinetic Curve

Fitting the complete kinetic curve extracts a maximum of information from the data and is an approach that is not limited to particular modes of tracer administration [2428]. It has been adopted for ^{15}O-water [1696, 2447, 2448], ^{18}F-fluoromethane [2449], bolus inhalation of ^{15}O-carbon dioxide [2450], ^{11}C-butanol [2437], and ^{15}O-butanol [2451, 2452]. Typical measurement times for a kinetic series are around 5 min. In addition to determination of blood flow F, the partition coefficient λ can also be calculated as the ratio of k_1 to k_2. Comparison of the kinetic curve in different brain regions allows conclusions to be drawn from the model on the common input function, which opens up an indirect way of achieving absolute quantitation of CBF without blood sampling [2453]. Another approach for absolute quantitation without arterial blood sampling involves the use of an average input function and calculation of absolute global CBF from the washout phase (k_2), which is then used for scaling of k_1-based CBF maps [1798].

4.9.1.2.3 The Steady State Technique

In a sense, this technique relies on the late part of the kinetic curve. By continuous inhalation of ^{15}O-carbon dioxide, which is converted to ^{15}O-water in the lung, a steady state of tracer delivery, washout, and isotope decay in tissue is reached [1713, 2454]. Historically, it was the first method for quantitative measurement of CBF with PET, mainly because it could also be implemented with the limited count-rate capabilities of first-generation PET scanners. It is closely related to and usually performed in conjunction with steady state inhalation of ^{15}O-oxygen gas for measurement of the cerebral metabolic rate of oxygen ($CMRO_2$). It has largely been abandoned for clinical studies because of the long measurement duration, relatively high radiation dose to airways, regulatory difficulties with the application of gaseous isotopes in some countries, and nonlinear signal to CBF relation [2440].

4.9.1.3   Mixed Effects of Transport and Blood Flow

Most tracers used for brain imaging show a high first-pass extraction from blood to tissue, which is necessary to provide a good tissue signal within a measurement time of 30–60 min, but are not entirely freely diffusible. In that case, Eqs. 4.1 and 4.2 apply for tracer transfer from blood to brain, but transfer from blood to brain (k_1) is now determined by blood flow, F, times the tracer extraction fraction, E, and k_2 reflects of low, extraction fraction, and partition coefficient. Thus, we now have:

$$k_1 = E \cdot F; k_2 = \frac{E \cdot F}{\lambda} \tag{4.4}$$

The extraction fraction is determined by blood flow, F, and the permeability-surface product, PS, of the tracer, according to the Renkin-Crone equation [2455]:

$$E = 1 - e^{-\frac{PS}{F}} \tag{4.5}$$

There are two important limiting cases: If permeability is very high, as is the case for freely diffusible tracers, then E = 1; if permeability (or surface) is very small, then EF $\cong$ PS, and k_1 is thus also approximately equal to the permeability surface product, PS. Therefore, in this case, we can use kinetic fits according to Eq. 4.2 to determine PS. This has been done for the measurement of BBB permeability with tracers ^{82}Rb [725, 2456] and ^{68}Ga-EDTA [729, 2457].

If independent measures of blood flow are available, Eqs. 4,4 and 4.5 can be used to measure PS for any substance from its k_1 by dynamic curve fitting. This is of particular interest in the study of pharmacokinetics. In practice, a blood flow measurement by ^{15}O-water can be done a few minutes before the kinetics of the test substance are measured in the same subject, recognizing of course, the modest diffusion limitation of ^{15}O-water.

4.9.1.4   Measurement of the Partition Coefficient

As mentioned in the discussion on CBF measurements, the partition coefficient (synonymous with distribution volume) λ can be determined from a curve fit of kinetic data as the ratio k_1/k_2. If there is relatively little change in tracer blood levels (compared with blood–brain transport kinetics), the ratio of tissue to blood activity, C_t/C_a, is also approximately equal to the partition coefficient. Thus, a single PET image taken in that phase of tracer kinetics may provide a reasonable representation of steady state tracer distribution. This approach has been used for instance for the estimation of benzodiazepine receptor binding capacity with ^{11}C-flumazenil in ischemic stroke and epilepsy [861, 2198].

4.9.1.5   Saturable Transport and Enzyme Reactions

The brain needs various biochemical substances that cannot cross the BBB by diffusion, such as glucose, amino acids, and nucleosides. They are transported by specific enzymatic carriers that can be saturated by these substrates [1872]. Since PET is operating with picomolar tracer concentrations far below the K_M values, which are in the micromolar range, we still have a linear relation between transfer rates and tracer concentration, and Eq. 4.1 can be applied. Rate constants k_1 and k_2 now depend on maximum velocity, V_{max}, and affinity of the carrier enzyme, and on the competing levels of unlabeled endogenous substrates (denoted by $^\circ$) in plasma, C_p°, and in tissue, C_t°. In general, enzyme affinity to competing unlabeled substrates, K_M°, does not need to be identical with that for the labeled tracer, K_M, particularly if the latter is an analogue tracer that is not chemically identical to the natural substrates. According to Michaelis-Menten kinetics, rate constants k_1 and k_2 are then related to these biochemical parameters by the following equations:

$$k_1 = \frac{V_{max}}{K_M + \frac{K_M}{K_M^\circ} C_p^\circ}$$

(4.6a)

and

$$k_2 = \frac{V_{max}}{K_M + \frac{K_M}{K_M^\circ} C_t^\circ}$$

(4.6b)

Related calculations have been used to study carrier-mediated blood-brain glucose transport by comparison of FDG and ^{11}C-O-methyl-glucose to natural glucose [2458, 2459].

4.9.2   Metabolism

Many tracers undergo metabolism or are bound to receptors in the brain and, quite often, metabolism and binding are the physiological processes of interest. An additional compartment is added to differentiate between free, C_f, and metabolized tracer, C_m, in tissue (Fig. 4.8). Differential equations describe the change of concentrations in these compartments over time (by analogy with Eq. 4.1):

$$\frac{dC_f}{dt} = k_1 \cdot C_a - (k_2 + k_3) \cdot C_f + k_4 \cdot C_m$$

(4.7a)

and

$$\frac{dC_m}{dt} = k_3 \cdot C_f - k_4 \cdot C_m$$

(4.7b)

Fig. 4.8. Model with two tissue compartments (*shaded areas*) to describe transport and metabolism

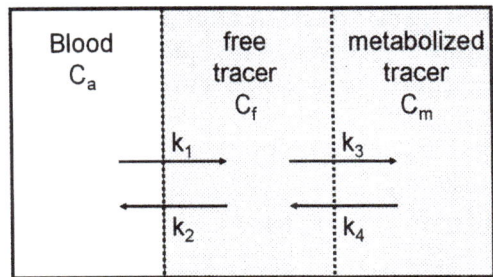

The corresponding solutions that describe the time–activity curves $C_f(t)$, $C_m(t)$, and total tissue activity $C_t(t)$ involve two exponential terms multiplied by time course of blood activity, $C_a(t)$ [2460]. In principle, rate constants k_1 to k_4 can be determined by curve fitting as described above, but the results are often not very accurate owing to a limited number of data points (should be more than 20) or too-short measurement time (90 min or more may necessary for a four-parameter fit). Unique determination of rate constants also requires that they are clearly different from each other, representing separable kinetics from fast (typically tracer delivery and transport at BBB) and slow (typically binding or metabolism) processes. Even more complications arise if tracer metabolites in blood occur that may also enter brain tissue (as is the case for FDOPA, for instance [2461]), or if we wish to measure processes in heterogeneous tissue, such as tumor or infarcted brain [2439]. Thus, this general model and the associated curve fitting are rarely applicable to clinical situations, and simplifications by approximation are needed.

4.9.2.1   Irreversible Metabolism (Metabolic Trapping)

Some tracers, e.g., FDG and FDOPA, are essentially trapped in tissue (at least during relevant measurement times) after metabolism. Irreversible trapping implies that k_4 is negligibly small, and the time course of total tissue tracer activity, $C_t(t)$, can be split into a first (reversible) term that represents free tracer, $C_f(t)$, and a second accumulating term that represents metabolized or bound tracer, $C_m(t)$:

$$C_t(t) = C_f(t) + C_m(t) \tag{4.8}$$

with

$$C_f(t) = \frac{k_1 k_2}{k_2 + k_3} \int_0^t C_a(\tau)\, e^{-(k_2+k_3)(t-\tau)}\, d\tau \text{ and } C_m(t) = \frac{k_1 k_3}{k_2 + k_3} \int_0^t C_a(\tau)\, d\tau$$

An important generalization of this approach refers is that if we are interested only in the metabolic rate, the exact time course of free tracer activity is not important when two conditions are fulfilled: (1) plasma activity declines to levels that are very small and change only slowly relative to metabolized tracer concentration in brain, and (2) free tracer activity follows plasma activity with transport kinetics that are more rapid than the metabolic rate. In that case the ratio of free tracer in tissue to plasma activity, $C_f(t)/C_t(t)$, approaches a constant value, B. The ratio of tissue activity to plasma activity can then be approximated by

$$\frac{C_t(t)}{C_a(t)} = \frac{k_1 k_3}{k_2+k_3} \frac{\int_0^t C_a(\tau)d\tau}{C_a(t)} + B \quad \text{with} \quad B = \frac{k_1 k_2}{(k_2+k_3)^2} \tag{4.9}$$

The validity of this approximation has been demonstrated by Patlak et al. [1936] even for irreversible accumulation preceded by an indefinite number of reversible transport steps. Thus, we can no longer use individual rate constants, and the rate of irreversible accumulation, $k_1k_3/(k_2+k_3)$ in Eq. 4.9 collapses into a single composite constant, which is called the influx constant, k_i. If C_t/C_a is plotted against the ratio containing C_a in the right-hand side of Eq. 4.9, the plot becomes levier with a slope of k_i. This convenient method of graphical analysis for irreversible processes is referred to as the Patlak plot [1936].

4.9.2.2 Measurement of Local Cerebral Glucose Metabolism

The technique of measuring local cerebral glucose metabolism (CMRglc) via measurement of FDG accumulation was based on the autoradiographic deoxyglucose technique for measuring CMRglc in laboratory animals [2462]. It has been adapted for PET in several variants. Unlike autoradiography, PET can deliver kinetic data and thus provide direct measurement of the kinetic constants of FDG, with k_1 and k_2 representing carrier-mediated transport at the BBB (with little influence from blood flow at an extraction rate of approx. 0.2) and k_3 representing phosphorylation by hexokinase yielding FDG phosphate that does not undergo further glycolysis. Dephosphorylation by brain phosphatase is slow. Thus, k_4 is nearly one order of magnitude smaller than k_3 and cannot be determined with short measurement times up to 60 min [2460, 2463]. Some FDG phosphate may undergo further metabolism in the pentose phosphate pathway, but also, in negligible amounts, at least in normal brain [2464].

Calculation of CMRglc requires measurement of plasma glucose, C_a^0, from which intracellular glucose is estimated (assuming a steady state of glucose levels and local glucose metabolism). This estimate and a correction for different enzyme affinities was put together in a "lumped constant", LC, which converts the metabolic rate of FDG to that of glucose [1740, 2465]. There has been some discussion about the actual value of the LC in human brain, with suggested values ranging from 0.42 to more recent, measured values of about 0.82 [1741, 1747, 2466, 2467]. Therefore, we need to know the actual value of the LC used for comparison of absolute values of CMRglc from different studies. Some older methods of synthesizing FDG included significant contaminations from ^{18}F-fluoro-2-deoxy-D-mannose, which can be accounted for by using slightly lower values for the LC [2468]. The lumped constant is increased by hypoglycemia [2469] and may also be higher in brain tumors and in regions of recent ischemic infarct than in normal brain [621, 2470]. With rate constants measured by dynamic curve fitting or by integration techniques, the equation for calculation of CMRglc is:

$$CMRglc = \frac{C_a^0}{LC} \frac{k_1 k_3}{k_2 + k_3} \qquad (4.10)$$

The term $k_1 k_3/(k_2+k_3)$, representing the metabolic rate of FDG, can be substituted by the influx rate constant k_i determined with the linear approximation of Patlak et al. [1936, 2471]. There have also been modifications that avoid the assumption of a fixed LC and refer instead to the Michaelis-Menten equation to account for the relations between enzyme affinities for FDG and glucose [2466].

Determination of individual rate constants is not very practical in many clinical applications, and methods are preferred that can be done with a single scan, a situation similar to the original development of the method for autoradiography. Then the deviation from population average CMRglc (given by the average rate constants) is estimated from a single scan. Actual measured FDG activity is compared with the activity that would have been expected at the time of the scan with the individual's blood activity–time course and average rate constants [2472]. The original approach [1740] underestimated CMRglc in brain areas with pathologically low FDG uptake, whereas later versions provide estimates of CMRglc that are in close correspondence with measurements based on individual determinations of rate constants [2473]. In that approach, local CMRglc is proportional to measured FDG activity, which implies that with scaling of whole-brain glucose metabolism to a population average, images of CMRglc and FDG distribution are identical.

Quantitation of CMRglc by dynamic fit or by single scan requires multiple arterial blood samples to provide the course of FDG activity in plasma, which is used as the input function $C_a(t)$. Short-term cannulation of the radial artery is a mildly invasive procedure that has a low complication rate [2474], but it is time consuming, can induce substantial discomfort, and is not possible in subjects with impaired blood coagulation. Therefore, alternatives are highly desirable. The peripheral extraction of FDG is rather low (approx. 10 % or less), which allows approximation of arterial concentration by sampling from an arterialized dorsal hand vein (achieved by heating the hand) [2465]. Another alternative is approximation of the input function by a population-based average curve that is scaled by one or two arterial samples [2433, 2475] or use of an image-derived input function from the region of the carotid bifurcation with scaling by a late venous blood sample [2476]. Even with these simplified approximations, calculation of CMRglc is preferable over less physiological values, such as the standardized uptake value (SUV, the tissue tracer concentration divided by the injected dose per unit body weight) that is frequently used to assess FDG uptake in other organs, because it takes due account of the influence and interindividual variation of plasma glucose and the amount of FDG delivered to the brain by the blood.

To avoid the conversion factor needed with the analogue tracer FDG, native glucose labeled with ^{11}C in the 1-position (1-^{11}C-D-glucose) has also been used for quantitation of CMRglc. Modeling is based on the same two-tissue-compartment model as with FDG, but an additional term is necessary to account for labeled metabolites (mainly lactate and other monocarboxylic acids and CO_2) [1746]. Metabolites occur in plasma and in brain, and loss of labeled CO_2 from the brain is dependent on CBF. Data indicate that there is a rapid loss of labeled lactate from brain, suggesting that it represents a significant nonoxidative part of glucose metabolism in brain (approx. 10 % of total CMRglc) [1746]. By comparison with FDG uptake, the method has been used to obtain estimates of the LC for FDG in human brain tumors [621].

4.9.2.3   The Reference Tissue Model for Dopamine Synthesis and Storage

Measurement of dopamine synthesis and storage with FDOPA also includes an essentially irreversible step, similar to phosphorylation of FDG by hexokinase. For FDOPA, this step is the decarboxylation by aromatic amino acid decarboxylase (AAAD) to ^{18}F-dopamine, which occurs only in dopaminergic cells and terminals (see section 3.10.1). Further metabolism occurs in only small quantities after synaptic release from vesicles. Thus, in principle, the relevant compartments are represented in Fig. 4.8, with k_1 and k_2 describing FDOPA transport at the BBB, which is mostly mediated by LNAA as for other large neutral amino acids, k_3 describing decarboxylation by AAAD, and k_4, which is negligibly small. There is a complication caused by the metabolite 3-O-methyl-6-^{18}F-fluoro-L-dopa (OMFD), which is present in plasma and in tissue, but is not a substrate for AAAD. Thus, in principle, it would be necessary to determine metabolites in plasma to determine plasma FDOPA activity and subtract the contribution of OMFD to measured tissue activity in the free tracer compartment before Eq. 4.8 can be applied [295, 2477].

Use of the Patlak approach (see section 4.9.2.9; Fig. 4.9) permits quantitation of AAAD activity in terms of the influx constant, k_i, on the assumption that the OMFD concentrations in plasma and in tissue parallel plasma FDOPA concentration and can thus be included in constant B of Eq. 4.9 after some time [293]. This simplifying approach has been driven even further by disregard of plasma activity while AAAD activity was estimated by comparison of local tracer accumulation with activity in occipital cortex, which is free of dopaminergic innervation [294]. Again, the assumption is made that the ratio of OMFD concentration to FDOPA concentration in tissue approaches a constant value after some time, which then allows the approximation

Fig. 4.9.
Patlak plot of
L-6-18F-fluoro-
3,4-dihydroxy-
phenylalanine
(FDOPA) uptake
using the refer-
ence tissue model

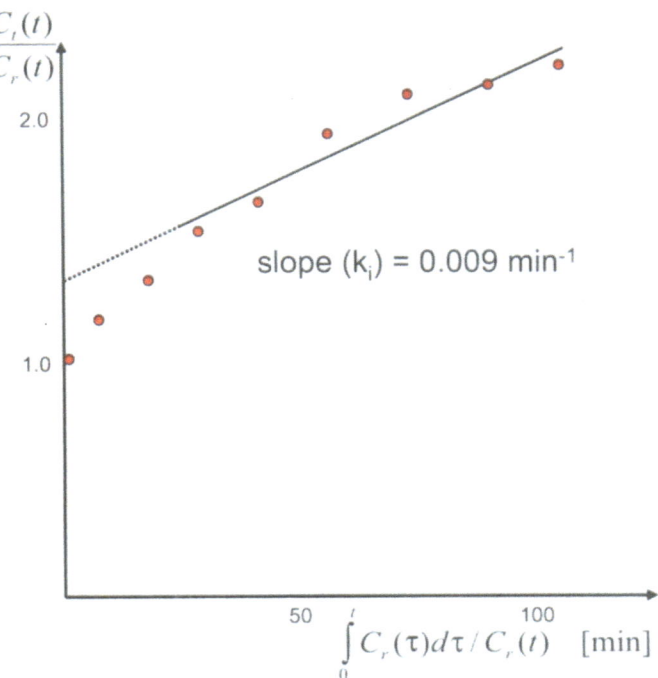

Fig. 4.9.
Patlak plot of L-6-^{18}F-fluoro-3,4-dihydroxy-phenylalanine (FDOPA) uptake using the reference tissue model

$$\frac{C_t(t)}{C_r(t)} = k_i \frac{\int\limits_0^t C_r(\tau)d\tau}{C_r(t)} + B \qquad (4.11)$$

where $C_r(t)$ is the tracer activity in occipital reference tissue, and the accumulation rate, k_i, is a measure of AAAD activity.

4.9.2.4   Reversible Metabolism or Binding

This is the most complicated situation that can be handled by dynamic curve fitting. It involves a k_4 to describe the reverse metabolic step (see Fig. 4.8), and it is usually necessary to have a measurement time extending over at least 90 min even with tracers with relatively fast kinetics to obtain a reasonable fit of k_4. Another important requirement is that rate constant k_1 is much faster than k_3, and k_2 is much faster than k_4, for it to be possible to separate these components of tracer kinetics by curve fitting. In this case, the total measured activity is related to plasma activity by two convolution terms [2460].

The accuracy of individual rate constants determined by curve fitting is usually rather low in this complicated situation, and convergence of fits needs to be checked. The procedure is therefore not suited for general clinical use. It has been noted that compound parameters derived from combinations of rate constants are more robust, because some errors tend to cancel each other out. There have been modifications to make the procedure more robust, for instance by including k_4 as a small fixed value, or by adjusting the Patlak plot to include a small k_4 [2478]. In the interpretation of such complex kinetic data, it should be kept in mind that they are sensitive to many causes of error, including tissue inhomogeneity, which is usually inevitable [2439].

4.9.3   Receptor Binding

The dynamics of receptor binding can be analyzed with reference to the same framework as used for measurement of metabolism (see Fig. 4.8). For intuitive understanding we replace the term used for metabolized tracer, C_m, by C_b to indicate specifically bound tracer. C_f comprises that part of tracer in tissue that is not specifically bound, which may include not only free but also nonspecifically bound tracer that cannot be displaced by specific competitors. The physiological interpretation of rate constants k_3 and k_4 now relates to receptor-binding parameters [2479]. Usually, B_{max} denotes the concentration of receptors that are available for binding, k_{on} the receptor-ligand association rate constant, k_{off} the corresponding dissociation rate constant, f_2 the fraction of free ligand that is available for specific binding (i.e., not bound unspecifically in tissue). Unlabeled ("cold") ligand is present in preparations of many radio ligands, and because some receptor systems may have rather small numbers of available binding sites, saturation effects cannot always be ignored. The amount of total injected ligand mass (labeled and unlabeled) is particularly critical for high-affinity substances. The proportion of labeled tracer relative to the total amount of receptor binding ligand is called specific activity, SA. The following equations apply:

$$k_3 = k_{on} f_2 \left(B_{max} - \frac{C_b}{SA} \right)$$ (4.12a)

and

$$k_4 = k_{off}$$ (4.12b)

If the amount of total receptor binding ligand, C_b/SA, is very small relative to the binding capacity, B_{max}, it can be disregarded, and k_3 is then simply the product of k_{on}, f_2, and B_{max}. In this situation only, k_3 can be assumed to be constant during measurement time, the linear model (Eqs. 4.7a, b) applies, and curve fitting procedures can be applied as usual. It should be noted that we do not get a separation of receptor density from the association kinetic constant in this case.

If receptor occupancy by the ligand cannot be disregarded, k_3 is not constant and it must be replaced by B_{max} (constant, determined by fit), C_b (time-dependent, part of total measured activity), SA (constant measured independently), and $k_{on} f_2$ (determined by fit) according to Eq. 4.12 in the fitting procedure. To extract k_{on}

from the product $k_{on}f_2$ that is determined by the fit, the relation $f_1/f_2 = k_1/k_2$, or $f_2 = f_1k_2/k_1$, must be used, where f_1 is the fraction of tracer in plasma that is available for transport across the BBB, which usually means that it is not bound to plasma protein. f_1 can be measured from plasma samples by chemical methods, although these measurements are cumbersome and may not accurately reflect the in vivo situation [2480]. Separate fitting of B_{max}, k_{on}, and k_{off} (which together with k_1 and k_2 now make five parameters to be determined by fitting) makes things very complicated, but it also offers the possibility of distinguishing between affinity and binding capacity, and thus providing quantitative measures of B_{max} and K_D (= k_{off}/k_{on}) in vivo [2011, 2203]. In order to get reliable results it is usually not sufficient to do just one fit; rather, multiple measurements with different specific activities are required.

In most clinical applications, no attempt is made to determine B_{max} and K_D separately, but the so-called binding potential, BP, is accepted as adequate [2481], which is B_{max} relative to K_D (BP = B_{max}/K_D). In most instances in this book, when a reference is made to receptor binding capacity, it is the binding potential or a closely related parameter (see section 4.9.3.2 for more details) that was measured. In the case of negligible occupancy of receptors by the tracer and co-injected cold ligand, the BP can be calculated from measured parameters in a kinetic study as

$$BP = \frac{k_1 k_3}{k_2 k_4 f_1} \tag{4.13}$$

Because of the additional work and problems that are involved in accurate measurement of f_1 in plasma by chemical means [2480], quite frequently a measure is used that is closely related to the binding potential as defined originally. It is denoted BP^* in this book, and is also known as V_3," (see reference [2044] for a more comprehensive review of common designations). It assumes that nonspecific binding (as taken into account by f_2) in tissue is a constant and can therefore be disregarded. It can be calculated directly from rate constants k_3 and k_4 by

$$BP^* = f_2 \cdot B_{max} / K_D = k_3 / k_4 \tag{4.14}$$

For reasons explained in the previous section, these full kinetic fits are unwieldy and require great care to make them reliable. Thus, approaches have been developed, in particular for the reversibly bound receptor ligands, to simplify calculations and to ensure that robust results are obtained. There approaches are described in section 4.9.3.2.

4.9.3.1   Irreversible Receptor Binding

In principle, the same situation as for metabolic trapping arises if a tracer is irreversibly bound to a high-affinity receptor, such as when ^{11}C-methylspiperone [2002] or ^{18}F-methylspiperone [2007] is bound to D2 receptors. In that situation, Eq. 4.8 provides, at least in theory, the basis for determination by curve fitting of k_3, which is proportional to B_{max} if receptor occupancy by tracer and coinjected cold ligand is negligible [2012, 2482]. In contrast to the situation with relatively slow metabolic processes, binding of spiperone to high-affinity receptors is usually rather fast, and therefore it is usually impossible to separate transport kinetics (k_1 and k_2) from binding kinetics (k_3) by curve fitting. Instead, the rate of tracer accumulation, $k_1k_3/(k_2+k_3)$, is determined. k_1 and k_2 are measured in a brain region devoid of specific receptors (for D2 receptor tracers usually in the cerebellum), assuming that BBB transport does not differ substantially between brain regions. Unfortunately, k_1 and k_2 depend on blood flow for tracers with high brain extraction, and thus local changes in blood flow cannot be separated from changes in receptor binding by this approach. For this reason, the use of receptor ligands with irreversible binding has largely been abandoned in spite of their favorable chemical properties.

4.9.3.2   Equilibrium Approaches for Quantification of Reversible Receptor Binding

Reversible receptor binding is best measured at equilibrium, because this eliminates possible sources of error from blood flow and transport kinetics. Equilibrium can be approximated in vivo by programmed infusion of the tracer starting with a bolus followed by a lower dose per unit of time to give a constant plasma level of free tracer. It is then necessary to wait until tissue compartments of free (or nonspecifically bound) and specifically bound tracer also approximate equilibrium, which may take up to several hours, depending on tracer kinetics [2479]. That approach has been adopted for ^{11}C-flumazenil [2194, 2200], ^{11}C-dihydrotetrabenazine [2483], RAC [2018, 2041], ^{18}F-altanserin [2175], and ^{18}F-cyclofoxy [2484]. Under such equilibrium conditions, the volumes of distribution can be obtained, typically by simple measurement of tissue activity, C_t, in two region types, and by comparison with plasma activity, C_p. The target region contains the receptor of interest and therefore comprises compartments with bound, C_b, and free (including nonspecifically bound) tracer, C_f. Thus, the tracer distribution volume there is defined by

$$DV_{rec} = C_t/C_p = \left(C_b + C_f\right)/C_p \tag{4.15a}$$

It is typically compared against a reference region where tissue activity is free of specific receptor binding, and thus defined by

$$DV_{ref} = C_f/C_p \tag{4.15b}$$

The modified binding potential, BP^*, is easily obtained from

$$BP^* = \left(DV_{rec}/DV_{ref}\right) - 1 = \left(C_t/C_f\right) - 1 \tag{4.15c}$$

Of course, the classic binding potential, BP, can easily be calculated from BP^* if f_2 is known, using $BP = BP^*/f_2$. Alternatively, if f_1 is known, BP can be obtained from

$$BP = \left(DV_{rec} - DV_{ref}\right)/f_1 \tag{4.15d}$$

This equilibrium method has solid foundations, but it may be difficult in practice to achieve actual equilibrium by programmed infusion, because transport kinetics and systemic tracer elimination may vary among individuals. Therefore, a substitute is often used that takes advantage of the property of many tracers with high brain extraction and substantial nonspecific binding, i.e., shows little change in plasma activity as opposed to relatively high tissue activity after initial tracer distribution following a bolus injection. Whether such an approximate equilibrium is actually achieved can be seen from a special graphical representation of kinetic data, the Logan plot [2485].

In the Logan plot (Fig. 4.10), the integral of regional activity over current regional activity is plotted versus the integral of plasma activity over current regional activity:

$$y(t) = \frac{\int_0^t C_t(\tau)d\tau}{C_t(t)} ; x(t) = \frac{\int_0^t C_p(\tau)d\tau}{C_t(t)} \tag{4.16a}$$

It has been shown that the slope of this curve approximates the regional tracer distribution volume [2485]. By comparing these slopes for the target (DV_{rec}) and the reference region (DV_{ref}), the binding potential is calculated according to Eq. 15c or 15d.

If a good estimate of average brain k_2 is available and if it can be assumed that it is not changed by the experiment in the target or the reference region, it is even

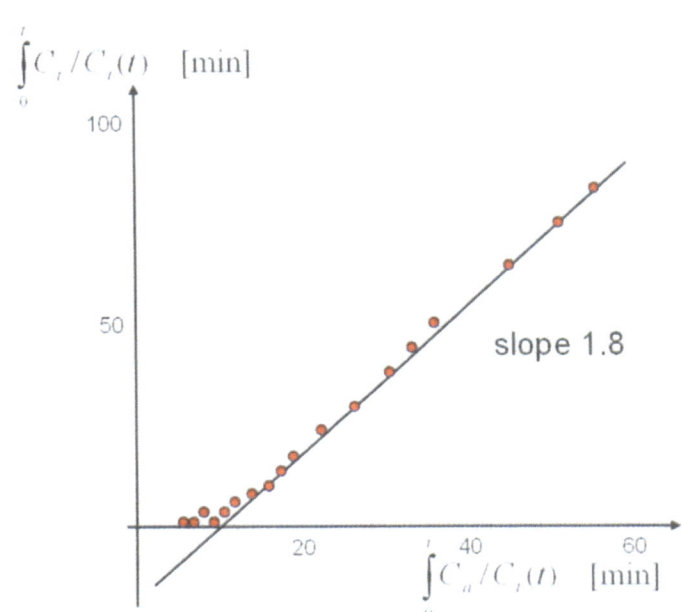

Fig. 4.10.
Logan plot
of striatal
^{11}C-raclopride
(RAC) binding

possible to avoid the use of a plasma input function [2486]. The integral of target region activity over current region activity is now plotted against a more complicated transformed time axis involving the integral of tissue activity in the reference region, which represents only free and nonspecifically bound tracer:

$$y(t) = \frac{\int\limits_0^t C_t(\tau)d\tau}{C_t(t)}; x(t) = \frac{\int\limits_0^t C_f(\tau)d\tau + \frac{C_f(t)}{k_2}}{C_t(t)} \qquad (4.16b)$$

The slope of the linear part of this plot approximates BP*. Although this variant of the Logan plot is very convenient, because all calculations can be done after a simple bolus injection without plasma sampling, it should be used with caution. In addition to changes in plasma binding or nonspecific tissue binding that may bias all estimates of BP*, it could also be biased by a change in blood flow because of its effect on k_2. The problem with possible CBF dependence is not so serious as with the irreversibly bound tracers. If measured data show that the ratio of target and reference tissue becomes rather constant for a sufficiently long period of time over the course of the experiment, the effect of k_2 (and any potential problem with CBF dependence) is so small as to be negligible and the correction term involving k_2 in Eq. 16b can be omitted (because it does not influence the slope).

Since the fitting of a slope in the Logan plot approach can be done by simple linear regression, it is also very well suited to generating parametric images [2487, 2488]. Refinements for the calculations have been suggested to test linearity and optimize goodness of approximation [2489], depending on kinetic properties of the tracer.

4.9.4   Extraction of Model Parameters from Kinetic Data

The standard approach applied to extract the model parameters from kinetic data is to fit the model to regional data (i.e., a time series measured at a particular location in the brain as defined by a volume of interest; see section 4.10.1) by minimization of the sum of squares (SSQ) of the residuals. Since models include linear and nonlinear (usually exponential) parameters, nonlinear fitting routines are required, which are provided by commonly used software packages. Users need to be aware of potential problems, such as poor definition of some parameters, lack of convergence, or convergence to a false-local SSQ minimum. Standard nonlinear SSQ minimization also allows for proper weighting to account for variable data variance, which originates from the Poisson statistics of radioactive decay and is modified by image reconstruction [2262]. It may also provide estimates of the accuracy of fitted parameters. Robustness and speed of multiple fits at different anatomical locations in the same subject can sometimes be improved by reducing the number of free parameters in individual fits [2490], e.g., by fixing the value of a parameter or of the ratio of two parameters.

Nonlinear SSQ minimization is time consuming and may not always converge properly. It is therefore hardly suitable for use on a voxel-by-voxel basis to create parametric images. Therefore, model equations are often transformed by weighted integration to linearize parameters that can then be estimated by linear regression, which can be solved noniteratively and do not have problems with multiple SSQ minima [2488, 2491–2496]. Weighted integration is equivalent to curve fitting of perfect data without noise. Integration does not preserve statistical independence of data points, and therefore weighted integration methods usually do not account properly for data error, but this is usually a minor problem. Special cases of linearizations are the Patlak and the Logan plots, which have been discussed above.

4.10   Quantitative Data Analysis

Quantitative image analysis is a key component of brain PET. Each PET image represents simultaneous quantitative and specific measurements at more than 1000 locations (resulting from a brain size of more than 1000 ml and a quantitative spatial resolution element of 1 ml or less). Presentation of this overwhelming amount of quantitative data as a functional image is convenient and efficient, but it limits data interpretation to effects that are easily visualized by the observer and also depends on his experience and expectations. Additional techniques are needed to detect and extract abnormalities in individual cases or differences between groups.

Extraction of relevant values can be based on regions or (preferably) on volumes of interest (ROIs, VOIs), thus concentrating analysis to anatomical locations that are known a priori to be particularly relevant (section 4.10.1). Determination of physiological parameters based on dynamic series, and blood samples or reference tissue is possible for VOIs, and can also be done for each voxel, producing a parametric image (section 4.10.4). Quantitative regional data from individuals then need to be compared by appropriate statistics against a reference group in order to detect significant abnormalities. In clinical and experimental studies, values observed in different groups or with different conditions are compared with each other (section 4.10.3). For values derived from a limited number of VOIs (typically under a hundred in number), this can best be achieved by standard statistical software packages. With prior spatial normalization (section 4.8.2) some statistics can even be handled on a voxel-by-voxel basis, producing statistical parametric images (also called statistical parametric maps, SPMs), as described in section 4.10.5.

Standard pathways for a diagnostic study start with images from a single individual. If this is a single image of tracer distribution or some calculated physiological parameter (parametric image), we can proceed either by ROI/VOI analysis (red arrow in Fig. 4.11) to obtain regional parameters or by spatial normalization (blue arrow). In both cases we need to compare the individual data with reference data. In the case of the regional data this is done by comparison with regional nor-

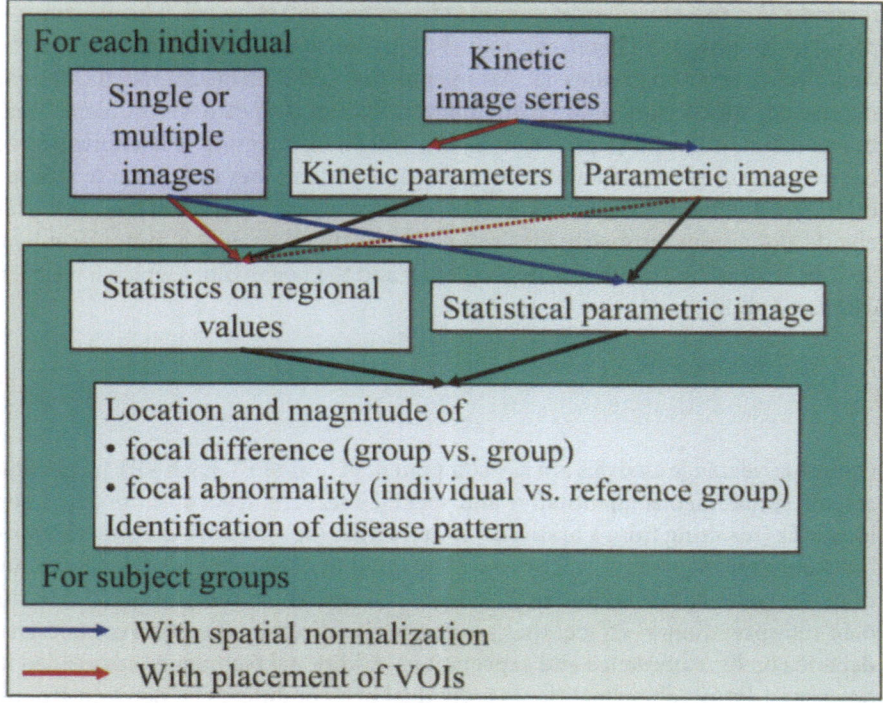

Fig. 4.11. Flowchart for quantitative data analysis, which starts for each individual either with a single image, or with multiple (coregistered) images, or with a kinetic image series

mative values (e.g., means and 95% confidence limits) obtained for the same regions in a reference group. The spatially normalized image can be compared with a set of spatially normalized reference images by statistical parametric mapping. The regional analysis will be more efficient if we are looking for an abnormality in few specific brain regions with known borders. Statistical parametric mapping (SPM) will be a reasonable option if we are searching in the whole brain for any abnormality. Unfortunately, this global search requires a massive statistical adjustment for multiple testing, as explained in sections 4.10.3 and 4.10.5, and may therefore be rather inefficient.

Standard pathways for a scientific study looking at differences between groups are very similar. Now, ROI/VOI analysis or spatial normalization is done for multiple subjects rather than for a single individual. Regional analysis will result in tabulated quantitative values that undergo standard statistical methods, as described in section 4.10.3. Normalized images will be subjected to SPM. Again, the optimum choice depends on whether there is a specific hypothesis about the values in certain brain regions. For an analysis of a few well-defined brain regions, regional analysis is more efficient, while for a global search SPM is usually more convenient.

4.10.1   Regions/Volumes of Interest

In older PET scanners that produced thick transaxial slices, it made good sense to place regions of interest (ROIs) on these slices. In more modern scanners with thin slices and near-isotropic voxels, ROIs defined on slices do not give good representation of brain structures. Volumes of interest (VOIs) extending over several slices, essentially without preference of one axis over another, are much more appropriate. However, it is a somewhat tedious task to define irregular VOIs on multiple slices manually, and therefore primitives, such as spheres or boxes, and special tools, including morphological 3D operations such as dilatation and erosion, are offered by dedicated software applications for that purpose [2377, 2497].

Placement of VOIs can be guided by specific abnormalities in the PET image, such as the hottest part of a brain tumor, or the core of an ischemic lesion. Nonetheless, with this concept, care must always be taken not to fall into the trap of circular reasoning (such as proving that such an area is abnormal compared with other regions by statistical tests, which would then just reflect the way it was selected). It could also be dangerous to pick extremes by means of small VOIs in noisy images (which could merely represent the extremes of a random variation). This caveat also applies to the use of thresholds for automatic definition of VOIs on PET images [2498]. For most purposes, it is advisable to guide VOI placement by external criteria, such as lesions or anatomical structures defined on a coregistered MRI image.

A common way of handling VOIs is by reference to templates and atlases. Like matching an image to an image template (see section 4.8.2), this usually involves scaling and shifting of VOIs to adapt to individual brain anatomy. In the past, such procedures have been developed to place ROIs on planes in standard orientation [2499]. Modern procedures for placement of VOIs involve various 3D transformations [2500–2502] and make reference to digital anatomical atlases (Fig. 4.12; see section 4.8.3). The corollary to this is that it is also possible to superimpose a fixed template of atlas VOIs onto functional brain images that have been spatially normalized [2395], but with this approach there is probably less flexibility in adaptation to situations with disease-related deformations of brain anatomy than is allowed by procedures with individual and interactive adjustment of VOIs.

In order to optimize sensitivity and specificity of VOI analysis, VOIs should represent structures that are as homogeneous as possible with regard to the parameter of interest. For instance, if D2 receptors in the caudate nucleus are of interest, there should be as little contamination from surrounding CSF and white matter as possible. This is to avoid undue bias of the signal and variation that stems from variable admixture of low signal from surrounding structures due to interindividual anatomical variation, rather than from changes in D2 receptors in the caudate nucleus.

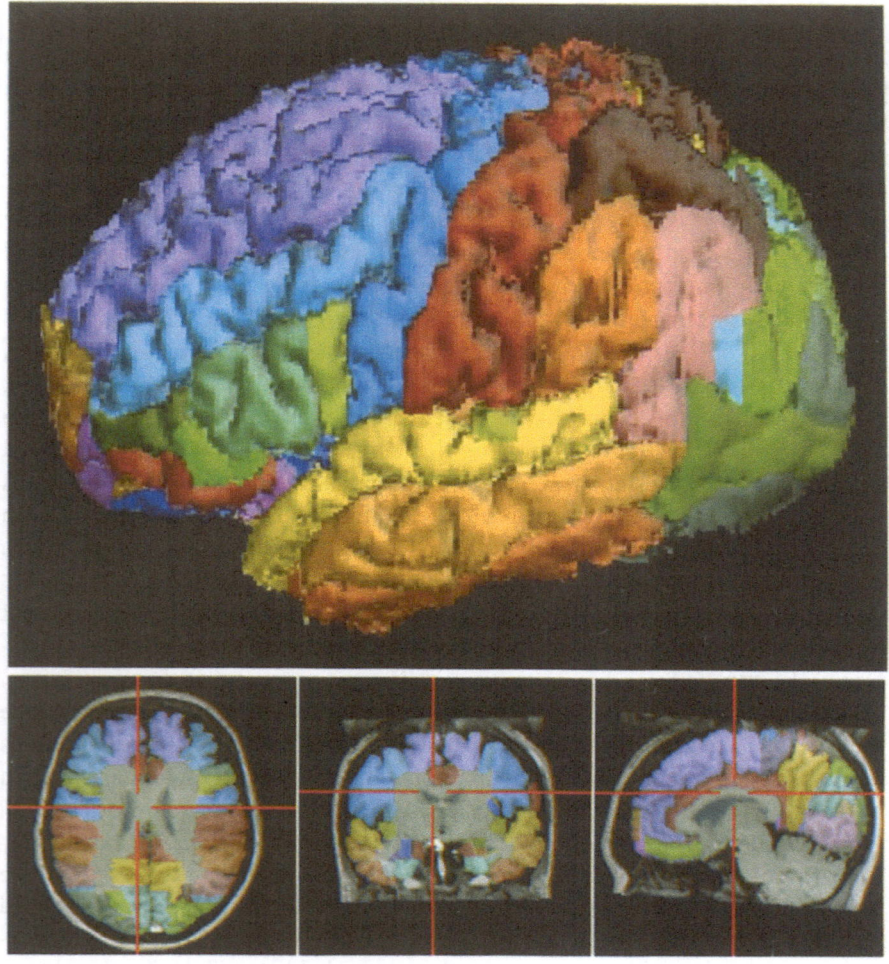

Fig. 4.12. Digital atlas of cortical gyri identified by different colors, that was created by manual segmentation of a normal brain MRI. It is shown as cortex rendering in left superior lateral view and three orthogonal cuts

4.10.2   Partial Volume Correction

For many physiological functions and receptor densities, there is a marked contrast between normal gray and white matter in the brain. Even in humans, many gray matter structures are rather small and cortical thickness is in the order of 3–5 mm. In animals, most structures are even smaller. With the spatial resolution of current PET scanners, it is often not possible to observe the gray matter activity in such small structures in a quantitatively accurate manner, because there is crosstalk between the gray and white matter signals, reducing the higher one (usually from gray matter) and increasing the lower one (usually from white matter) (Fig. 4.13). This contribution of signal from another tissue compartment is called the partial volume effect. As a general rule, better than 95 % quantitative accuracy is achieved at the center of a region only if the diameter of the structure that is measured is larger than twice the FWHM resolution of the scanner [125].

The loss of gray matter signal due to the smallness of structures is particularly disturbing if we want to compare metabolism of a structure between subjects that may also be affected by volume changes, e.g., due to atrophy. Thus, a reduction of apparent local PET signal in degenerative disease could also be due to loss of volume, rather than function [126, 283]. Therefore, there have been attempts to correct for size-dependent partial volume effects. This is possible if certain assumptions are made. Most frequently, it is assumed that the low signal from white matter and CSF is homogeneous and constant within these compartments. Thus, these values can be obtained from pure and sufficiently large white matter (centrum semiovale) and CSF regions. For several tracers, CSF signal may be approximated as zero. If we then know the extent of the gray matter region that we wish to measure (usually from coregistered MRI), the reduction of signal due to crosstalk from white matter and CSF can be estimated.

Several algorithms for partial volume correction (PVC) have been proposed and evaluated [2317–2319, 2503]. Corrections are relatively straightforward and robust for homogeneous VOIs, but are more sensitive to severe bias and artifacts if applied voxel by voxel, mainly because voxels with small contributions from gray matter undergo huge corrections that are very sensitive to errors in the estimation of relative compartment volume contributions. Even minor errors in MRI geometry, coregistration and segmentation can introduce large PVC errors. Thus, PVC cannot currently be considered a routine tool for quantitative PET studies. Better accuracy is achieved by the use of scanners with 2.5-mm resolution (see section 4.2).

4.10.3   Statistical Models for Quantitative Data

Statistical analysis is necessary to progress from plain observation of individual cases to detection of abnormalities in individual scans and to detection of significant differences between subject groups to test scientific hypotheses. Standard approaches as described in textbooks (e.g., [2504]) and provided by software packages (e.g., those provided by SPSS Inc. or SAS Institute Inc., Cary, NC) are available for regional data extracted from images by ROIs or VOIs. Some of these procedures have also been transferred (at least to some extent) to the image domain by statistical parametric mapping (see section 4.10.5).

First, a decision has to be made as to whether data on an absolute scale (such as CBF in units of milliliters per 100 grams per minute, CMRglc in units of micromoles per 100 grams per minute, or B_{max} in nanomoles per liter) are available and

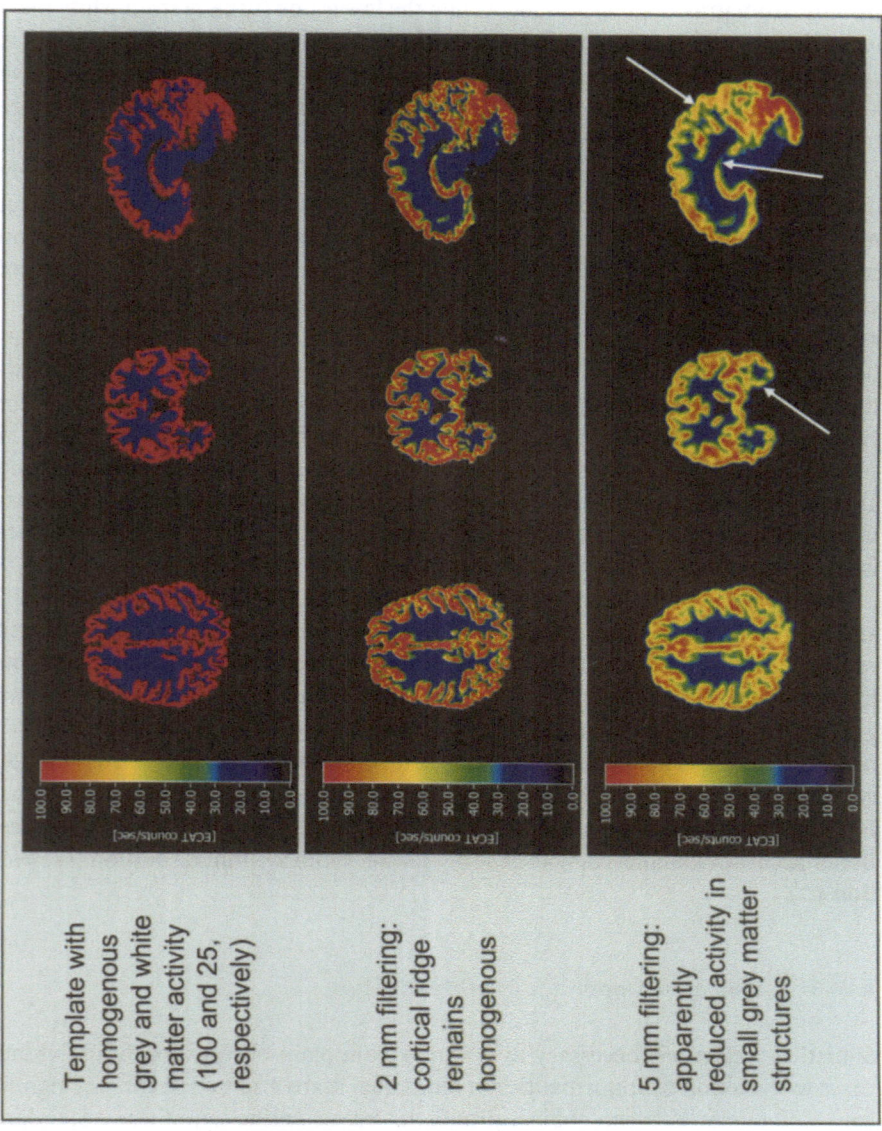

Fig. 4.13. Simulation of the effect of spatial resolution on image quantitation by a template with homogeneous activity in gray and white matter. At 5-mm spatial resolution (as in current standard clinical scanners) cortical glucose metabolism appears inhomogeneous, with values below the original values in many areas (pronounced effects in tail of caudate, mesial temporal and parietal cortex marked by *white arrows*). At 2-mm spatial resolution (now reached by ultra-high-resolution clinical PET scanners) quantitative accuracy is maintained in most structures

should be analyzed. Otherwise, regional data are often considered relative to some reference region or whole-brain average. The latter may vary arbitrarily, depending on some irrelevant factor (such as injected dose, systemic tracer elimination, cardiac output) that is not of interest for the study and would introduce unnecessary variance. Even if measurements are available on some absolute physiological scale, we may decide to look at only relative values because they frequently have a lower variance in normals than the absolute values (see also sections on CBF [3.2] and CMRglc [3.4]), and therefore usually have higher power to detect focal changes.

Nonetheless, results obtained from relative values may be misleading if there is a physiologically relevant or disease-related change in the reference value. It may mask a regional change (if it is of the same magnitude as the change in the reference value) or even cause a regional change to point in the wrong direction (if it is smaller than the change in the reference region). Therefore, absolute values should be checked whenever possible, even if most of the statistical analysis is based on regional values.

The choice of a proper reference region to scale relative values is not a trivial task. If in a receptor-binding study a region devoid of specific receptors is chosen (which may happen as early as the kinetic modeling stage; see section 4.9.3.2), stochastic fluctuations in that region can be large if tracer accumulation is low (meaning that the respective VOI needs to be large to allow collection of as many counts as possible), and small changes in nonspecific binding may introduce a significant bias [2505]. With whole-brain average reference, it is necessary to specify what exactly is meant by "whole brain". It could either be everything, including the ventricles (which may be biased by enlargement of ventricles with age or in a patient group), both gray and white matter (which may not make much sense if we want to analyze functional changes in gray matter only and could again be biased by atrophy of either gray or white matter), or gray matter only. In the last case, a procedure is needed that extracts gray matter values. This could be done by thresholding [1822], by use of a gray matter template that has been adapted to individual anatomy, or by using gray matter segmented on a coregistered MRI (which is probably the most accurate method, but also the most complicated). These details are rarely described in the methods sections of scientific papers, even though in certain cases the selection may have an impact on the interpretation of results. The potential bias from global estimators introduced by focal changes, and means of minimizing them, have been discussed in detail in the context of activation studies [2506].

In the case of a single image from each individual, regional image data may directly enter statistical tests. Sometimes, we may be interested in statistics on values derived from regional data rather than on the original data. For instance, differences between corresponding regions from the two brain hemispheres may be analyzed in the case of unilateral disorders (such as stroke, focal epilepsy, or brain tumors). Alternatively, in follow-up studies, the difference in corresponding values between the second (or other subsequent) measurement and the first measurement may be of interest. Instead of absolute differences, we could also use relative differences, such as (B–A)/A or (B–A)/(B+A), where B indicates the target value and A, the reference value. Such relative differences are frequently expressed as percentage changes: [100(B–A)/A, or 50(B–A)/(B+A)]. Their use is appropriate whenever there is reason to believe that the magnitude of the change is related to the magnitude of the reference value, and when it is desired to control for this, but such relative values may become unstable and deviate considerably from normal distribution if some of the reference values (in the denominator) are small or have high variance.

It is also necessary to consider the issue of data distribution, because standard parametric statistical methods are generally based on the assumption of normal distribution. Normal data distribution appears to be a reasonable assumption for most functional values determined by PET, but has rarely been tested explicitly. Scatter plots in most studies show no major abnormalities suggesting significant deviation from normality, but deviations have been noted in certain instances and may require the use of appropriate transformations or nonparametric methods [2507]. Data distribution in PET studies should be checked at least by scatter plots to exclude major deviations and detect outliers that might distort parametric statistics.

In a diagnostic setting, it is usually important to determine whether a regional value is abnormal compared to a reference sample. In the case of a normal data distribution this is easily done by t-statistics to determine confidence limits of reference values and the probability that the test value lies within the distribution of reference values. In scientific studies, t-tests can be used to compare values between two groups, and analysis of variance (ANOVA) is a standard procedure for comparisons of regional values among two or more groups. ANOVA may also be appropriate for complex experimental designs and can easily be extended as a general linear model to include continuous independent parameters (e.g., age), usually called covariates, for regression. If normal data distribution is not present and cannot easily be achieved by data transformation, nonparametric tests should be used.

A common issue with image-derived data is that data are not usually analyzed for a single region only; rather, images provide data for multiple regions that are all to be tested statistically. Thus, multiple testing is frequently encountered. This requires adjustment of statistical thresholds because multiple tests are likely to yield some false-positive results. With 100 independent tests performed at the 5% error level on completely random data lacking any significant signal, an average of five falsely significant results would be expected. A commonly used procedure to guard against this is the Bonferroni correction, which basically sets the critical significance threshold at a value calculated from the overall (experiment-wise) error probability threshold (usually 0.05) divided by the number of regions to be tested.

To complicate the situation even further, however, multiple regional values obtained in the same individuals are not really independent of each other, but we expect correlations among these regional values. As a consequence, the Bonferroni correction is too conservative and is likely to miss significant findings. This situation can be handled by treating these multiple regional values as "repeated measures" in ANOVA and general linear models [2507], which then provide statistics allowing us to decide whether there is a significant overall difference between groups and whether regional differences affect some regions significantly more than others.

4.10.4   Parametric Imaging

Construction of parametric images of physiological values can be regarded as an extension of regional kinetic parameter extraction (see section 4.9.4), but now we look at very many small "VOIs" that consist of only single voxels, and we do not tabulate the resulting parameters but display them as images. There are various obvious consequences. Because of the large number of voxels, the calculation demands are much higher than for VOIs. It is usually not practical to do iterative nonlinear least-squares fit for each voxel and faster methods are needed. Individual voxels usually represent very few counts that have been actually measured, and consequently kinetic curves derived from individual voxels are usually subject to exceedingly large noise. The usual remedy for this problem is substantial smoothing prior to the generation of parametric images.

In most instances, smoothing is done by Gaussian filtering of the original reconstructed images with typical filter widths between 6 and 12 mm, but other low-pass filters (that could be applied at the time of image reconstruction) can also be used. The reduction in high-frequency noise is achieved at the cost of a loss of spatial resolution. The use of nonstationary filters that extract information from the time domain so as to select a locally optimal filter width that maximizes counts in homogeneous regions while preserving the contrast between such regions [2508] has been suggested, but results are difficult to handle, in particular with respect to subsequent statistical procedures, which usually assume homogenous variance. A nonlinear ridge regression algorithm with spatial constraint has been suggested to integrate both spatial and temporal information for improvement of parametric image quality [2509].

As mentioned in section 4.9.4, algorithms for the generation of parametric images are often based on integration of kinetic equations to linearize parameters that can then be estimated by linear regression. Parametric imaging procedures have been developed for CBF [2449, 2510], CMRglc [2492, 2511, 2512], and receptor binding [2488, 2513], especially for benzodiazepine receptors with FMZ [2514, 2515].

Generalized linear approaches that are not restricted to specific compartmental models have also been developed [2491]. One of these approaches, dubbed spectral analysis, extracts the spectrum of exponential functions that would describe tracer kinetics in brain after a sharp bolus input [2516]. Because of their model independence, however, biological interpretation in physiologically mean-ingful terms is not always clear. Owing to their linear properties, most parametric imaging procedures can be applied to projection data (after correction for scatter and attenuation) before reconstruction [2517].

Parametric imaging may involve a reference region that has to be selected by placement of a reference VOI, as described for instance for RAC (D2 receptors), ^{11}C-SCH 23390 (D1 receptors), ^{11}C-CFT (dopamine transporters), AChE activity, and FDOPA [334, 2513, 2518].

4.10.5   Statistical Parametric Mapping

With the relatively straightforward extension of the general linear statistical model that accommodates both discrete and continuous independent variables (see section 4.10.3) on voxels (rather than VOI-derived values), a freely available software package for statistical parametric mapping (SPM) developed and dis-tributed by the Wellcome Institute (London, U.K) [2519, 2520] has become very popular, especially in the context of functional activation studies in normal sub-jects [1825]. It is also useful to study other biological effects (such as aging or phar-macological interventions) in normals [31, 2521]. It offers some resources that are useful for the study of patient groups, e.g., in neurodegenerative disease and epilepsy [87, 122, 124, 210, 390, 1012], but limitations still apply that are mainly due to unsolved issues with respect to spatial normalization in disease (see sec-tion 4.8.2) and some restrictions with respect to model building for different sub-ject groups. The statistical concepts related to SPM have reached a high degree of sophistication that tends to prevent a broad appreciation of its limitations, as indicated by recent reviews [2522, 2523].

Significance of local effects is determined either by the maximum t-value (peak height) or by the extent of the effect (cluster size) [1822, 2524, 2525], recog-nizing that in contrast to random noise, which has little low-frequency spatial cor-relation in PET images, the effects with the greatest biological relevance tend to affect larger brain areas. As in the statistical analysis of VOI-derived values (see section 4.10.3), it is necessary to determine the proper adjustment of significance levels for peak height and cluster size, also considering the covariance among vox-els [2526]. This has been done (assuming homogeneous covariance) on the basis of Gaussian random fields [2527–2529]. The algorithm provides an estimate of the size of appropriate resolution elements ("resels"), and the number of these resels in the total volume of analysis (usually the whole brain, but restriction to a par-ticular VOI is also possible) is used for a Bonferroni correction of statistical thresholds.

Expected peak height and cluster size depend on prior smoothing of images [2530]. Selection of the proper filter size is still often done empirically to find some pragmatic optimum between spatial resolution and sensitivity for detection of small but extended peaks. Variation of filter size for maximizing power across the brain has been suggested to incorporate underlying neuroanatomy of functional units in the a priori selection of filter size [2531]. Because a linear statistical model is used, this can also be applied to PET images that have been transformed into wavelet space [2532], which offers integration of spatial filtering into statistical estimation but is difficult to understand intuitively.

SPM has also been applied to parametric images to study changes in local receptor occupancy [2533], opiate and D2 receptors in Huntington and Parkinson disease [534, 541, 2534], benzodiazepine receptors in epilepsy [1071, 2535], 5-HT$_2$ receptors in psychiatric disorders [1652, 1685], and to develop a multicenter database of 5-HT$_{1A}$ receptor binding [2153].

Use of multivariate methods on a voxel-by-voxel basis is limited by the same considerations as in the standard statistical setting and, compared with the very large number of variables (voxels), the relative number of subjects is now even smaller, precluding confirmatory statistical testing. The potential and limitations of such methods in the study of brain connectivity have been pointed out in a recent review [2536]. Principal-components analysis has been used to extract covariance patterns [33, 2537–2539], and function connectivity has been studied by multiple regression and correlation analysis [2540, 2541] and by neuronal net modeling [2542]. Multivariate methods also appear to be useful for pattern extraction and recognition (see section 4.10.6).

A few nonparametric statistical tests that avoid the sometimes uncertain assumption of normal data distribution have also been made available for statistical parametric mapping. This assumption is particularly difficult to maintain for the extremes of the statistical distribution that are used with Bonferroni-corrected thresholds in SPM. Significance of activations can be determined accurately without any assumptions about data distribution by a randomization algorithm [2543] that has been shown to be less prone to false-positive results in individual subjects than parametric SPM procedures [2544].

4.10.6   Lesion Detection and Pattern Recognition

Detection of lesions (meaning any significant abnormality) is the ultimate goal of diagnostic imaging. Up to now, the standard procedure has been visual evaluation by an experienced observer. The level of experience varies between observers, however, and accepted standards for reporting abnormalities are lacking. The overwhelming number of possible abnormalities and normal variations of human brain, both anatomical and physiological, makes it extremely difficult to establish reproducible standards. These difficulties are further increased by the relatively small number of examinations that are typically performed with each tracer at most laboratories, and the correspondingly small number of reference scans that are available to observers. This situation is probably one of the main limitations preventing full development of the clinical potential of PET in neurology and psychiatry.

Quantitative comparison of an individual clinical PET scan against a reference sample by means of image processing and statistical parametric mapping (see earlier sections) may provide substantial help. Yet, in a global brain search for abnormal voxels, the level of statistical significance needs to be corrected for the large number of multiple comparisons that are performed in this search. That correction, if performed rigorously, usually requires a very large abnormality to be accepted as significant, and thus reduces sensitivity to a degree that may make the investigation clinically useless. It is the characteristic pattern of specific diseases, rather than abnormalities in individual voxels or an abnormality at very specific anatomical location predicted by clinical findings, that we are looking for in a typical clinical situation.

Few algorithms have so far been developed to meet the challenges of pattern recognition in brain PET. Principal-components analysis [33, 2539, 2545] is a powerful multivariate technique designed to extract covariance patterns from functional imaging data and can be applied to disease detection [2546–2548]. Other approaches employ discriminant analysis [2549], neural networks [2550], and fuzzy clustering [2551]. The sum of abnormal t-values in regions that are typically hypometabolic in AD has been used as an indicator of disease [32]. Such techniques need to be developed further in connection with large image databases, to approach the goal of detecting disease-related quantitative abnormalities that cannot be adequately described qualitatively as focal lesions. Recognition of patterns in 3D PET data sets is still in a very early phase of development but promises to increase our diagnostic abilities in an objective and reproducible way.

References

1. Sung NS, Crowley WF, Jr et al. JAMA 2003; 289(10):1278–1287.
2. Chalmers TC. JAMA 1988; 260(18):2713–2715.
3. Powers WJ, Berg L et al. Neurology 1991; 41(9):1339–1340.
4. Robert G, Milne R. Health Technol Assess 1999; 3(16):1–54.
5. Kuwert T, Bartenstein P et al. Nervenarzt 1998; 69:1045–1060.
6. Reid MC, Lachs MS et al. JAMA 1995; 274(8):645–651.
7. FDA. Federal Register 2000; 65:12999–13010.
8. Gambhir SS, Schwimmer J. Q J Nucl Med 2000; 44(2):121–137.
9. Gambhir SS. J Nucl Med 1999; 40(9):1570–1581.
10. Gee AD. Br Med Bull 2003; 65:169–177.
11. Eckelman WC. Drug Discov Today 2003; 8(9):404–410.
12. Chatziioannou A, Tai YC et al. Phys Med Biol 2001; 46(11):2899–2910.
13. Benson DF, Kuhl DE et al. Arch Neurol 1983; 40(12):711–714.
14. Friedland RP, Budinger TF et al. J Comput Assist Tomogr 1983; 7(4):590–598.
15. DeLeon MJ, Ferris SH et al. J Cereb Blood Flow Metab 1983; 3(3):391–394.
16. Foster NL, Chase TN et al. Neurology 1983; 33(8):961–965.
17. Frackowiak RS, Pozzilli C et al. Brain 1981; 104(Pt 4):753–778.
18. Knopman DS, Dekosky ST et al. Neurology 2001; 56(9):1143–1153.
19. Deutsche Gesellschaft für Psychiatrie PuND. Behandlungsleitlinie Demenz. Steinkopff, Darmstadt, 2000.
20. Deutsche Gesellschaft für Neurologie. Diagnose und Therapie der Alzheimer-Demenz (AD) und der Demenz mit Lewy-Körperchen (DLB). 2002.
21. Halliday G, Ng T et al. Acta Neuropathol (Berl) 2002; 104(1):72–78.
22. Reiman EM, Caselli RJ et al. Proc Natl Acad Sci U S A 2001; 98(6):3334–3339.
23. Petersen RC, Stevens JC et al. Neurology 2001; 56(9):1133–1142.
24. Wolf H, Jelic V et al. Acta Neurol Scand Suppl 2003; 179:52–76.
25. Salmon E, Sadzot B et al. J Nucl Med 1994; 35(3):391–398.
26. Chase TN, Foster NL et al. Annals of Neurology 1984; 15 Suppl:S170-4.
27. Haxby JV, Grady CL et al. Neurology 1988; 38(12):1853–1863.
28. Grady CL, Haxby JV et al. Journal of Neuropsychiatry & Clinical Neurosciences 1990; 2(4):373–384.
29. Kuhl DE, Metter EJ et al. J Cereb Blood Flow Metab 1982; 2(2):163–171.
30. Moeller JR, Ishikawa T et al. J Cereb Blood Flow Metab 1996; 16(3):385–398.
31. Petit-Taboue MC, Landeau B et al. Neuroimage 1998; 7(3):176–184.
32. Herholz K, Salmon E et al. Neuroimage 2002; 17:302–316.
33. Zündorf G, Kerrouche N et al. Hum Brain Mapp 2003; 18:13–21.
34. Minoshima S, Giordani B et al. Ann Neurol 1997; 42(1):85–94.
35. Gusnard DA, Raichle ME et al. Nat Rev Neurosci 2001; 2(10):685–694.
36. Chetelat G, Baron JC. Neuroimage 2003; 18(2):525–541.
37. de Leon MJ, Convit A et al. Proc Natl Acad Sci U S A 2001; 98(19):10966–10971.
38. Yamaguchi S, Meguro K et al. J Neurol Neurosurg Psychiatry 1997; 62(6):596–600.
39. Haxby JV, Duara R et al. J Cereb Blood Flow Metab 1985; 5(2):193–200.
40. Akiyama H, Harrop R et al. Neurology 1989; 39(4):541–548.

41. Mielke R, Herholz K et al. Neurobiology of Aging 1992; 13(1):93–98.
42. Small GW, Kuhl DE et al. Arch Gen Psychiatry 1989; 46(6):527–532.
43. Azari NP, Pettigrew KD et al. Dementia 1994; 5(2):69–78.
44. Schapiro MB, Ball MJ et al. Neurology 1988; 38(6):938–942.
45. Piert M, Koeppe RA et al. J Nucl Med 1996; 37(2):201–208.
46. Jagust WJ, Seab JP et al. J Cereb Blood Flow Metab 1991; 11(2):323–330.
47. Friedland RP, Jagust WJ et al. Neurology 1989; 39(11):1427–1434.
48. Simpson IA, Davies P. Ann Neurol 1994; 36(5):800–801.
49. Hoyer S. J Neural Transm 2002; 109(3):341–360.
50. Heiss WD, Hebold I et al. J Cereb Blood Flow Metab 1988; 8(4):613–617.
51. Mielke R, Kittner B et al. Journal of the Neurological Sciences 1996; 141:59–64.
52. Mega MS, Cummings JL et al. Neuropsychiatry Neuropsychol Behav Neurol 2001; 14(1):63–68.
53. Blin J, Ivanoiu A et al. Psychopharmacology (Berl) 1998; 136(3):256–263.
54. Jagust WJ, Friedland RP et al. Neurology 1988; 38(6):909–912.
55. Smith GS, de Leon MJ et al. Arch Neurol 1992; 49(11):1142–1150.
56. Grady CL, Haxby JV et al. Neurology 1986; 36(10):1390–1392.
57. Haxby JV, Grady CL et al. Arch Neurol 1990; 47(7):753–760.
58. Slansky I, Herholz K et al. Neuroradiology 1995; 37(4):270–277.
59. Reed BR, Eberling JL et al. Ann Neurol 2000; 48(3):275–284.
60. Desgranges B, Baron JC et al. Brain 2002; 125(Pt 5):1116–1124.
61. Haxby JV, Grady CL et al. Arch Neurol 1986; 43(9):882–885.
62. Haxby JV, Grady CL et al. Journal of Neural Transmission 1994; Supplementum 1987;24:49–53.
63. Hirono N, Mori E et al. Dement Geriatr Cogn Disord 1998; 9(2):68–73.
64. Kessler J, Mielke R et al. International Journal of Neuroscience 2000; 104:1–15.
65. Sultzer DL, Brown CV et al. Am J Psychiatry 2003; 160(2):341–349.
66. Salmon E, Garraux G et al. Neuroimage 2003; 20(1):435–440.
67. Kumar A, Schapiro MB et al. J Psychiatr Res 1990; 24(2):97–109.
68. Holthoff VA, Herholz K et al. J Cereb Blood Flow Metab 2003; 23 (Suppl. 1):631.
69. Hoffman JM, Welsh-Bohmer KA et al. J Nucl Med 2000; 41(11):1920–1928.
70. Mielke R, Schröder R et al. Acta Neuropathologica 1996; 91(2):174–179.
71. Silverman DH, Small GW et al. JAMA 2001; 286(17):2120–2127.
72. Terry RD, Masliah E et al. Ann Neurol 1991; 30(4):572–580.
73. Mega MS, Chen SS et al. Neuroimage 1997; 5(2):147–153.
74. Mega MS, Chu T et al. Neuroreport 1999; 10(14):2911–2917.
75. DeCarli C, Atack JR et al. Neurodegeneration 1992; 1:113–121.
76. Braak H, Braak E. Acta Neuropathologica 1991; 82(4):239–259.
77. Selkoe DJ. Science 2002; 298(5594):789–791.
78. Thal DR, Rub U et al. Neurology 2002; 58(12):1791–1800.
79. Petersen RC. Arch Neurol 2000; 57(5):643–644.
80. Thompson SA, Hodges JR. Neurocase 2002; 8(6):405–410.
81. Nestor PJ, Fryer TD et al. Ann Neurol 2003; 54(3):343–351.
82. Small GW, Ercoli LM et al. Proc Natl Acad Sci U S A 2000; 97(11):6037–6042.
83. Grigoletto F, Zappala G et al. Neurology 1999; 53(2):315–320.
84. Herholz K, Nordberg A et al. European Journal of Neurology 1998; 5(Suppl. 3):S24.
85. Berent S, Giordani B et al. J Psychiatr Res 1999; 33(1):7–16.
86. Arnaiz E, Jelic V et al. Neuroreport 2001; 12(4):851–855.
87. Drzezga A, Lautenschlager N et al. Eur J Nucl Med Mol Imaging 2003; 30:1104–1113.
88. Chetelat G, Desgranges B et al. Neurology 2003; 60(8):1374–1377.
89. Reiman EM, Caselli RJ et al. New England Journal of Medicine 1996; 334(12):752–758.
90. Small GW, Mazziotta JC et al. JAMA 1995; 273(12):942–947.
91. Alexander GE, Chen K et al. Am J Psychiatry 2002; 159(5):738–745.
92. Fukuyama H, Ogawa M et al. J Nucl Med 1994; 35(1):1–6.

93. Johnson KA, Mueller ST et al. Arch Neurol 1987; 44(2):165–168.
94. Powers WJ, Perlmutter JS et al. Neurology 1992; 42(4):765–770.
95. Herholz K, Schopphoff H et al. J Nucl Med 2002; 43(1):21–26.
96. Mielke R, Pietrzyk U et al. Eur J Nucl Med 1994; 21(10):1052–1060.
97. Messa C, Perani D et al. J Nucl Med 1994; 35(2):210–216.
98. Silverman DHS, Cummings JL et al. Molecular Imaging & Biology 2002; 4(4):283–293.
99. Okamura N, Arai H et al. Am J Psychiatry 2002; 159(3):474–476.
100. Huang C, Wahlund LO et al. BMC Neurol 2002; 2(1):9.
101. Gonzalez RG, Fischman AJ et al. AJNR Am J Neuroradiol 1995; 16(9):1763–1770.
102. Harris GJ, Lewis RF et al. AJNR Am J Neuroradiol 1998; 19(9):1727–1732.
103. Bancher C, Braak H et al. Neuroscience Letters 1993; 162(1–2):179–182.
104. McKhann G, Drachman D et al. Neurology 1984; 34:939–944.
105. Galasko D, Hansen LA et al. Arch Neurol 1994; 51(9):888–895.
106. Rutschmann OT, Matchar DB. JAMA 2002; 287(8):985–986.
107. Gill SS, Rochon PA et al. J Am Geriatr Soc 2003; 51(2):258–264.
108. Kulasingham SL, Samsa GP et al. Value in Health 2003; 6(5):542–550.
109. Kent DL, Larson EB. Invest Radiol 1992; 27(3):245–254.
110. McMahon PM, Araki SS et al. Radiology 2003;2282020915.
111. Kuhl DE, Metter EJ et al. Annals of Neurology 1984; 15(5):419–424.
112. Schapiro MB, Pietrini P et al. J Neurol Neurosurg Psychiatry 1993; 56(8):859–864.
113. Eidelberg D, Moeller JR et al. J Nucl Med 1995; 36(3):378–383.
114. Hilker R, Voges J et al. Journal of Neural Transmission 2002; 10:1257–1264.
115. Grady CL, Haxby JV et al. Neuropsychologia 1987; 25(5):807–816.
116. Foster NL, VanDerSpek AF et al. J Cereb Blood Flow Metab 1987; 7(4):415–420.
117. Minoshima S, Frey KA et al. Journal of Computer Assisted Tomography 1995; 19(4):541–547.
118. McNamara D, Horwitz B et al. Int J Neurosci 1987; 36(1–2):89–97.
119. Kuhl DE, Koeppe RA et al. Neurology 1999; 52(4):691–699.
120. Herholz K, Perani D et al. J Nucl Med 1993; 34(9):1460–1466.
121. Burdette JH, Minoshima S et al. Radiology 1996; 198(3):837–843.
122. Signorini M, Paulesu E et al. Neuroimage 1999; 9(1):63–80.
123. Ishii K, Willoch F et al. J Nucl Med 2001; 42(4):548–557.
124. Salmon E, Collette F et al. Hum Brain Mapp 2000; 10(1):39–48.
125. Mazziotta JC, Phelps ME et al. Journal of Computer Assisted Tomography 1981; 5(5):734–743.
126. Meltzer CC, Zubieta JK et al. Neurology 1996; 47:454–461.
127. Alavi A, Newberg AB et al. J Nucl Med 1993; 34(10):1681–1687.
128. Ibanez V, Pietrini P et al. Neurology 1998; 50(6):1585–1593.
129. Perry EK, Haroutunian V et al. Neuroreport 1994; 5(7):747–749.
130. Candy JM, Perry RH et al. J Neurol Sci 1983; 59:277–289.
131. Mann DM, Yates PO et al. Acta Neuropathol (Berl) 1986; 71(3–4):332–337.
132. Shinotoh H, Aotsuka A et al. J Cereb Blood Flow Metab 2003; 23 (Suppl. 1):S598.
133. Irie T, Fukushi K et al. J Nucl Med 1996; 37(4):649–655.
134. Kilbourn MR, Snyder SE et al. Synapse 1996; 22(2):123–131.
135. Shinotoh H, Namba H et al. Ann Neurol 2000; 48(2):194–200.
136. Herholz K, Bauer B et al. J Neural Transm 2000; 12:1457–1468.
137. Herholz K, Weisenbach S et al. Neuroimage 2004; 29(9):136–143.
138. Weisenbach S, Hilker R et al. Neurology 2002; 58 (Suppl. 3):A144.
139. Kuhl DE, Minoshima S et al. Ann Neurol 2000; 48(3):391–395.
140. Shinotoh H, Aotsuka A et al. Neurology 2001; 56(3):408–410.
141. Blomqvist G, Tavitian B et al. J Cereb Blood Flow Metab 2001; 21:114–131.
142. Bencherif B, Endres CJ et al. Synapse 2002; 45(1):1–9.
143. Traykov L, Tavitian B et al. Eur J Neurol 1999; 6(3):273–278.
144. Mulholland GK, Wieland DM et al. Synapse 1998; 30(3):263–274.
145. Kuhl DE, Minoshima S et al. Annals of Neurology 1996; 40(3):399–410.

146. Paterson D, Nordberg A. Prog Neurobiol 2000; 61(1):75–111.
147. Sihver W, Langstrom B et al. Acta Neurol Scand Suppl 2000; 176:27–33.
148. Gatley SJ, Ding YS et al. Nucl Med Biol 1998; 25(5):449–454.
149. Ding YS, Molina PE et al. Nucl Med Biol 1999; 26(1):139–148.
150. Dolci L, Dolle F et al. Bioorg Med Chem 1999; 7(3):467–479.
151. Horti AG, Scheffel U et al. J Med Chem 1998; 41(22):4199–4206.
152. Villemagne VL, Horti A et al. J Nucl Med 1997; 38(11):1737–1741.
153. Valette H, Bottlaender M et al. J Nucl Med 1999; 40(8):1374–1380.
154. Valette H, Bottlaender M et al. J Neurochem 2003; 84(1):105–111.
155. Scheffel U, Horti AG et al. Nucl Med Biol 2000; 27(1):51–56.
156. Horti AG, Chefer SI et al. Life Sci 2000; 67(4):463–469.
157. Nordberg A, Lundqvist H et al. Alzheimer Disease & Associated Disorders 1995; 9(1):21–27.
158. Zubieta JK, Koeppe RA et al. Synapse 2001; 39(4):275–287.
159. Skaddan MB, Kilbourn MR et al. J Cereb Blood Flow Metab 2001; 21(2):144–148.
160. Hartvig P, Nordberg A et al. Dement Geriatr Cogn Disord 2002; 13(4):199–204.
161. Cohen RM, Podruchny TA et al. Synapse 2003; 49(3):150–156.
162. Meltzer CC, Price JC et al. Am J Psychiatry 1999; 156(12):1871–1878.
163. Nacmias B, Tedde A et al. Biol Psychiatry 2001; 50(6):472–475.
164. Tyrrell PJ, Sawle GV et al. Arch Neurol 1990; 47(12):1318–1323.
165. Kemppainen N, Ruottinen H et al. Neurology 2000; 55(2):205–209.
166. Kemppainen N, Laine M et al. Eur J Neurosci 2003; 18(1):149–154.
167. Meyer M, Koeppe RA et al. Arch Neurol 1995; 52(3):314–317.
168. Banati RB. GLIA 2002; 40(2):206–217.
169. Cagnin A, Brooks DJ et al. Lancet 2001; 358(9280):461–467.
170. Volkow ND, Ding YS et al. Biol Psychiatry 2001; 49(3):211–220.
171. Bacskai BJ, Klunk WE et al. J Cereb Blood Flow Metab 2002; 22(9):1035–1041.
172. Shoghi-Jadid K, Small GW et al. Am J Geriatr Psychiatry 2002; 10(1):24–35.
173. Klunk W, Debnath M et al. Neurobiology of Aging 2002; 23(1):1149.
174. Agdeppa ED, Kepe V et al. J Neurosci 2001; 21(24):RC189.
175. Agdeppa ED, Kepe V et al. Neuroscience 2003; 117(3):723–730.
176. Bresjanac M, Smid LM et al. J Neurosci 2003; 23(22):8029–8033.
177. Mathis CA, Bacskai BJ et al. Bioorg Med Chem Lett 2002; 12(3):295–298.
178. Kung MP, Skovronsky DM et al. Journal of Molecular Neuroscience 2003; 20(1):15–23.
179. Zhuang ZP, Kung MP et al. Journal of Medicinal Chemistry 2003; 46(2):237–243.
180. Kung MP, Hou C et al. Brain Research 2002; 956(2):202–210.
181. Klunk WE, Wang Y et al. J Neurosci 2003; 23(6):2086–2092.
182. Mathis CA, Wang YM et al. Journal of Medicinal Chemistry 2003; 46(13):2740–2754.
183. Bacskai BJ, Hickey GA et al. PNAS 2003; 100(21):12462–12467.
184. Klunk WE, Engler H et al. Imaging brain amyloid in Alzheimer's disease with Pittsburgh Compound-B. Ann Neurol 2004; 55(3):306–319.
185. Ono M, Wilson A et al. Nucl Med Biol 2003; 30(6):565–571.
186. Pietrini P, Dani A et al. American Journal of Psychiatry 1997; 154(8):1063–1069.
187. Duara R, Barker WW et al. J Cereb Blood Flow Metab 1992; 12(6):927–934.
188. Grady CL, Haxby JV et al. Neurobiology of Aging 1993; 14(1):35–44.
189. Pietrini P, Furey ML et al. Am J Psychiatry 1999; 156(3):470–473.
190. Kessler J, Herholz K et al. Neuropsychologia 1991; 29(3):229–243.
191. Buchsbaum MS, Kesslak JP et al. Arch Gen Psychiatry 1991; 48(9):840–847.
192. Backman L, Almkvist O et al. J Cogn Neurosci 2000; 12(1):134–141.
193. Bäckman L, Andersson JL et al. Neurology 1999; 52(9):1861–1870.
194. Grady CL, Furey ML et al. Brain 2001; 124(Pt 4):739–756.
195. Perry RH, Irving D et al. J Neurol Sci 1990; 95(2):119–139.
196. Verghese J, Crystal HA et al. Neurology 1999; 53(9):1974–1982.
197. Hurtig HI, Trojanowski JQ et al. Neurology 2000; 54(10):1916–1921.

198. Jellinger KA. J Neural Transm Suppl 2000; 59(1–2):185–212.
199. Ishii K, Imamura T et al. Neurology 1998; 51(1):125–130.
200. Higuchi M, Tashiro M et al. Exp Neurol 2000; 162(2):247–256.
201. Minoshima S, Foster NL et al. Ann Neurol 2001; 50(3):358–365.
202. Hu XS, Okamura N et al. Neurology 2000; 55(10):1575–1577.
203. Hisanaga K, Suzuki H et al. J Neurol 2001; 248(10):905–906.
204. Neary D, Snowden JS et al. Neurology 1998; 51(6):1546–1554.
205. Kertesz A, Munoz D. Arch Neurol 1998; 55(3):302–304.
206. Lindau M, Almkvist O et al. Dement Geriatr Cogn Disord 2000; 11(5):286–293.
207. van Swieten JC, Stevens M et al. Ann Neurol 1999; 46(4):617–626.
208. Friedland RP, Koss E et al. Dementia 1993; 4(3–4):192–203.
209. Kamo H, McGeer PL et al. Neurology 1987; 37(3):439–445.
210. Hosaka K, Ishii K et al. J Neurol Sci 2002; 199(1–2):67–71.
211. Feany MB, Mattiace LA et al. Journal of Neuropathology & Experimental Neurology 1996; 55(1):53–67.
212. Karbe H, Grond M et al. J Neurol 1992; 239(2):98–102.
213. Soong B, Liu R et al. Arch Neurol 2001; 58(2):300–304.
214. Volkow ND, Hitzemann R et al. Synapse 1992; 11(3):184–190.
215. Hodges JR, Patterson K et al. Brain 1992; 115 (Pt 6):1783–1806.
216. Mummery CJ, Patterson K et al. Brain 1999; 122(Pt 1):61–73.
217. Kempler D, Metter EJ et al. J Neurol Neurosurg Psychiatry 1990; 53(11):987–993.
218. De Oliveira SA, Castro MJ et al. Arq Neuropsiquiatr 1989; 47(1):72–75.
219. Nagy TG, Jelencsik I et al. Eur J Neurol 1999; 6(4):515–519.
220. Cappa SF, Perani D et al. Annals of the New York Academy of Sciences 1996; 777:243–248.
221. Chawluk JB, Mesulam MM et al. Annals of Neurology 1986; 19(1):68–74.
222. Tyrrell PJ, Warrington EK et al. Journal of Neurology, Neurosurgery & Psychiatry 1990; 53(12):1046–1050.
223. Ludolph AC, Langen KJ et al. Acta Neurol Scand 1992; 85(2):81–89.
224. Abrahams S, Goldstein LH et al. Brain 1996; 119 (Pt 6):2105–2120.
225. Abrahams S, Leigh PN et al. J Neurol Sci 1995; 129 Suppl:44–46.
226. Erkinjuntti T, Ostbye T et al. N Engl J Med 1997; 337(23):1667–1674.
227. Chui HC, Mack W et al. Arch Neurol 2000; 57(2):191–196.
228. Jellinger KA. Journal of the Neurological Sciences 2002; 203–204:153–157.
229. Ikeda M, Hokoishi K et al. Neurology 2001; 57(5):839–844.
230. Mielke R, Herholz K et al. Arch Neurol 1992; 49(9):909–913.
231. Sultzer DL, Mahler ME et al. Arch Neurol 1995; 52:773–780.
232. Mendez MF, Ottowitz W et al. Dement Geriatr Cogn Disord 1999; 10(6):518–525.
233. Ogawa T, Inugami A et al. Ajnr: American Journal of Neuroradiology 1995; 16(4 Suppl):978–981.
234. Matochik JA, Molchan SE et al. Acta Neurologica Scandinavica 1995; 91(2):153–157.
235. Goldman S, Laird A et al. Neurology 1993; 43(9):1828–1830.
236. Holthoff VA, Sandmann J et al. Arch Neurol 1990; 47(9):1035–1038.
237. Engler H, Lundberg PO et al. Eur J Nucl Med Mol Imaging 2003; 30(1):85–95.
238. Perani D, Cortelli P et al. Neurology 1993; 43(12):2565–2569.
239. Cortelli P, Perani D et al. Neurology 1997; 49(1):126–133.
240. Heiss WD, Pawlik G et al. Cerebrovascular & Brain Metabolism Reviews 1992; 4(1):1–27.
241. Perani D, Bressi S et al. Brain 1993; 116(Pt 4):903–919.
242. Schaafsma A, de Jong BM et al. Journal of the Neurological Sciences 2003; 210(1–2):23–30.
243. Aupee AM, Desgranges B et al. Neuroimage 2001; 13(6 Pt 1):1164–1173.
244. Fazio F, Perani D et al. J Cereb Blood Flow Metab 1992; 12(3):353–358.
245. Kuwert T, Homberg V et al. J Neurol Sci 1993; 118(1):10–16.
246. Reed LJ, Marsden P et al. Memory 1999; 7(5–6):599–612.
247. Szelies B, Herholz K et al. Arch Neurol 1991; 48(2):178–182.

248. Levasseur M, Baron JC et al. Brain 1992; 115(Pt 3):795–807.
249. Jacobs A, Neveling M et al. Stroke 1998; 29(3):660–667.
250. Welch LW, Nimmerrichter A et al. Psychol Med 1996; 26(2):421–425.
251. Matsuda K, Yamaji S et al. Ann Nucl Med 1997; 11(1):33–35.
252. Joyce EM, Rio DE et al. Psychiatry Res 1994; 54(3):225–239.
253. Paller KA, Acharya A et al. Journal of Cognitive Neuroscience 1997; 9(2):277–293.
254. Pappata S, Chabriat H et al. J Neural Transm Park Dis Dement Sect 1994; 8(1–2):131–137.
255. Teitelbaum JS, Zatorre RJ et al. N Engl J Med 1990; 322(25):1781–1787.
256. Guillery B, Desgranges B et al. Neurosci Lett 2002; 325(1):62–66.
257. Fujii K, Sadoshima S et al. J Neurol Neurosurg Psychiatry 1989; 52(5):622–630.
258. Markowitsch HJ, Kessler J et al. Neuropsychiatry Neuropsychol Behav Neurol 2000; 13(1):60–66.
259. Yasuno F, Nishikawa T et al. Psychiatry Res 2000; 99(1):43–57.
260. Riachi NJ, Arora PK et al. J Neurochem 1988; 50(4):1319–1321.
261. Calne DB, Langston JW et al. Nature 1985; 317(6034):246–248.
262. Gilman S, Low PA et al. J Neurol Sci 1999; 163(1):94–98.
263. Sawle GV, Playford ED et al. Arch Neurol 1994; 51(3):237–243.
264. Stoof JC, Winogrodzka A et al. Eur J Pharmacol 1999; 375(1–3):75–86.
265. Hughes AJ, Daniel SE et al. Brain 2002; 125(Pt 4):861–870.
266. Antonini A, Kazumata K et al. Mov Disord 1998; 13(2):268–274.
267. Otsuka M, Kuwabara Y et al. Ann Nucl Med 1997; 11(3):251–257.
268. Ghaemi M, Hilker R et al. J Neurol Neurosurg Psychiatry 2002; 73(5):517–523.
269. Boecker H, Weindl A et al. Acta Neurol Scand 1996; 93(6):387–392.
270. Karbe H, Holthoff V et al. J Neural Transm Park Dis Dement Sect 1992; 4(2):121–130.
271. Kume A, Shiratori M et al. J Neurol Sci 1992; 110(1–2):37–45.
272. Burn DJ, Sawle GV et al. Journal of Neurology, Neurosurgery & Psychiatry 1994; 57(3):278–284.
273. Eidelberg D, Moeller JR et al. Neurology 1995; 45(11):1995–2004.
274. Hardy J, Cookson MR et al. The Lancet Neurology 2003; 2(4):221–228.
275. Garnett ES, Firnau G et al. Nature 1983; 305(5930):137–138.
276. Pate BD, Kawamata T et al. Annals of Neurology 1993; 34(3):331–338.
277. Snow BJ, Tooyama I et al. Annals of Neurology 1993; 34(3):324–330.
278. Yee RE, Irwin I et al. Mov Disord 2001; 16(5):838–848.
279. Nahmias C, Garnett ES et al. Journal of the Neurological Sciences 1985; 69(3):223–230.
280. Leenders KL, Palmer AJ et al. Journal of Neurology, Neurosurgery & Psychiatry 1986; 49(8):853–860.
281. Damier P, Hirsch EC et al. Brain 1999; 122(Pt 8):1437–1448.
282. Garnett ES, Lang AE et al. Can J Neurol Sci 1987; 14(3 Suppl):444–447.
283. Rousset OG, Deep P et al. Synapse 2000; 37(2):81–89.
284. Nurmi E, Ruottinen HM et al. Mov Disord 2001; 16(4):608–615.
285. Antonini A, Vontobel P et al. Arch Neurol 1995; 52(12):1183–1190.
286. Pal PK, Lee CS et al. Parkinsonism Relat Disord 2001; 7(4):305–309.
287. Bruck A, Portin R et al. Neurosci Lett 2001; 311(2):81–84.
288. Holthoff-Detto VA, Kessler J et al. Arch Neurol 1997; 54:145–150.
289. Rinne JO, Portin R et al. Arch Neurol 2000; 57(4):470–475.
290. Brooks DJ, Playford ED et al. Neurology 1992; 42(8):1554–1560.
291. Ghaemi M, Raethjen J et al. Mov Disord 2002; 17(4):782–788.
292. Antonini A, Moresco RM et al. Neurol Sci 2001; 22(1):47–48.
293. Martin WR, Palmer MR et al. Annals of Neurology 1989; 26(4):535–542.
294. Sawle GV, Colebatch JG et al. Annals of Neurology 1990; 28(6):799–804.
295. Dhawan V, Ishikawa T et al. J Nucl Med 1996; 37(2):209–216.
296. Vingerhoets FJ, Schulzer M et al. J Nucl Med 1996; 37(3):421–426.
297. Whone AL, Moore RY et al. Ann Neurol 2003; 53(2):206–213.

298. Ghaemi M, Rudolf J et al. J Pineal Res 2001; 30(4):213–219.
299. Rakshi JS, Uema T et al. Brain 1999; 122(Pt 9):1637–1650.
300. Ghaemi M, Hilker R et al. J Neural Transm 1999; 106:X.
301. Doder M, Rabiner EA et al. Neurology 2003; 60(4):601–605.
302. Ouchi Y, Yoshikawa E et al. J Cereb Blood Flow Metab 2002; 22(6):746–752.
303. Goerendt IK, Messa C et al. Brain 2003; 126(2):312–325.
304. Piccini P, Pavese N et al. Ann Neurol 2003; 53(5):647–653.
305. Tedroff J, Pedersen M et al. Neurology 1996; 46:1430–1436.
306. Doudet DJ, Chan GL et al. Synapse 1998; 29(3):225–232.
307. Sossi V, Fuente-Fernandez R et al. J Cereb Blood Flow Metab 2002; 22(2):232–239.
308. Fuente-Fernandez R, Lim AS et al. Synapse 2003; 47(2):152–158.
309. Barrio JR, Huang SC et al. Biochem Pharmacol 1997; 54(3):341–348.
310. Aquilonius SM, Bergstrom K et al. Acta Neurol Scand 1987; 76(4):283–287.
311. Tedroff J, Aquilonius SM et al. Acta Neurol Scand 1988; 77(3):192–201.
312. Leenders KL, Salmon EP et al. Arch Neurol 1990; 47(12):1290–1298.
313. Brooks DJ, Salmon EP et al. Brain 1990; 113(Pt 5):1539–1552.
314. Danielsen EH, Cumming P et al. Cell Transplant 2000; 9(2):247–259.
315. Poyot T, Conde F et al. J Cereb Blood Flow Metab 2001; 21(7):782–792.
316. Rinne JO, Ruottinen H et al. J Neurol Neurosurg Psychiatry 1999; 67(6):737–741.
317. Frost JJ, Rosier AJ et al. Ann Neurol 1993; 34(3):423–431.
318. Guttman M, Burkholder J et al. Neurology 1997; 48(6):1578–1583.
319. Ribeiro MJ, Vidailhet M et al. Arch Neurol 2002; 59(4):580–586.
320. Ilgin N, Zubieta J et al. Neurology 1999; 52(6):1221–1226.
321. Wullner U, Pakzaban P et al. Exp Neurol 1994; 126(2):305–309.
322. Ouchi Y, Yoshikawa E et al. Ann Neurol 1999; 45(5):601–610.
323. Frey KA, Koeppe RA et al. Ann Neurol 1996; 40(6):873–884.
324. Kumar A, Mann S et al. Brain 2003; 126(12):2648–2655.
325. Lee CS, Samii A et al. Ann Neurol 2000; 47(4):493–503.
326. Kaasinen V, Ruottinen HM et al. J Nucl Med 2000; 41(1):65–70.
327. Hume SP, Opacka-Juffry J et al. Synapse 1995; 21(1):45–53.
328. Ginovart N, Farde L et al. Synapse 1997; 25(4):321–325.
329. Perlmutter JS, Kilbourn MR et al. Neurology 1987; 37(10):1575–1579.
330. Antonini A, Schwarz J et al. Neurology 1994; 44(7):1325–1329.
331. Ouchi Y, Kanno T et al. Ann Neurol 1999; 46(5):723–731.
332. Ouchi Y, Kanno T et al. Brain 2001; 124(4):784–792.
333. Ouchi Y, Yoshikawa E et al. Ann Neurol 1999; 45(5):608–610.
334. Ito K, Nagano-Saito A et al. Brain 2002; 125(Pt 6):1358–1365.
335. Kaasinen V, Nurmi E et al. Brain 2001; 124(Pt 6):1125–1130.
336. Kerenyi L, Ricaurte GA et al. Arch Neurol 2003; 60(9):1223–1229.
337. Asahina M, Shinotoh H et al. Acta Neurol Scand 1995; 91(6):437–443.
338. Asahina M, Suhara T et al. J Neurol Neurosurg Psychiatry 1998; 65(2):155–163.
339. Shinotoh H, Namba H et al. Ann Neurol 1999; 46(1):62–69.
340. Bohnen NI, Kaufer DI et al. J Nucl Med 2002; 43 (Suppl.)(5):63P.
341. Eidelberg D, Moeller JR et al. Movement Disorders 1990; 5(3):203–213.
342. Rougemont D, Baron JC et al. J Neurol Neurosurg Psychiatry 1984; 47(8):824–830.
343. Antonini A, Moeller JR et al. Neurology 1998; 51(3):803–810.
344. Arahata Y, Hirayama M et al. J Neurol Sci 1999; 163(2):119–126.
345. Berding G, Odin P et al. Mov Disord 2001; 16(6):1014–1022.
346. Hu MT, Taylor-Robinson SD et al. Brain 2000; 123(Pt 2):340–352.
347. Metter EJ, Kuhl DE et al. Advances in Neurology 1990; 53:135–139.
348. Wu JC, Iacono R et al. Neuroreport 2000; 11(10):2139–2144.
349. Bohnen NI, Minoshima S et al. Neurology 1999; 52(3):541–546.
350. Peppard RF, Martin WR et al. Arch Neurol 1992; 49(12):1262–1268.

351. Turjanski N, Brooks DJ. J Neural Transm Suppl 1997; 51:37–48.
352. Mentis MJ, McIntosh AR et al. Am J Psychiatry 2002; 159(5):746–754.
353. Playford ED, Jenkins IH et al. Ann Neurol 1992; 32(2):151–161.
354. Jahanshahi M, Jenkins IH et al. Brain 1995; 118(Pt 4):913–933.
355. Jenkins IH, Fernandez W et al. Ann Neurol 1992; 32(6):749–757.
356. Samuel M, Ceballos-Baumann AO et al. Neuroreport 2001; 12(4):821–828.
357. Thobois S, Dominey PF et al. Neurology 2000; 55(7):996–1002.
358. Samuel M, Ceballos-Baumann AO et al. Brain 1997; 120(Part 6):963–976.
359. Rascol O, Sabatini U et al. Brain 1997; 120(Part 1):103–110.
360. Catalan MJ, Ishii K et al. Brain 1999; 122 (Pt 3):483–495.
361. Albani G, Kunig G et al. Neurol Sci 2001; 22(1):43–44.
362. Brooks DJ. J Neural Transm 2001; 108(11):1283–1298.
363. Thiel A, Hilker R et al. J Cereb Blood Flow Metab 2001; 21 (Suppl. 1):S301.
364. Holthoff VA, Vieregge P et al. Ann Neurol 1994; 36(2):176–182.
365. Piccini P, Burn DJ et al. Ann Neurol 1999; 45(5):577–582.
366. Piccini P, Morrish PK et al. Ann Neurol 1997; 41(2):222–229.
367. Wu RM, Shan DE et al. Mov Disord 2002; 17(4):670–675.
368. Hilker R, Klein C et al. Annals of Neurology 2001; 49:367–376.
369. Khan NL, Valente EM et al. Ann Neurol 2002; 52(6):849–853.
370. Morrish PK, Rakshi JS et al. J Neurol Neurosurg Psychiatry 1998; 64(3):314–319.
371. Vingerhoets FJ, Snow BJ et al. Ann Neurol 1994; 36(5):759–764.
372. Nurmi E, Ruottinen HM et al. Ann Neurol 2000; 47(6):804–808.
373. Hilker R, Schweitzer K et al. European Journal of Neurology 2003; 10 (Suppl. 1):SC202.
374. Whone AL, Watts RL et al. Ann Neurol 2003; 54(1):93–101.
375. Hilker R, Schweitzer K et al. Movement Disorders 2002; 17:131.
376. Vingerhoets FJ, Snow BJ et al. Ann Neurol 1994; 36(5):765–770.
377. Antonini A, Schwarz J et al. Mov Disord 1997; 12(1):33–38.
378. Kaasinen V, Aalto S et al. J Neural Transm 2003; 110(6):591–601.
379. Mayberg HS, Starkstein SE et al. Ann Neurol 1990; 28(1):57–64.
380. Torstenson R, Hartvig P et al. Ann Neurol 1997; 41(3):334–340.
381. Hershey T, Black KJ et al. J Neurol Neurosurg Psychiatry 2003; 74(7):844.
382. Ruottinen HM, Rinne JO et al. Synapse 1997; 27(4):336–346.
383. Ruottinen HM, Rinne JO et al. Journal of Neural Transmission - Parkinsons Disease & Dementia Section 1995; 10(2–3):91–106.
384. Sawle GV, Burn DJ et al. Neurology 1994; 44(7):1292–1297.
385. Doudet DJ, Chan GL et al. Neuropharmacology 1997; 36(3):363–371.
386. Ceravolo R, Piccini P et al. Synapse 2002; 43(3):201–207.
387. Moresco RM, Volonte MA et al. J Neural Transm 2002; 109(10):1265–1274.
388. Deep P, Dagher A et al. Synapse 1999; 34(4):313–318.
389. Su PC, Ma Y et al. Ann Neurol 2001; 50(4):514–520.
390. Henselmans JM, de Jong BM et al. Clin Neurol Neurosurg 2000; 102(2):84–90.
391. Boecker H, Wills AJ et al. Ann Neurol 1997; 41(1):108–111.
392. Katayama Y, Tsubokawa T et al. Appl Neurophysiol 1986; 49(1–2):76–85.
393. Kazumata K, Antonini A et al. Neurology 1997; 49(4):1083–1090.
394. Eidelberg D, Moeller JR et al. Annals of Neurology 1996; 39(4):450–459.
395. Grafton ST, Waters C et al. Ann Neurol 1995; 37(6):776–783.
396. Samuel M, Ceballos-Baumann AO et al. Brain 1997; 120(Part 8):1301–1313.
397. Nakajima T, Nimura T et al. J Neurosurg 2003; 98(1):57–63.
398. Ghika J, Ghika-Schmid F et al. J Neurosurg 1999; 91(2):313–321.
399. Meissner W, Harnack D et al. Neuroscience Letters 2002; 328(2):105–108.
400. Strafella AP, Sadikot AF et al. Neuroreport 2003; 14(9):1287–1289.
401. Ceballos-Baumann AO, Boecker H et al. Arch Neurol 1999; 56(8):997–1003.
402. Strafella AP, Dagher A et al. Neurology 2003; 60(6):1039–1042.

403. Hilker R, Voges J et al. J Cereb Blood Flow Metab 2004; 24:7-16
404. Fukuda M, Mentis MJ et al. Brain 2001; 124(Pt 8):1601–1609.
405. Fukuda M, Mentis M et al. Ann Neurol 2001; 49(2):155–164.
406. Fukuda M, Ghilardi MF et al. Ann Neurol 2002; 52(2):144–152.
407. Hilker R, Voges J et al. Mov Disord 2003; 18(1):41–48.
408. Davis KD, Taub E et al. Nat Med 1997; 3(6):671–674.
409. Kordower JH, Emborg ME et al. Science 2000; 290(5492):767–773.
410. Bankiewicz KS, Eberling JL et al. Exp Neurol 2000; 164(1):2–14.
411. Subramanian T, Emerich DF et al. Cell Transplant 1997; 6(5):469–477.
412. Sawle GV, Bloomfield PM et al. Ann Neurol 1992; 31(2):166–173.
413. Widner H, Tetrud J et al. N Engl J Med 1992; 327(22):1556–1563.
414. Kordower JH, Freeman TB et al. N Engl J Med 1995; 332(17):1118–1124.
415. Iacono RP, Tang ZS et al. Stereotactic & Functional Neurosurgery 1992; 58:84–87.
416. Brundin P, Pogarell O et al. Brain 2000; 123(Pt 7):1380–1390.
417. Hagell P, Schrag A et al. Brain 1999; 122(Pt 6):1121–1132.
418. Freed CR, Greene PE et al. N Engl J Med 2001; 344(10):710–719.
419. Freeman TB, Olanow CW et al. Ann Neurol 1995; 38(3):379–388.
420. Piccini P, Brooks DJ et al. Nat Neurosci 1999; 2(12):1137–1140.
421. Piccini P, Lindvall O et al. Ann Neurol 2000; 48(5):689–695.
422. Cochen V, Ribeiro MJ et al. Mov Disord 2003; 18(8):928–932.
423. Cumming P, Danielsen EH et al. Acta Neurol Scand 2001; 103(5):309–315.
424. Brownell AL, Livni E et al. Ann Neurol 1998; 43(3):387–390.
425. Hu MT, Chaudhuri KR et al. Mov Disord 2002; 17(6):1321–1328.
426. Brooks DJ. Neurology 1993; 43(12 Suppl 6):S6–16.
427. Perani D, Bressi S et al. Arch Neurol 1995; 52(2):179–185.
428. Gilman S, Koeppe RA et al. Ann Neurol 1995; 38(2):176–185.
429. Rinne JO, Burn DJ et al. Ann Neurol 1995; 37(5):568–573.
430. Gilman S, Markel DS et al. Ann Neurol 1988; 23(3):223–230.
431. Gilman S, Koeppe RA et al. Ann Neurol 1994; 36(2):166–175.
432. Rosenthal G, Gilman S et al. Ann Neurol 1988; 24(3):414–419.
433. Kluin KJ, Gilman S et al. Ann Neurol 1988; 23(6):547–554.
434. Matthew E, Nordahl T et al. J Neurol Sci 1993; 119(2):134–140.
435. Yamaguchi S, Fukuyama H et al. J Neurol Sci 1994; 125(1):56–61.
436. Mishina M, Senda M et al. Acta Neurol Scand 1999; 100(6):369–376.
437. Brooks DJ, Ibanez V et al. Ann Neurol 1990; 28(4):547–555.
438. De Volder AG, Francart J et al. Annals of Neurology 1989; 26(2):239–247.
439. Eidelberg D, Moeller JR et al. J Cereb Blood Flow Metab 1994; 14(5):783–801.
440. Eidelberg D, Takikawa S et al. Annals of Neurology 1993; 33(5):518–527.
441. Dethy S, Van Blercom N et al. Mov Disord 1998; 13(2):275–280.
442. Otsuka M, Ichiya Y et al. J Neurol Sci 1996; 144(1–2):77–83.
443. Gilman S, Koeppe RA et al. Ann Neurol 1999; 45(6):769–777.
444. Brooks DJ, Ibanez V et al. Ann Neurol 1992; 31(2):184–192.
445. Shinotoh H, Inoue O et al. J Neurol Neurosurg Psychiatry 1993; 56(5):467–472.
446. Burn DJ, Rinne JO et al. Brain 1995; 118(Pt 4):951–958.
447. Schatz IJ. Ann Intern Med 1996; 125(1):74–75.
448. Bhatt MH, Snow BJ et al. Ann Neurol 1990; 28(1):101–103.
449. Ogawa M, Fukuyama H et al. J Neurol Sci 1998; 158(2):173–179.
450. Goldstein DS, Holmes C et al. N Engl J Med 1997; 336(10):696–702.
451. D'Antona R, Baron JC et al. Brain 1985; 108(Pt 3):785–799.
452. Foster NL, Gilman S et al. Ann Neurol 1988; 24(3):399–406.
453. Goffinet AM, De Volder AG et al. Ann Neurol 1989; 25(2):131–139.
454. Blin J, Baron JC et al. Arch Neurol 1990; 47(7):747–752.
455. Yamauchi H, Fukuyama H et al. Ann Neurol 1997; 41(5):606–614.

456. Garraux G, Salmon E et al. Mov Disord 2000; 15(5):894–904.
457. Salmon E, Vanderlinden M et al. Neuroimage 1997; 5(3):173–178.
458. Garraux G, Salmon E et al. Neuroimage 1999; 10(2):149–162.
459. Bhatt MH, Snow BJ et al. Arch Neurol 1991; 48(4):389–391.
460. Leenders KL, Frackowiak RS et al. Brain 1988; 111(Pt 3):615–630.
461. Cordes M, Snow BJ et al. Neuroradiology 1993; 35(6):404–409.
462. Hornykiewicz O, Shannak K. Journal of Neural Transmission 1994; Supplementum. 42:219–227.
463. Baron JC, Maziere B et al. J Cereb Blood Flow Metab 1986; 6(2):131–136.
464. Blin J, Mazetti P et al. Brain 1995; 118(Pt 6):1485–1495.
465. Otsuka M, Ichiya Y et al. Ann Nucl Med 1989; 3(3):111–118.
466. Santens P, De Reuck J et al. Eur Neurol 1997; 37(1):18–22.
467. Piccini P, de Yebenez J et al. Arch Neurol 2001; 58(11):1846–1851.
468. Foster NL, Gilman S et al. J Neurol Neurosurg Psychiatry 1992; 55(8):707–713.
469. Laureys S, Salmon E et al. J Neurol 1999; 246(12):1151–1158.
470. Sawle GV, Brooks DJ et al. Brain 1991; 114(Pt 1B):541–556.
471. Nagasawa H, Tanji H et al. Journal of the Neurological Sciences 1996; 139(2):210–217.
472. Frasson E, Moretto G et al. Ital J Neurol Sci 1998; 19(5):321–328.
473. Blin J, Vidailhet MJ et al. Movement Disorders 1992; 7(4):348–354.
474. Eidelberg D, Dhawan V et al. Journal of Neurology, Neurosurgery & Psychiatry 1991; 54(10):856–862.
475. Nagahama Y, Fukuyama H et al. Mov Disord 1997; 12(5):691–696.
476. Hayflick SJ, Westaway SK et al. N Engl J Med 2003; 348(1):33–40.
477. Castelnau P, Zilbovicius M et al. Pediatr Neurol 2001; 25(2):170–174.
478. Cooper GE, Rizzo M et al. Alzheimer Dis Assoc Disord 2000; 14(2):120–126.
479. Snow BJ, Bhatt M et al. J Neurol Neurosurg Psychiatry 1991; 54(1):12–17.
480. Westermark K, Tedroff J et al. Mov Disord 1995; 10(5):596–603.
481. Schlaug G, Hefter H et al. J Neurol Sci 1996; 136(1–2):129–139.
482. Hawkins RA, Mazziotta JC et al. Neurology 1987; 37(11):1707–1711.
483. Kuwert T, Hefter H et al. Eur J Nucl Med 1992; 19(2):96–101.
484. Schwarz J, Antonini A et al. Neurology 1994; 44(6):1079–1082.
485. Schlaug G, Hefter H et al. J Neurol 1994; 241(10):577–584.
486. Cordato DJ, Fulham MJ et al. Mov Disord 1998; 13(1):162–166.
487. De Volder A, Sindic CJ et al. J Neurol Neurosurg Psychiatry 1988; 51(7):947–949.
488. Tanner CM. Occupational Medicine 1992; 7(3):503–513.
489. Frey KA. Neurol Clin 2000; 18(3):615–630.
490. Herholz K. Der medizinische Sachverständige 2001; 97(5):181–184.
491. Yoshii F, Kozuma R et al. J Neurol Sci 1998; 160(1):87–91.
492. Rosenow F, Herholz K et al. Ann Neurol 1995; 38(5):825–828.
493. Pezzoli G, Antonini A et al. Mov Disord 1995; 10(3):279–282.
494. Hageman G, van der Hoek JAF et al. J Neurol 1999; 246:198–206.
495. Tetrud JW, Langston JW et al. Neurology 1994; 44(6):1051–1054.
496. Uitti RJ, Snow BJ et al. Ann Neurol 1994; 35(5):616–619.
497. Pal PK, Samii A et al. Neurotoxicology 1999; 20(2–3):227–238.
498. Wolters EC, Huang CC et al. Ann Neurol 1989; 26(5):647–651.
499. Shinotoh H, Snow BJ et al. Neurology 1997; 48(4):1053–1056.
500. Eriksson H, Tedroff J et al. Archives of Toxicology 1992; 66(6):403–407.
501. Ghaemi M, Rudolf J et al. J Neural Transm 2000; 107(11):1289–1295.
502. Picard F, Saint-Martin A et al. Mov Disord 1996; 11(5):567–570.
503. Caparros-Lefebvre D, Cabaret M et al. J Neural Transm 1998; 105(4–5):489–495.
504. Kew JJ, Brooks DJ et al. Neurology 1994; 44(6):1101–1110.
505. Tanaka M, Kondo S et al. J Neurol Sci 1993; 120(1):22–28.
506. Hatazawa J, Brooks RA et al. J Comput Assist Tomogr 1988; 12(4):630–636.

507. Dalakas MC, Hatazawa J et al. Ann Neurol 1987; 22(5):580–586.
508. Kew JJ, Goldstein LH et al. Brain 1993; 116(Pt 6):1399–1423.
509. Przedborski S, Dhawan V et al. Neurology 1996; 47(6):1546–1551.
510. Minami T, Otsuka M et al. Brain Dev 1994; 16(4):335–338.
511. Wszolek ZK, Pfeiffer RF et al. Annals of Neurology 1992; 32(3):312–320.
512. Brashear A, Mulholland GK et al. Mov Disord 1999; 14(1):132–137.
513. Mazziotta JC. Seminars in Neurology 1989; 9(4):360–369.
514. Otsuka M, Ichiya Y et al. J Neurol Sci 1993; 115(2):153–157.
515. Martin WR, Clark C et al. Neurology 1992; 42(1):223–229.
516. Kuwert T, Lange HW et al. Brain 1990; 113 (Pt 5):1405–1423.
517. De Volder A, Bol A et al. Brain Dev 1988; 10(1):47–50.
518. Hayden MR, Martin WR et al. Neurology 1986; 36(7):888–894.
519. Young AB, Penney JB et al. Ann Neurol 1986; 20(3):296–303.
520. Garnett ES, Firnau G et al. J Neurol Sci 1984; 65(2):231–237.
521. Kuhl DE, Phelps ME et al. Ann Neurol 1982; 12(5):425–434.
522. Dierks T, Linden DE et al. Psychiatry Res 1999; 90(1):67–75.
523. Matthews PM, Evans AC et al. Pediatr Neurol 1989; 5(6):353–356.
524. Grafton ST, Mazziotta JC et al. Arch Neurol 1992; 49(11):1161–1167.
525. Kuwert T, Noth J et al. Mov Disord 1993; 8(1):98–106.
526. Grafton ST, Mazziotta JC et al. Ann Neurol 1990; 28(5):614–621.
527. Hayden MR, Hewitt J et al. Neurology 1987; 37(9):1441–1447.
528. Mazziotta JC, Phelps ME et al. N Engl J Med 1987; 316(7):357–362.
529. Young AB, Penney JB et al. Arch Neurol 1987; 44(3):254–257.
530. Weeks RA, Ceballos-Baumann A et al. Brain 1997; 120 (Pt 9):1569–1578.
531. Ginovart N, Lundin A et al. Brain 1997; 120(Part 3):503–514.
532. Turjanski N, Weeks R et al. Brain 1995; 118(Pt 3):689–696.
533. Sedvall G, Karlsson P et al. Eur Arch Psychiatry Clin Neurosci 1994; 243(5):249–255.
534. Pavese N, Andrews TC et al. Brain 2003; 126(5):1127.
535. Andrews TC, Weeks RA et al. Brain 1999; 122(Pt 12):2353–2363.
536. Leenders KL, Frackowiak RS et al. Mov Disord 1986; 1(1):69–77.
537. Antonini A, Leenders KL et al. Ann Neurol 1998; 43(2):253–255.
538. Weeks RA, Piccini P et al. Ann Neurol 1996; 40(1):49–54.
539. Berent S, Giordani B et al. Ann Neurol 1988; 23(6):541–546.
540. Brandt J, Folstein SE et al. J Neuropsychiatry Clin Neurosci 1990; 2(1):20–27.
541. Weeks RA, Cunningham VJ et al. J Cereb Blood Flow Metab 1997; 17(9):943–949.
542. Bohnen NI, Koeppe RA et al. Neurology 2000; 54(9):1753–1759.
543. Holthoff VA, Koeppe RA et al. Ann Neurol 1993; 34(1):76–81.
544. Kunig G, Leenders KL et al. Ann Neurol 2000; 47(5):644–648.
545. Bachoud-Levi AC, Remy P et al. Lancet 2000; 356(9246):1975–1979.
546. Brownell AL, Hantraye P et al. Exp Neurol 1994; 125(1):41–51.
547. Burns LH, Pakzaban P et al. Neuroscience 1995; 64(4):1007–1017.
548. Hume SP, Lammertsma AA et al. J Neurosci Methods 1996; 67(2):103–112.
549. Goldman S, Amrom D et al. Mov Disord 1993; 8(3):355–358.
550. Sunden-Cullberg J, Tedroff J et al. Mov Disord 1998; 13(1):147–149.
551. Guttman M, Lang AE et al. Mov Disord 1987; 2(3):201–210.
552. Hosokawa S, Ichiya Y et al. J Neurol Neurosurg Psychiatry 1987; 50(10):1284–1287.
553. Tanaka M, Hirai S et al. Mov Disord 1998; 13(1):100–107.
554. Kuwert T, Lange HW et al. J Neurol 1990; 237(2):80–84.
555. Suchowersky O, Hayden MR et al. Mov Disord 1986; 1(1):33–44.
556. Turjanski N, Lees AJ et al. Neurology 1999; 52(5):932–937.
557. Hallett M, Dubinsky RM. J Neurol Sci 1993; 114(1):45–48.
558. Jenkins IH, Bain PG et al. Ann Neurol 1993; 34(1):82–90.
559. Wills AJ, Thompson PD et al. Neurology 1996; 46(3):747–752.

560. Wills AJ, Jenkins IH et al. Ann Neurol 1994; 36(4):636–642.
561. Boecker H, Wills AJ et al. Ann Neurol 1996; 39(5):650–658.
562. Ceballos-Baumann AO, Boecker H et al. Neurology 2001; 56(10):1347–1354.
563. Haslinger B, Boecker H et al. Neuroimage 2003; 18(2):517–524.
564. Braun AR, Stoetter B et al. Neuropsychopharmacology 1993; 9(4):277–291.
565. Braun AR, Randolph C et al. Neuropsychopharmacology 1995; 13(2):151–168.
566. Eidelberg D, Moeller JR et al. Neurology 1997; 48(4):927–934.
567. Jeffries KJ, Schooler C et al. Neuropsychopharmacology 2002; 27(1):92–104.
568. Wong DF, Singer HS et al. J Nucl Med 1997; 38(8):1243–1247.
569. Meyer P, Bohnen NI et al. Neurology 1999; 53(2):371–374.
570. Ernst M, Zametkin AJ et al. J Am Acad Child Adolesc Psychiatry 1999; 38(1):86–94.
571. Krause KH, Dresel S et al. J Neurol 2002; 249(8):1116–1118.
572. Singer HS, Szymanski S et al. Am J Psychiatry 2002; 159(8):1329–1336.
573. Stern E, Silberzweig DA et al. Arch Gen Psychiatry 2000; 57(8):741–748.
574. Eidelberg D, Moeller JR et al. Brain 1995; 118(Pt 6):1473–1484.
575. Galardi G, Perani D et al. Acta Neurol Scand 1996; 94(3):172–176.
576. Magyar-Lehmann S, Antonini A et al. Mov Disord 1997; 12(5):704–708.
577. Chase TN, Tamminga CA et al. Adv Neurol 1988; 50:237–241.
578. Eidelberg D, Moeller JR et al. Ann Neurol 1998; 44(3):303–312.
579. Gilman S, Junck L et al. Adv Neurol 1988; 50:231–236.
580. Karbe H, Holthoff VA et al. Neurology 1992; 42(8):1540–1544.
581. Perlmutter JS, Raichle ME. Ann Neurol 1984; 15(3):228–233.
582. Ceballos-Baumann AO, Passingham RE et al. Ann Neurol 1995; 37(3):363–372.
583. Playford ED, Passingham RE et al. Mov Disord 1998; 13(2):309–318.
584. Naumann M, Magyar-Lehmann S et al. Ann Neurol 2000; 47(3):322–328.
585. Kumar R, Dagher A et al. Neurology 1999; 53(4):871–874.
586. Ceballos-Baumann AO, Brooks DJ. Adv Neurol 1998; 78:135–152.
587. Perlmutter JS, Stambuk MK et al. J Neurosci 1997; 17(2):843–850.
588. Dhawan V, Ma Y et al. J Cereb Blood Flow Metab 2003; 23 (Suppl. 1):S643.
589. Leenders K, Hartvig P et al. J Neural Transm Park Dis Dement Sect 1993; 5(2):79–87.
590. Jarman PR, Bhatia KP et al. Mov Disord 2000; 15(4):648–657.
591. Calne DB, Fuente-Fernandez R et al. J Neural Transm Suppl 1997; 50:47–52.
592. Turjanski N, Bhatia K et al. Neurology 1993; 43(8):1563–1568.
593. Otsuka M, Ichiya Y et al. J Neurol Sci 1992; 111(2):195–199.
594. Takahashi H, Levine RA et al. Ann Neurol 1994; 35(3):354–356.
595. Snow BJ, Nygaard TG et al. Ann Neurol 1993; 34(5):733–738.
596. Nygaard TG, Takahashi H et al. Ann Neurol 1992; 32(5):603–608.
597. Sawle GV, Leenders KL et al. Ann Neurol 1991; 30(1):24–30.
598. Fuente-Fernandez R, Furtado S et al. Synapse 2003; 49(1):20–28.
599. Kishore A, Nygaard TG et al. Neurology 1998; 50(4):1028–1032.
600. Kunig G, Leenders KL et al. Ann Neurol 1998; 44(5):758–762.
601. Lang AE. Mov Disord 1995; 10(4):489–495.
602. Przedborski S, Goldman S et al. Mov Disord 1993; 8(3):391–395.
603. Vidailhet M, Dupel C et al. Arch Neurol 1999; 56(8):982–989.
604. Wong DF, Harris JC et al. Proc Natl Acad Sci U S A 1996; 93(11):5539–5543.
605. Gilman S, Sima AAF et al. Annals of Neurology 1996; 39(2):241–255.
606. Gilman S, Heumann M et al.In: Brain mapping: The disorders 2000;(17):417–455.
607. Soong BW, Liu RS. J Neurol Neurosurg Psychiatry 1998; 64(4):499–504.
608. Taniwaki T, Sakai T et al. J Neurol Sci 1997; 145(1):63–67.
609. Shinotoh H, Thiessen B et al. Neurology 1997; 49(4):1133–1136.
610. Furtado S, Farrer M et al. Neurology 2002; 59(10):1625–1627.
611. Gilman S, Junck L et al. Annals of Neurology 1990; 28(6):750–757.
612. Anderson NE, Posner JB et al. Ann Neurol 1988; 23(6):533–540.

613. Gilman S, Adams K et al. Annals of Neurology 1990; 28(6):775-785.
614. Kleihues P, Burger PC et al. Histological typing of tumors of the central nervous system. Springer, Berlin, 1993.
615. Paulus W, Peiffer J. Cancer 1989; 64(2):442-447.
616. Warburg O. Science 1956; 123:309-314.
617. Reske SN, Kotzerke J. Eur J Nucl Med 2001; 28(11):1707-1723.
618. Rottenberg DA, Strother SC et al. J Nucl Med 1989; 30(4):564.
619. Di Chiro G, Brooks RA. J Nucl Med 1988; 29(9):1603-1604.
620. Herholz K, Heindel W et al. Annals of Neurology 1992; 31(3):319-327.
621. Spence AM, Muzi M et al. J Nucl Med 1998; 39(3):440-448.
622. Tyler JL, Diksic M et al. J Nucl Med 1987; 28(7):1123-1133.
623. Kapoor R, Spence AM et al. Journal of Neurochemistry 1989; 53(1):37-44.
624. Hawkins RA, Phelps ME et al. J Cereb Blood Flow Metab 1986; 6(2):170-183.
625. Meyer PT, Schreckenberger M et al. Eur J Nucl Med 2001; 28(2):165-174.
626. Kim CK, Alavi JB et al. Journal of Neuro-Oncology 1991; 10(1):85-91.
627. Alavi JB, Alavi A et al. Cancer 1988; 62(6):1074-1078.
628. Di Chiro G, DeLaPaz RL et al. Neurology 1982; 32(12):1323-1329.
629. Barker FG, Chang SM et al. Cancer 1997; 79(1):115-126.
630. Patronas NJ, Di Chiro G et al. Journal of Neurosurgery 1985; 62(6):816-822.
631. Delbeke D, Meyerowitz C et al. Radiology 1995; 195(1):47-52.
632. Holthoff VA, Herholz K et al. Cancer 1993; 72(4):1394-1403.
633. Okada J, Yoshikawa K et al. J Nucl Med 1991; 32(4):686-691.
634. Basso U, Brandes AA. Eur J Cancer 2002; 38(10):1298-1312.
635. Hoffman JM, Waskin HA et al. J Nucl Med 1993; 34(4):567-575.
636. Rohren EM, Provenzale JM et al. Radiology 2003; 226(1):181-187.
637. Di Chiro G, Hatazawa J et al. Radiology 1987; 164(2):521-526.
638. Borbely K, Fulham MJ et al. J Nucl Med 1992; 33(11):1931-1934.
639. De Souza B, Brunetti A et al. Radiology 1990; 177(1):39-44.
640. Komori T, Martin WH et al. Clin Nucl Med 2002; 27(3):176-178.
641. Daemen BJ, Zwertbroek R et al. Eur J Nucl Med 1991; 18(7):453-460.
642. Wilkinson MD, Fulham MJ et al. J Comput Assist Tomogr 2003; 27(1):26-29.
643. Roelcke U, Radu EW et al. J Neurooncol 1998; 36(3):279-283.
644. Fulham MJ, Melisi JW et al. Radiology 1993; 189(1):221-225.
645. DeLaPaz RL, Patronas NJ et al. Ajnr: American Journal of Neuroradiology 1983; 4(3):826-829.
646. Hölzer T, Herholz K et al. Journal of Computer Assisted Tomography 1993; 17(5):681-687.
647. Roelcke U, Blasberg RG et al. J Nucl Med 1998; 39(5):879-884.
648. Fulham MJ, Brunetti A et al. Journal of Neurosurgery 1995; 83(4):657-664.
649. Ishizu K, Nishizawa S et al. J Nucl Med 1994; 35(7):1104-1109.
650. Kaschten B, Stevenaert A et al. J Nucl Med 1998; 39(5):778-785.
651. Chung JK, Kim YK et al. Eur J Nucl Med Mol Imaging 2002; 29(2):176-182.
652. Ogawa T, Inugami A et al. Ajnr: American Journal of Neuroradiology 1996; 17(2):345-353.
653. Bustany P, Chatel M et al. Journal of Neuro-Oncology 1986; 3(4):397-404.
654. Sasaki M, Kuwabara Y et al. Eur J Nucl Med 1998; 25(9):1261-1269.
655. Herholz K, Hölzer T et al. Neurology 1998; 50(5):1316-1322.
656. De Witte O, Goldberg I et al. J Neurosurg 2001; 95(5):746-750.
657. Kracht LW, Friese M et al. Eur J Nucl Med Mol Imaging 2003; 30:868-873.
658. Ericson K, Lilja A et al. J Comput Assist Tomogr 1985; 9(4):683-689.
659. Ogawa T, Shishido F et al. Radiology 1993; 186(1):45-53.
660. Derlon JM, Petit-Taboue MC et al. Neurosurgery 1997; 40(2):276-288.
661. Ogawa T, Kanno I et al. Radiographics 1994; 14(1):101-110.
662. Sawataishi J, Mineura K et al. Neuroradiology 1992; 34(6):517-519.
663. Padma MV, Jacobs M et al. J Neurooncol 2001; 55(1):39-44.

664. Nyberg G, Bergstrom M et al. Acta Otolaryngol 1997; 117(4):482–489.
665. Kracht L-W, Bauer A et al. Journal of Computer Assisted Tomography 1999; 23:365–368.
666. Nuutinen J, Sonninen P et al. Int J Radiat Oncol Biol Phys 2000; 48(1):43–52.
667. Utriainen M, Metsahonkala L et al. Cancer 2002; 95(6):1376–1386.
668. Ribom D, Eriksson A et al. Cancer 2001; 92(6):1541–1549.
669. Sato N, Inoue T et al. Neuroradiology 2003; epub.
670. Mineura K, Sasajima T et al. J Nucl Med 1991; 32(4):726–728.
671. Bergstrom M, Muhr C et al. J Comput Assist Tomogr 1987; 11(3):384–389.
672. Bergstrom M, Muhr C et al. Neuroradiology 1987; 29(3):221–225.
673. Mankoff DA, Shields AF et al. J Nucl Med 1999; 40(4):614–624.
674. Shields AF, Grierson JR et al. Nat Med 1998; 4(11):1334–1336.
675. Vesselle H, Grierson J et al. Clin Cancer Res 2002; 8(11):3315–3323.
676. Blasberg RG, Roelcke U et al. Cancer Res 2000; 60(3):624–635.
677. Shapiro WR. Annals of Neurology 1992; 31(4):437–438.
678. Jackson RJ, Fuller GN et al. Neuro -oncol 2001; 3(3):193–200.
679. Hanson MW, Glantz MJ et al. J Comput Assist Tomogr 1991; 15(5):796–801.
680. Levivier M, Goldman S et al. Journal of Neurosurgery 1995; 82(3):445–452.
681. Goldman S, Levivier M et al. J Nucl Med 1997; 38(9):1459–1462.
682. Pirotte B, Goldman S et al. Pediatr Neurosurg 2003; 38(3):146–155.
683. Di Chiro G. Investigative Radiology 1987; 22:360–371.
684. Francavilla TL, Miletich RS et al. Neurosurgery 1989; 24(1):1–5.
685. Forsyth PA, Kelly PJ et al. J Neurosurg 1995; 82(3):436–444.
686. Bryan RN. AJNR Am J Neuroradiol 1998; 19(3):590.
687. Schiffer D, Giordana MT et al. Acta Neuropathol (Berl) 1982; 58(4):291–299.
688. Patronas NJ, Di Chiro G et al. Radiology 1982; 144(4):885–889.
689. Di Chiro G, Oldfield E et al. AJR Am J Roentgenol 1988; 150(1):189–197.
690. Doyle WK, Budinger TF et al. Journal of Computer Assisted Tomography 1987; 11(4):563–570.
691. Levivier M, Becerra A et al. Journal of Neurosurgery 1996; 84(1):148–149.
692. Reinhardt MJ, Kubota K et al. J Nucl Med 1997; 38(2):280–287.
693. Kracht LW, Miletic H et al. Eur J Nucl Med 2003; 30 (Suppl.2):190.
694. Tsuyuguchi N, Sunada I et al. J Neurosurg 2003; 98(5):1056–1064.
695. Ogawa T, Kanno I et al. Acta Radiologica 1991; 32(3):197–202.
696. Thiel A, Pietrzyk U et al. Neurosurgery 2000; 46:232–234.
697. Tralins KS, Douglas JG et al. J Nucl Med 2002; 43(12):1667–1673.
698. Jacobs AH, Dittmar C et al. J Cereb Blood Flow Metab 2003; 23 (Suppl. 1):376.
699. Takeda N, Diksic M et al. Cancer 1996; 77(6):1167–1179.
700. Brock CS, Young H et al. Br J Cancer 2000; 82(3):608–615.
701. Yamamoto T, Nishizawa S et al. Ann Nucl Med 2001; 15(2):103–109.
702. De Witte O, Hildebrand J et al. Cancer 1994; 74(10):2836–2842.
703. Rozental JM, Levine RL et al. Arch Neurol 1989; 46(12):1302–1307.
704. Daemen BJ, Elsinga PH et al. J Nucl Med 1992; 33(3):373–379.
705. Würker M, Herholz K et al. Eur J Nucl Med 1996; 23(5):583–586.
706. Ribom D, Engler H et al. Eur J Nucl Med 2002; 29(5):632–640.
707. Heesters MA, Go KG et al. Neuroradiology 1998; 40(2):103–108.
708. Francavilla TL, Miletich RS et al. Neurosurgery 1991; 28(6):826–833.
709. Bergstrom M, Muhr C et al. J Nucl Med 1991; 32(4):610–615.
710. Bergstrom M, Muhr C et al. J Comput Assist Tomogr 1987; 11(5):815–819.
711. Parodi K, Enghardt W et al. Phys Med Biol 2002; 47(1):21–36.
712. Enghardt W, Debus J et al. Strahlenther Onkol 1999; 175 Suppl 2:33–36.
713. Rhodes CG, Wise RJ et al. Annals of Neurology 1983; 14(6):614–626.
714. Brooks DJ, Beaney RP et al. J Cereb Blood Flow Metab 1986; 6(5):529–535.
715. Rottenberg DA, Ginos JZ et al. Annals of Neurology 1985; 17(1):70–79.
716. Horowitz M, Blasberg R et al. Cancer Res 1983; 43(8):3800–3807.

717. Valk PE, Mathis CA et al. J Nucl Med 1992; 33(12):2133–2137.
718. Yamamoto F, Aoki M et al. Biol Pharm Bull 2002; 25(5):616–621.
719. Ziemer S, Evans M et al. Eur J Nucl Med Mol Imaging 2003; 30(2):259–266.
720. Braun V, Dempf S et al. Neurosurgery 2001; 48(5):1178–1181.
721. Lammertsma AA, Wise RJ et al. Br J Radiol 1985; 58(692):725–734.
722. Blasberg RG, Kobayashi T et al. Annals of Neurology 1983; 14(2):189–201.
723. Iannotti F, Fieschi C et al. J Comput Assist Tomogr 1987; 11(3):390–397.
724. Hawkins RA, Phelps ME et al. J Cereb Blood Flow Metab 1984; 4(4):507–515.
725. Yen CK, Yano Y et al. J Nucl Med 1982; 23(6):532–537.
726. Roelcke U, Radu EW et al. Journal of the Neurological Sciences 1995; 132(1):20–27.
727. Jarden JO. Acta Neurologica Scandinavica 1994; Supplementum. 156:1–35.
728. Jarden JO, Dhawan V et al. Annals of Neurology 1985; 18(6):636–646.
729. Kessler RM, Goble JC et al. J Cereb Blood Flow Metab 1984; 4(3):323–328.
730. Zunkeler B, Carson RE et al. Journal of Neurosurgery 1996; 84(3):494–502.
731. Zunkeler B, Carson RE et al. Journal of Neurosurgery 1996; 85(6):1056–1065.
732. Mitsuki S, Diksic M et al. J Neurooncol 1991; 10(1):47–55.
733. Diksic M, Sako K et al. Cancer Res 1984; 44(7):3120–3124.
734. Meikle SR, Matthews JC et al. Cancer Chemother Pharmacol 1998; 42(3):183–193.
735. Tyler JL, Yamamoto YL et al. J Nucl Med 1986; 27(6):775–780.
736. Saleem A, Brown GD et al. Cancer Res 2003; 63(10):2409.
737. Hendrikse NH, Vaalburg W. Methods 2002; 27(3):228–233.
738. Levchenko A, Mehta BM et al. J Nucl Med 2000; 41(3):493–501.
739. Hendrikse NH, Franssen EJ et al. Eur J Nucl Med 1999; 26(3):283–293.
740. Kurdziel KA, Kiesewetter DO et al. J Nucl Med 2003; 44(8):1330–1339.
741. Levivier M, Wikler D, Jr. et al. J Neurosurg 2002; 97(5 Suppl):542–550.
742. Levivier M, Wikier D et al. J Neurosurg 2000; 93 Suppl 3:233–238.
743. Gross MW, Weber WA et al. Int J Radiat Oncol Biol Phys 1998; 41(5):989–995.
744. Imahori Y, Ueda S et al. Clin Cancer Res 1998; 4(8):1825–1832.
745. Imahori Y, Ueda S et al. Clin Cancer Res 1998; 4(8):1833–1841.
746. Voges J, Reszka R et al. Ann Neurol 2003; 54(4):479–487.
747. Tjuvajev JG, Chen SH et al. Cancer Res 1999; 59(20):5186–5193.
748. Jacobs A, Tjuvajev JG et al. Cancer Res 2001; 61(7):2983–2995.
749. Jacobs A, Voges J et al. Lancet 2001; 358(9283):727–729.
750. Blasberg R. Eur J Cancer 2002; 38(16):2137–2146.
751. Jayson GC, Zweit J et al. J Natl Cancer Inst 2002; 94(19):1484–1493.
752. Herbst RS, Mullani NA et al. J Clin Oncol 2002; 20(18):3804–3814.
753. Ben David I, Rozen Y et al. Appl Radiat Isot 2003; 58(2):209–217.
754. Brunetti A, Blasberg RG et al. International Journal of Radiation Applications & Instrumentation – Part B, Nuclear Medicine & Biology 1988; 15(6):665–672.
755. Mintun MA, Dennis DR et al. J Nucl Med 1987; 28(11):1704–1716.
756. Aloj L, Jogoda E et al. J Nucl Med 1999; 40(9):1547–1555.
757. Li JY, Boado RJ et al. J Cereb Blood Flow Metab 2001; 21(1):61–68.
758. Shinoura N, Nishijima M et al. Radiology 1997; 202(2):497–503.
759. DeGrado TR, Baldwin SW et al. J Nucl Med 2001; 42(12):1805–1814.
760. Guhlke S, Wester HJ et al. Nucl Med Biol 1994; 21(6):819–825.
761. Smith-Jones PM, Stolz B et al. J Nucl Med 1994; 35(2):317–325.
762. Wester HJ, Brockmann J et al. Nucl Med Biol 1997; 24(4):275–286.
763. Henze M, Schuhmacher J et al. J Nucl Med 2001; 42(7):1053–1056.
764. Jamar F, Barone R et al. Eur J Nucl Med Mol Imaging 2003.
765. Blankenstein MA, Verheijen FM et al. Steroids 2000; 65(10–11):795–800.
766. Katzenellenbogen JA, Welch MJ et al. Anticancer Res 1997; 17(3B):1573–1576.
767. Bergstrom M, Muhr C et al. Neurosurgery 1992; 30(6):855–861.
768. Muhr C, Bergstrom M et al. Acta Radiol Suppl 1986; 369:406–408.

769. Lucignani G, Losa M et al. Eur J Nucl Med 1997; 24(9):1149–1155.
770. Muhr C, Bergstrom M et al. J Comput Assist Tomogr 1986; 10(2):175–180.
771. Ahmed AR, Watanabe H et al. Eur J Nucl Med 2001; 28(10):1541–1551.
772. Shah N, Sibtain A et al. J Comput Assist Tomogr 2000; 24(1):55–56.
773. Hirai K, Umesaki N et al. Oncol Rep 2001; 8(4):773–775.
774. Cardona S, Schwarzbach M et al. Eur J Surg Oncol 2003; 29(6):536–541.
775. Shulkin BL, Wieland DM et al. J Nucl Med 1992; 33(6):1125–1131.
776. Gourgiotis L, Sarlis NJ et al. J Clin Endocrinol Metab 2003; 88(2):637–641.
777. Hwang JJ, Uchio EM et al. J Urol 2003; 169(1):274–275.
778. Pacak K, Eisenhofer G et al. Ann N Y Acad Sci 2002; 970:170–176.
779. Berry CR, DeGrado TR et al. Vet Radiol Ultrasound 2002; 43(2):183–186.
780. Taniguchi K, Ishizu K et al. Eur J Surg 2001; 167(11):866–870.
781. Shulkin BL, Thompson NW et al. Radiology 1999; 212(1):35–41.
782. Hoegerle S, Nitzsche E et al. Radiology 2002; 222(2):507–512.
783. Neumann DR, Basile KE et al. J Comput Assist Tomogr 1996; 20(2):312–316.
784. Vaidyanathan G, Affleck DJ et al. Journal of Medicinal Chemistry 1994; 37(21):3655–3662.
785. Vaidyanathan G, Affleck DJ et al. J Nucl Med 1995; 36(4):644–650.
786. Higano S, Shishido F et al. Journal of Computer Assisted Tomography 1990; 14(2):297–299.
787. Giannopoulou C. Eur J Nucl Med Mol Imaging 2003; 30(3):333–338.
788. Antoine JC, Cinotti L et al. Ann Neurol 2000; 48(1):105–108.
789. Rees JH, Hain SF et al. Brain 2001; 124(Pt 11):2223–2231.
790. Shinohara Y, Ohnuki Y et al. Ann Neurol 1998; 43(5):684.
791. Klee B, Law I et al. Eur J Neurol 2002; 9(6):657–662.
792. Ojemann JG, Miller JW et al. Neurosurgery 1996; 39(2):253–258.
793. Krings T, Schreckenberger M et al. J Neurol Neurosurg Psychiatry 2001; 71(6):762–771.
794. Kaplan AM, Bandy DJ et al. J Neurosurg 1999; 91(5):797–803.
795. Herholz K, Reulen HJ et al. Neurosurgery 1997; 41:1253–1262.
796. Vinas FC, Zamorano L et al. Neurol Res 1997; 19(6):601–608.
797. Chang JY, Duara R et al. J Nucl Med 1987; 28(5):852–860.
798. Steinmetz H, Huang Y et al. J Cereb Blood Flow Metab 1992; 12(6):919–926.
799. Pietrzyk U, Herholz K et al. J Nucl Med 1994; 35(12):2011–2018.
800. Pietrzyk U, Herholz K et al. Journal of Computer Assisted Tomography 1990; 14(1):51–59.
801. Schreckenberger M, Spetzger U et al. Surg Neurol 1999; 52(1):24–29.
802. Ojemann JG, Neil JM et al. J Cereb Blood Flow Metab 1998; 18(2):148–153.
803. Herholz K, Thiel A et al. Neuroimage 1999; 9 (Supplement):S569.
804. Thiel A, Herholz K et al. Annals of Neurology 2001; 50:620–629.
805. Seitz RJ, Huang Y et al. Neuroreport 1995; 6(5):742–744.
806. Müller RA, Rothermel RD et al. Neuroreport 1997; 8(14):3103–3108.
807. Knorr U, Weder B et al. Journal of Computer Assisted Tomography 1993; 17(4):517–528.
808. Walter H, Kristeva R et al. Brain Topography 1992; 5(2):183–187.
809. Seitz RJ, Schlaug G et al. Advances in Neurology 1996; 70:167–175.
810. Müller RA, Rothermel RD et al. J Child Neurol 1998; 13(11):550–556.
811. Wunderlich G, Knorr U et al. Neurosurgery 1998; 42(1):18–26.
812. Herholz K, Heiss W-D. J Cereb Blood Flow Metab 2000; 20:1619–1631.
813. Thiel A, Herholz K et al. Neuroimage 1998; 7:284–295.
814. Ojemann G, Ojemann J et al. Journal of Neurosurgery 1989; 71:316–326.
815. Herholz K, Thiel A et al. Neuroimage 1996; 3(3):185–194.
816. Xiong J, Rao S et al. Neuroimage 2000; 12(3):326–339.
817. Hunter KE, Blaxton TA et al. Ann Neurol 1999; 45(5):662–665.
818. Müller RA, Rothermel RD et al. Journal of Epilepsy 1998; 11:152–161.
819. Pardo JV, Fox PT. Human Brain Mapping 1993; 1:57–68.
820. Bookheimer SY, Zeffiro TA et al. Neurology 1997; 48(4):1056–1065.
821. Nariai T, Senda M et al. J Nucl Med 1997; 38(10):1563–1568.

822. Papathanassiou D, Etard O et al. Neuroimage 2000; 11(4):347–357.
823. Karbe H, Herholz K et al. Brain & Language 1998; 63:108–121.
824. Buckner RL, Corbetta M et al. Proc Natl Acad Sci USA 1996; 93(3):1249–1253.
825. Leblanc R, Meyer E et al. Neurosurgery 1992; 31(2):369–373.
826. Bookheimer SY, Zeffiro TA et al. Neurology 2000; 55(8):1151–1157.
827. Herholz K, Thiel A et al. In: Quantitative functional brain imaging with positron emission tomography 1998;(23):159–163.
828. Müller RA, Behen ME et al. Prog Neuropsychopharmacol Biol Psychiatry 1999; 23(4):657–668.
829. Tierney MC, Varga M et al. Neuropsychologia 2001; 39:114–121.
830. Müller RA, Rothermel RD et al. Neuropsychologia 1999; 37(5):545–557.
831. Miyamoto RT, Wong D. J Commun Disord 2001; 34(6):473–478.
832. Fujiki N, Naito Y et al. Auris Nasus Larynx 1999; 26(3):229–236.
833. Giraud AL, Price CJ et al. Brain 2001; 124(Pt 7):1307–1316.
834. Chao ST, Suh JH et al. Int J Cancer 2001; 96(3):191–197.
835. Thompson TP, Lunsford LD et al. Stereotact Funct Neurosurg 1999; 73(1–4):9–14.
836. Ricci PE, Karis JP et al. AJNR Am J Neuroradiol 1998; 19(3):407–413.
837. Valk PE, Budinger TF et al. J Neurosurg 1988; 69(6):830–838.
838. Kahn D, Follett KA et al. AJR 1994; 163(6):1459–1465.
839. Ericson K, Kihlstrom L et al. Stereotact Funct Neurosurg 1996; 66 Suppl 1:214–224.
840. Belohlavek O, Simonova G et al. Eur J Nucl Med Mol Imaging 2003; 30(1):96–100.
841. Sonoda Y, Kumabe T et al. Neurol Med Chir (Tokyo) 1998; 38(6):342–347.
842. Herholz K, Hölzer T et al. Journal of Neuro-Oncology 1997; 35 (Suppl.1):S20.
843. Heiss WD. Stroke 1992; 23(11):1668–1672.
844. Heiss WD, Graf R. Curr Opin Neurol 1994; 7(1):11–19.
845. Lyden PD, Grotta JC et al. Neurology 1997; 49(1):14–20.
846. Baron JC, Delattre JY et al. Ajnr: American Journal of Neuroradiology 1983; 4(3):536–540.
847. Powers WJ, Grubb RL, Jr. et al. J Cereb Blood Flow Metab 1985; 5(4):600–608.
848. Marchal G, Beaudouin V et al. Stroke 1996; 27(4):599–606.
849. Furlan M, Marchal G et al. Annals of Neurology 1996; 40:216–226.
850. Heiss WD, Graf R et al. J Cereb Blood Flow Metab 1994; 14(6):892–902.
851. Sakoh M, Ostergaard L et al. J Neurosurg 2000; 93(4):647–657.
852. Baron JC, Bousser MG et al. Stroke 1981; 12(4):454–459.
853. Wise RJ, Bernardi S et al. Brain 1983; 106(Pt 1):197–222.
854. Heiss WD, Huber M et al. J Cereb Blood Flow Metab 1992; 12(2):193–203.
855. De Ley G, Weyne J et al. J Cereb Blood Flow Metab 1988; 8(4):539–545.
856. Heiss WD, Graf R et al. J Cereb Blood Flow Metab 1997; 17(4):388–400.
857. Hakim AM, Evans AC et al. J Cereb Blood Flow Metab 1989; 9(4):523–534.
858. Sette G, Baron JC et al. Stroke 1993; 24(12):2046–2057.
859. Heiss WD, Graf R et al. Stroke 1997; 28(10):2045–2051.
860. Heiss WD, Grond M et al. Stroke 1998; 29(2):454–461.
861. Heiss WD, Kracht LW et al. Brain 2001; 124(Pt 1):20–29.
862. Markus R, Donnan GA et al. Neuroimage 2002; 16(2):425–433.
863. Read SJ, Hirano T et al. Neurology 1998; 51(6):1617–1621.
864. Read SJ, Hirano T et al. Ann Neurol 2000; 48(2):228–235.
865. Ackerman RH, Correia JA et al. Arch Neurol 1981; 38(9):537–543.
866. Hakim AM, Pokrupa RP et al. Ann Neurol 1987; 21(3):279–289.
867. Marchal G, Young AR et al. J Cereb Blood Flow Metab 1999; 19(5):467–482.
868. Marchal G, Serrati C et al. Lancet 1993; 341:925–927.
869. Fink GR, Herholz K et al. Journal of Stroke and Cerebrovascular Disease 1993; 3:123–131.
870. Lassen NA. Lancet 1966; II:1113–1115.
871. Gerhard A, Neumaier B et al. Neuroreport 2000; 11(13):2957–2960.
872. Kuhl DE, Phelps ME et al. Ann Neurol 1980; 8(1):47–60.

873. Kushner M, Reivich M et al. Neurology 1987; 37(7):1103–1110.
874. Baron JC, Rougemont D et al. J Cereb Blood Flow Metab 1984; 4(2):140–149.
875. Wise RJ, Rhodes CG et al. Ann Neurol 1983; 14(6):627–637.
876. Senda M, Alpert NM et al. J Cereb Blood Flow Metab 1989; 9(6):859–873.
877. Syrota A, Castaing M et al. Ann Neurol 1983; 14(4):419–428.
878. Syrota A, Samson Y et al. J Cereb Blood Flow Metab 1985; 5(3):358–368.
879. Heiss WD, Herholz K et al. J Comput Assist Tomogr 1986; 10(6):903–911.
880. Uemura K, Shishido F et al. Acta Radiol Suppl 1986; 369:426–428.
881. Ishikawa T, Kawamura S et al. J Neurol Neurosurg Psychiatry 1992; 55(5):401–403.
882. Ishikawa T, Kawamura S et al. Surg Neurol 1995; 43(2):130–136.
883. Zazulia AR, Diringer MN et al. J Cereb Blood Flow Metab 2001; 21(7):804–810.
884. Dethy S, Goldman S et al. J Nucl Med 1994; 35(7):1162–1166.
885. Ogawa T, Hatazawa J et al. J Nucl Med 1995; 36(12):2175–2179.
886. Hino A, Mizukawa N et al. Stroke 1989; 20(11):1504–1510.
887. De Ley G, Eechaute W et al. Stroke 1993; 24(3):400–405.
888. Carpenter DA, Grubb RL, Jr. et al. J Cereb Blood Flow Metab 1991; 11(5):837–844.
889. Hayashi T, Suzuki A et al. J Neurosurg 2000; 93(6):1014–1018.
890. Feeney DM, Baron JC. Stroke 1986; 17(5):817–830.
891. Baron JC, Bousser MG et al. Transactions of the American Neurological Association 1980; 105:459–461.
892. De Reuck J, Decoo D et al. Clin Neurol Neurosurg 1997; 99(1):11–16.
893. Jacobs A, Herholz K et al. J Neurol 1996; 243(2):131–136.
894. Kushner M, Alavi A et al. Ann Neurol 1984; 15(5):425–434.
895. Miura H, Nagata K et al. J Neuroimaging 1994; 4(2):91–96.
896. Yamauchi H, Fukuyama H et al. Stroke 1999; 30(7):1429–1431.
897. Fulham MJ, Brooks RA et al. Neurology 1992; 42(12):2267–2273.
898. Shamoto H, Chugani HT. J Child Neurol 1997; 12(7):407–414.
899. Ito H, Kanno I et al. Ann Nucl Med 2002; 16(4):249–254.
900. Brunberg JA, Frey KA et al. Ajnr: American Journal of Neuroradiology 1992; 13:58–61.
901. Dettmers C, Hartmann A et al. Neurological Research 1995; 17(2):109–112.
902. Pawlik G, Herholz K et al.In: Functional mapping of the brain in vascular disorders 1985; 59–84.
903. Pantano P, Formisano R et al. Cerebrovascular Diseases 1993; 3(2):80–85.
904. Pantano P, Baron JC et al. Brain 1986; 109(Pt 4):677–694.
905. Tanaka M, Kondo S et al. J Neurol Neurosurg Psychiatry 1992; 55(2):121–125.
906. Tecco JM, Wuilmart P et al. J Neuroimaging 1998; 8(2):115–116.
907. Niimura K, Chugani DC et al. Neurology 1999; 52(4):792–797.
908. Baron JC, D'Antona R et al. Brain 1986; 109(Pt 6):1243–1259.
909. Chabriat H, Pappata S et al. Acta Neurol Scand 1992; 86(3):285–290.
910. Baron JC, Levasseur M et al. J Neurol Neurosurg Psychiatry 1992; 55(10):935–942.
911. Pappata S, Mazoyer B et al. Stroke 1990; 21(4):519–524.
912. Iglesias S, Marchal G et al. Cerebrovasc Dis 2000; 10(5):391–402.
913. Dobkin JA, Levine RL et al. Arch Neurol 1989; 46(12):1333–1336.
914. Lagreze HL, Levine RL et al. Stroke 1987; 18(5):882–886.
915. Shishido F, Uemura K et al. Radiat Med 1987; 5(2):36–41.
916. Pawlik G, Wienhard K et al. Prog Brain Res 1984; 62:253–262.
917. Cappa SF, Perani D et al. Brain & Language 1997; 56(1):55–67.
918. Seitz RJ, Azari NP et al. Stroke 1999; 30(9):1844–1850.
919. Baron JC. Semin Neurol 1989; 9(4):281–285.
920. Fink GR, Pietrzyk U et al. Neurol Res 1992; 14(2 Suppl):139–142.
921. Powers WJ, Press GA et al. Ann Intern Med 1987; 106(1):27–34.
922. Gibbs JM, Wise RJS et al. Lancet 1984; I:310–314.

923. Yamauchi H, Fukuyama H et al. Journal of Neurology, Neurosurgery & Psychiatry 1996; 61(1):18–25.
924. Pozzilli C, Itoh M et al. J Cereb Blood Flow Metab 1987; 7(2):137–142.
925. Powers WJ, Derdeyn CP et al. Neurology 2000; 54(4):878–882.
926. Derdeyn CP, Videen TO et al. Brain 2002; 125(Pt 3):595–607.
927. Kawamura S, Sayama I et al. Acta Neurochir (Wien) 1994; 126(2–4):135–139.
928. Samson Y, Baron JC et al. Stroke 1985; 16(4):609–616.
929. Gibbs JM, Wise RJ et al. J Neurol Neurosurg Psychiatry 1987; 50(2):140–150.
930. Ishikawa T, Yasui N et al. Neurol Med Chir (Tokyo) 1992; 32(1):5–9.
931. Ogawa A, Kameyama M et al. J Neurosurg 1992; 76(6):955–960.
932. Powers WJ, Grubb RL, Jr. et al. J Neurosurg 1989; 70(1):61–67.
933. Adams HP, Jr., Powers WJ et al. Neurosurg Clin N Am 2001; 12(3):613–624.
934. Nariai T, Suzuki R et al. AJNR Am J Neuroradiol 1995; 16(3):563–570.
935. Hirano T, Minematsu K et al. J Cereb Blood Flow Metab 1994; 14(5):763–770.
936. Arigoni M, Kneifel S et al. Eur J Nucl Med 2000; 27(10):1557–1563.
937. Kuwabara Y, Ichiya Y et al. J Nucl Med 1998; 39(5):760–765.
938. Gambhir S, Inao S et al. Neurol Res 1997; 19(2):139–144.
939. Hayashida K, Tanaka Y et al. Nucl Med Commun 1996; 17(12):1047–1051.
940. Kuwabara Y, Ichiya Y et al. Stroke 1995; 26(10):1825–1829.
941. Taki W, Yonekawa Y et al. Childs Nerv Syst 1988; 4(5):259–262.
942. Matsushima T, Inoue T et al. Acta Neurochir (Wien) 1994; 131(3–4):196–202.
943. Taki W, Yonekawa Y et al. Acta Neurochir (Wien) 1989; 100(3–4):150–154.
944. Kuwabara Y, Ichiya Y et al. Stroke 1990; 21(2):272–277.
945. Ikezaki K, Matsushima T et al. J Neurosurg 1994; 81(6):843–850.
946. Kuwabara Y, Ichiya Y et al. Clin Neurol Neurosurg 1997; 99 Suppl 2:S74–S78.
947. Morimoto M, Iwama T et al. Acta Neurochir (Wien) 1999; 141(4):377–384.
948. Takahashi S, Tohgi H et al. J Neurol Sci 1998; 158(1):58–64.
949. Nariai T, Ohno K et al. J Neuroimaging 2001; 11(3):325–329.
950. Ooiwa Y, Uematsu Y et al. Stroke 1993; 24(2):304–309.
951. Sano M, Ishii K et al. Acta Neurol Scand 1995; 92(6):497–502.
952. Tuominen S, Juvonen V et al. Stroke 2001; 32(8):1767–1774.
953. Tatsch K, Koch W et al. J Nucl Med 2003; 44(6):862–869.
954. Heiss WD, Emunds HG et al. Stroke 1993; 24(12):1784–1788.
955. Müller RA, Chugani HT et al. J Child Neurol 1998; 13(1):16–22.
956. Chugani HT, Jacobs B. Annals of Neurology 1994; 36(5):794–797.
957. Mies G, Auer LM et al. Stroke 1983; 14(1):22–27.
958. Nakayama H, Jorgensen HS et al. Stroke 1994; 25(4):808–813.
959. Metter EJ, Hanson WR et al. Arch Neurol 1990; 47(11):1235–1238.
960. Karbe H, Herholz K et al. Neurology 1989; 39(8):1083–1088.
961. Metter EJ, Riege WH et al. Arch Neurol 1988; 45(11):1229–1234.
962. Kumar R, Masih AK et al. Archives of Physical Medicine & Rehabilitation 1996; 77(12):1312–1315.
963. Heiss WD, Kessler J et al. Arch Neurol 1993; 50(9):958–964.
964. Karbe H, Kessler J et al. Arch Neurol 1995; 52(2):186–190.
965. Heiss WD, Kessler J et al. Ann Neurol 1999; 45(4):430–438.
966. Warburton E, Price CJ et al. J Neurol Neurosurg Psychiatry 1999; 66(2):155–161.
967. Belin P, Vaneeckhout P et al. Neurology 1996; 47(6):1504–1511.
968. Berthier ML, Starkstein SE et al. Brain 1991; 114 (Pt 3):1409–1427.
969. Kessler J, Thiel A et al. Stroke 2000; 31(9):2112–2116.
970. Ohyama M, Senda M et al. Stroke 1996; 27(5):897–903.
971. Weiller C, Isensee C et al. Annals of Neurology 1995; 37(6):723–732.
972. Musso M, Weiller C et al. Brain 1999; 122(Pt 9):1781–1790.
973. Karbe H, Thiel A et al. Brain Lang 1998; 64(2):215–230.

974. Bakar M, Kirshner HS et al. Arch Neurol 1996; 53(10):1026–1032.
975. Tzourio N, Crivello F et al. Neuroimage 1998; 8(1):1–16.
976. Weiller C, Chollet F et al. Annals of Neurology 1992; 31(5):463–472.
977. Weder B, Knorr U et al. Brain 1994; 117(Pt 3):593–605.
978. Nelles G, Spiekermann G et al. Stroke 1999; 30(8):1510–1516.
979. Seitz RJ, Hoflich P et al. Arch Neurol 1998; 55(8):1081–1088.
980. Jenkins IH, Brooks DJ et al. Journal of Neuroscience 1994; 14(6):3775–3790.
981. Schlaug G, Knorr U et al. Experimental Brain Research 1994; 98(3):523–534.
982. Rauch SL, Savage CR et al. Human Brain Mapping 1995; 3(4):271–286.
983. Chollet F, DiPiero V et al. Annals of Neurology 1991; 29(1):63–71.
984. Honda M, Nagamine T et al. J Neurol Sci 1997; 146(2):117–126.
985. Cramer SC, Nelles G et al. Stroke 1997; 28(12):2518–2527.
986. Weiller C, Ramsay SC et al. Ann Neurol 1993; 33(2):181–189.
987. Calautti C, Baron JC. Stroke 2003; 34(6):1553–1566.
988. Calautti C, Leroy F et al. Neuroreport 2001; 12(18):3883–3886.
989. Weder B, Seitz RJ. Neuroreport 1994; 5(4):457–460.
990. Iglesias S, Marchal G et al. Stroke 1996; 27(7):1192–1199.
991. Binkofski F, Seitz RJ et al. Annals of Neurology 1996; 39(4):460–470.
992. Azari NP, Binkofski F et al. Human Brain Mapping 1996; 4(4):240–253.
993. Nirkko AC, Rosler KM et al. Neurology 1997; 48(4):1090–1093.
994. von Giesen HJ, Schlaug G et al. Journal of the Neurological Sciences 1994; 125(1):29–38.
995. Fiorelli M, Blin J et al. Journal of the Neurological Sciences 1991; 104(2):135–142.
996. Perani D, Vallar G et al. Neuropsychologia 1993; 31:115–125.
997. Pizzamiglio L, Perani D et al. Arch Neurol 1998; 55(4):561–568.
998. Gardiner RM. Epilepsy Res 1999; 36(2–3):91–95.
999. Kuhl DE, Engel J, Jr. et al. Annals of Neurology 1980; 8(4):348–360.
1000. European Federation of Neurological Societies Task Force. European Journal of Neurology 2000; 7:119–122.
1001. Kuzniecky RI, Knowlton RC. Semin Neurol 2002; 22(3):279–288.
1002. Theodore WH. Curr Science 2002; 82:1–9.
1003. Richardson MP. Br Med Bull 2003; 65:179–192.
1004. Neuroimaging Subcommission of the International League Against Epilepsy. Epilepsia 2000; 41(10):1350–1356.
1005. Duncan JS. Rev Neurol (Paris) 1999; 155(6–7):482–488.
1006. Henry TR, Pennell PB. Q J Nucl Med 1998; 42(3):199–210.
1007. Engel J, Jr., Kuhl DE et al. Science 1982; 218(4567):64–66.
1008. Henry TR. Epilepsia 1996; 37(12):1141–1154.
1009. Engel J, Jr., Brown WJ et al. Annals of Neurology 1982; 12(6):518–528.
1010. Henry TR, Mazziotta JC et al. Arch Neurol 1993; 50(6):582–589.
1011. Sackellares JC, Siegel GJ et al. Neurology 1990; 40(9):1420–1426.
1012. Kim YK, Lee DS et al. J Nucl Med 2003; 44(7):1006–1012.
1013. Diehl B, LaPresto E et al. Epilepsia 2003; 44(4):559–564.
1014. Vielhaber S, Von Oertzen JH et al. Epilepsia 2003; 44(2):193–199.
1015. Pfund Z, Chugani DC et al. J Cereb Blood Flow Metab 2000; 20(5):871–878.
1016. Peyron R, Cinotti L et al. Epilepsy Research 1994; 19(1):55–62.
1017. Cornford EM, Hyman S et al. J Cereb Blood Flow Metab 1998; 18(1):26–42.
1018. Knowlton RC, Laxer KD et al. Neurology 2001; 57(7):1184–1190.
1019. Theodore WH, Gaillard WD et al. Epilepsia 2001; 42(1):130–132.
1020. Foldvary N, Lee N et al. Epilepsia 1999; 40(1):26–29.
1021. Lamusuo S, Jutila L et al. Arch Neurol 2001; 58(6):933–939.
1022. Knowlton RC, Abou-Khalil B et al. Arch Neurol 2002; 59(12):1882–1886.
1023. Breier JI, Mullani NA et al. Neurology 1997; 48(4):1047–1053.
1024. Rausch R, Henry TR et al. Arch Neurol 1994; 51(2):139–144.

1025. Bittar RG, Andermann F et al. Epilepsia 1999; 40(2):170–178.
1026. Sperling MR, Gur RC et al. Epilepsia 1990; 31(2):145–155.
1027. Bouilleret V, Dupont S et al. Ann Neurol 2002; 51(2):202–208.
1028. Dupont S, Semah F et al. Neurology 1998; 51(5):1289–1292.
1029. Dlugos DJ, Jaggi J et al. Epilepsia 1999; 40(4):408–413.
1030. Pawlik G, Fink GR et al. Acta Neurologica Scandinavica 1994; Supplementum 1994;15:150–156.
1031. Pawlik G, Holthoff VA et al. Acta Neurochirurgica 1990; 50 (Suppl.):84–87.
1032. Bromfield EB, Ludlow CL et al. Epilepsy Research 1991; 9(1):49–58.
1033. Leiderman DB, Balish M et al. Epilepsy Research 1992; 13(2):153–157.
1034. Theodore WH, Gaillard WD et al. Annals of Neurology 1994; 36(2):241–244.
1035. Engel J, Jr., Henry TR et al. Epilepsy Research – Supplement 1992; 5:111–120.
1036. Debets RM, Sadzot B et al. J Neurol Neurosurg Psychiatry 1997; 62(2):141–150.
1037. Lamusuo S, Ruottinen HM et al. J Neurol Neurosurg Psychiatry 1997; 63(6):743–748.
1038. DellaBadia J, Jr., Bell WL et al. Seizure 2002; 11(5):303–309.
1039. Theodore WH. Curr Opin Neurol 2002; 15(2):191–195.
1040. Meyer PT, Cortes-Blanco A et al. Eur J Nucl Med 2001; 28(10):1529–1540.
1041. Hwang SI, Kim JH et al. AJNR Am J Neuroradiol 2001; 22(5):937–946.
1042. Stefan H, Pawlik G et al. J Neurol 1987; 234(6):377–384.
1043. O'Brien TJ, Hicks RJ et al. J Nucl Med 2001; 42(8):1158–1165.
1044. Franck G, Maquet P et al. Advances in Neurology 1992; 57:471–485.
1045. Muzik O, Da Silva EA et al. Neurology 2000; 54:171–179.
1046. Theodore WH, Gaillard WD. Adv Neurol 2000; 84:435–446.
1047. Swartz BE, Halgren E et al. Epilepsia 1989; 30(5):547–558.
1048. Henry TR, Sutherling WW et al. Epilepsy Research 1991; 10(2–3):174–182.
1049. Ryvlin P, Bouvard S et al. Brain 1998; 121(Pt 11):2067–2081.
1050. Swartz BW, Khonsari A et al. Epilepsia 1995; 36(4):388–395.
1051. Da Silva EA, Chugani DC et al. Epilepsia 1997; 38(11):1198–1208.
1052. Theodore WH, Balish M et al. Epilepsia 1996; 37(8):796–802.
1053. Meltzer CC, Adelson PD et al. Epilepsia 2000; 41(2):193–200.
1054. Park YD, Hoffman JM et al. J Child Neurol 1994; 9(2):139–143.
1055. Leiderman DB, Albert P et al. Arch Neurol 1994; 51(9):932–936.
1056. Franck G, Sadzot B et al. Adv Neurol 1986; 44:935–948.
1057. Barrington SF, Koutroumanidis M et al. Epilepsia 1998; 39(7):753–766.
1058. Stefan H, Schneider S et al. Electroencephalography & Clinical Neurophysiology 1991; 79(1):1–10.
1059. Arnold S, Berthele A et al. Epilepsia 2000; 41(7):818–824.
1060. Kaminska A, Chiron C et al. Brain 2003; 126(1):248–260.
1061. Avery RA, Spencer SS et al. Eur J Nucl Med 1999; 26(8):830–836.
1062. Baumgartner C, Serles W et al. J Nucl Med 1998; 39(6):978–982.
1063. Kahane P, Merlet I et al. Brain 1999; 122(Pt 10):1851–1865.
1064. Tatlidil R. Epilepsy Res 2000; 42(2–3):83–88.
1065. Engel J, Jr., Kuhl DE et al. Neurology 1983; 33(4):400–413.
1066. Hajek M, Antonini A et al. Epilepsy Research 1991; 9(1):44–48.
1067. Handforth A, Cheng JT et al. Epilepsia 1994; 35(4):876–881.
1068. Savic I, Persson A et al. Lancet 1988; 2(8616):863–866.
1069. Koepp MJ, Richardson MP et al. Neurology 1997; 49(3):764–773.
1070. Juhasz C, Nagy F et al. Epilepsia 1999; 40(5):566–574.
1071. Koepp MJ, Hammers A et al. Neurology 2000; 54(2):332–339.
1072. Szelies B, Weber-Luxenburger G et al. Eur J Neurol 2000; 7(4):393–400.
1073. Lamusuo S, Pitkanen A et al. Neurology 2000; 54(12):2252–2260.
1074. Savic I, Ingvar M et al. Journal of Neurology, Neurosurgery & Psychiatry 1993; 56(6):615–621.

1075. Hammers A, Koepp MJ et al. Neurology 2001; 56(7):897–906.
1076. Szelies B, Weber-Luxenburger G et al. Neuroimage 1996; 3:109–286.
1077. Savic I, Blomqvist G et al. Acta Neurol Scand 1998; 97(5):279–286.
1078. Savic I, Svanborg E et al. Epilepsia 1996; 37(3):236–244.
1079. Savic I, Thorell JO. Arch Neurol 1996; 53(7):656–662.
1080. Koepp MJ, Hand KS et al. Ann Neurol 1998; 43(5):618–626.
1081. Hand KS, Baird VH et al. Br J Pharmacol 1997; 122(2):358–364.
1082. Nagy F, Chugani DC et al. J Cereb Blood Flow Metab 1999; 19(9):939–947.
1083. Hammers A, Koepp MJ et al. Brain 2002; 125(10):2257–2271.
1084. Juhasz C, Chugani DC et al. Neurology 2001; 56(12):1650–1658.
1085. Savic I, Thorell JO et al. Epilepsia 1995; 36(12):1225–1232.
1086. Juhasz C, Chugani DC et al. Neurology 2000; 55(6):825–835.
1087. Richardson MP, Hammers A et al. Epilepsia 2001; 42(10):1327–1334.
1088. Frost JJ, Mayberg HS et al. Annals of Neurology 1988; 23(3):231–237.
1089. Mayberg HS, Sadzot B et al. Annals of Neurology 1991; 30(1):3–11.
1090. Madar I, Lesser RP et al. Ann Neurol 1997; 41(3):358–367.
1091. Bartenstein PA, Prevett MC et al. Epilepsy Res 1994; 18(2):119–125.
1092. Theodore WH, Carson RE et al. Epilepsy Research 1992; 13(2):129–139.
1093. Koepp MJ, Richardson MP et al. Lancet 1998; 352(9132):952–955.
1094. Kumlien E, Hartvig P et al. Epilepsia 1999; 40(1):30–37.
1095. Toczek MT, Carson RE et al. Neurology 2003; 60(5):749–756.
1096. Iinuma K, Yokoyama H et al. Lancet 1993; 341(8839):238.
1097. Asano E, Chugani DC et al. Neurology 2000; 54(10):1976–1984.
1098. Fedi M, Reutens DC et al. Epilepsy Research 2003; 52(3):203–213.
1099. Szelies B, Herholz K et al. Journal of Computer Assisted Tomography 1983; 7(6):946–953.
1100. Banati RB, Goerres GW et al. Neurology 1999; 53(9):2199–2203.
1101. Goerres GW, Revesz T et al. AJR Am J Roentgenol 2001; 176(4):1016–1018.
1102. Bergstrom M, Kumlien E et al. Acta Neurol Scand 1998; 98(4):224–231.
1103. Kumlien E, Nilsson A et al. Acta Neurol Scand 2001; 103(6):360–366.
1104. Fakhoury T, Abou-Khalil B et al. Seizure 1999; 8(7):427–431.
1105. Theodore WH, Gaillard WD. Prog Brain Res 2002; 135:305–313.
1106. Gaillard WD, Kopylev L et al. Neurology 2002; 58(5):717–722.
1107. Theodore WH, Bhatia S et al. Neurology 1999; 52(1):132–136.
1108. Radtke RA, Hanson MW et al. Neurology 1993; 43(6):1088–1092.
1109. Delbeke D, Lawrence SK et al. Invest Radiol 1996; 31(5):261–266.
1110. Theodore WH, Sato S et al. Annals of Neurology 1992; 32(6):789–794.
1111. Choi JY, Kim SJ et al. Eur J Nucl Med Mol Imaging 2003; 30:581–587.
1112. Manno EM, Sperling MR et al. Neurology 1994; 44(12):2331–2336.
1113. Blum DE, Ehsan T et al. Epilepsia 1998; 39(6):651–659.
1114. Newberg AB, Alavi A et al. J Nucl Med 2000; 41(12):1964–1968.
1115. Hammers A, Koepp MJ et al. Brain 2003; 126(6):1300–1318.
1116. Hammers A, Koepp MJ et al. Brain 2001; 124(Pt 8):1555–1565.
1117. Miura K, Watanabe K et al. Brain & Development 1993; 15(4):288–290.
1118. De Volder AG, Gadisseux JF et al. Pediatric Neurology 1994; 11(4):290–294.
1119. Bairamian D, Di Chiro G et al. J Comput Assist Tomogr 1985; 9(6):1137–1139.
1120. Falconer J, Wada JA et al. Can J Neurol Sci 1990; 17:35–39.
1121. Richardson MP, Koepp MJ et al. Brain 1998; 121(Pt 7):1295–1304.
1122. Chugani HT, Chugani DC. Adv Neurol 1999; 79:883–891.
1123. Juhasz C, Chugani HT et al. Brain Dev 2001; 23(7):488–495.
1124. Theodore WH, Rose D et al. Annals of Neurology 1987; 21(1):14–21.
1125. Chugani HT, Mazziotta JC et al. Annals of Neurology 1987; 21(1):4–13.
1126. Ferrie CD, Marsden PK et al. J Neurol Neurosurg Psychiatry 1997; 63(2):181–187.
1127. Olson DM, Chugani HT et al. Epilepsia 1990; 31(6):731–739.

1128. Maquet P, Hirsch E et al. Epilepsia 1990; 31(6):778–783.
1129. Da Silva EA, Chugani DC et al. J Child Neurol 1997; 12(8):489–495.
1130. Rintahaka PJ, Chugani HT et al. Journal of Child Neurology 1995; 10(2):127–133.
1131. Majerus S, Laureys S et al. Hum Brain Mapp 2003; 19(3):133–144.
1132. Maquet P, Hirsch E et al. Brain 1995; 118(Part 6):1497–1520.
1133. Muller RA, Chugani HT et al. J Child Neurol 1998; 13(1):16–22.
1134. Duncan JD, Moss SD et al. Pediatr Neurosurg 1997; 26(3):144–156.
1135. Tatlidil R, Xiong J et al. Acta Neurol Scand 2000; 102(2):73–80.
1136. Bahn MM, Lin W et al. AJR Am J Roentgenol 1997; 169(2):575–579.
1137. van der Kallen BF, Morris GL et al. AJNR Am J Neuroradiol 1998; 19(1):73–77.
1138. Yetkin FZ, Swanson S et al. AJNR Am J Neuroradiol 1998; 19(6):1095–1098.
1139. Springer JA, Binder JR et al. Brain 1999; 122(Pt 11):2033–2046.
1140. Lehericy S, Cohen L et al. Neurology 2000; 54(8):1625–1633.
1141. Rutten GJ, Ramsey NF et al. Ann Neurol 2002; 51(3):350–360.
1142. Sabbah P, Chassoux F et al. Neuroimage 2003; 18(2):460–467.
1143. Carpentier A, Pugh KR et al. Epilepsia 2001; 42(10):1241–1254.
1144. Stoeckel MC, Kleinschmidt A et al. J Neuroimaging 2002; 12(3):276–281.
1145. Spanaki MV, Kopylev L et al. Arch Neurol 2000; 57(10):1447–1452.
1146. Hajek M, Wieser HG et al. Neurology 1994; 44(11):2125–2132.
1147. Dasheiff RM, Rosenbek J et al. J Neurol 1987; 234(5):283–288.
1148. Gur RC, Sussman NM et al. Neurology 1982; 32(10):1191–1194.
1149. Theodore WH, Bairamian D et al. J Cereb Blood Flow Metab 1986; 6(3):315–320.
1150. Theodore WH, Di Chiro G et al. Neurology 1986; 36(1):60–64.
1151. Theodore WH, Bromfield E et al. Ann Neurol 1989; 25(5):516–520.
1152. Leiderman DB, Balish M et al. Epilepsia 1991; 32(3):417–422.
1153. Gaillard WD, Zeffiro T et al. Epilepsia 1996; 37(6):515–521.
1154. Spanaki MV, Siegel H et al. Neurology 1999; 53(7):1518–1522.
1155. Baron JC, Roeda D et al. Neurology 1983; 33(5):580–585.
1156. Engel J, Jr., Lubens P et al. Annals of Neurology 1985; 17(2):121–128.
1157. Savic I, Pauli S et al. Journal of Neurology, Neurosurgery & Psychiatry 1994; 57(7):797–804.
1158. Bartenstein PA, Duncan JS et al. J Neurol Neurosurg Psychiatry 1993; 56(12):1295–1302.
1159. Prevett MC, Cunningham VJ et al. Epilepsy Res 1994; 19(1):71–77.
1160. Langfitt TW, Obrist WD et al. J Neurosurg 1986; 64(5):760–767.
1161. Ricker JH, Zafonte RD. J Head Trauma Rehabil 2000; 15(2):859–868.
1162. Moore TH, Osteen TL et al. J Cereb Blood Flow Metab 2000; 20(10):1492–1501.
1163. Umile EM, Sandel ME et al. Arch Phys Med Rehabil 2002; 83(11):1506–1513.
1164. Worley G, Hoffman JM et al. Dev Med Child Neurol 1995; 37(3):213–220.
1165. Bergsneider M, Hovda DA et al. J Head Trauma Rehabil 2001; 16(2):135–148.
1166. Bergsneider M, Hovda DA et al. J Neurosurg 1997; 86(2):241–251.
1167. Yamaki T, Imahori Y et al. J Nucl Med 1996; 37(7):1166–1170.
1168. Bergsneider M, Hovda DA et al. J Neurotrauma 2000; 17(5):389–401.
1169. Steiner LA, Coles JP et al. J Neurol Neurosurg Psychiatry 2003; 74(6):765.
1170. Diringer MN, Videen TO et al. J Neurosurg 2002; 96(1):103–108.
1171. Steiner LA, Coles JP et al. J Cereb Blood Flow Metab 2003; 23(11):1371–1377.
1172. Levine B, Cabeza R et al. J Neurol Neurosurg Psychiatry 2002; 73(2):173–181.
1173. McAllister TW, Saykin AJ et al. Neurology 1999; 53(6):1300–1308.
1174. Ricker JH, Muller RA et al. J Clin Exp Neuropsychol 2001; 23(2):196–206.
1175. Mattioli F, Grassi F et al. Cortex 1996; 32(1):121–129.
1176. Roberts MA, Manshadi FF et al. Brain Inj 1995; 9(5):427–436.
1177. Bicik I, Radanov BP et al. Neurology 1998; 51(2):345–350.
1178. Radanov BP, Bicik I et al. J Neurol Neurosurg Psychiatry 1999; 66(4):485–489.
1179. Plum F, Schiff N et al. Philos Trans R Soc Lond B Biol Sci 1998; 353(1377):1929–1933.
1180. De Volder AG, Michel C et al. Acta Neurol Belg 1994; 94(3):183–189.

1181. Rudolf J, Sobesky J et al. Lancet 2000; 355(9198):115–116.
1182. Larsen PD, Gupta NC et al. Pediatr Neurol 1993; 9(4):323–326.
1183. Schiff ND, Ribary U et al. Brain 2002; 125(Pt 6):1210–1234.
1184. Rudolf J, Sobesky J et al. Eur J Neurol 2002; 9(6):671–677.
1185. Kassubek J, Juengling FD et al. J Neurol Sci 2003; 212(1–2):85–91.
1186. Laureys S, Antoine S et al. Acta Neurol Belg 2002; 102(4):177–185.
1187. Owen AM, Menon DK et al. Neurocase 2002; 8(5):394–403.
1188. Powers WJ, Rosenbaum JL et al. J Cereb Blood Flow Metab 1998; 18(6):632–638.
1189. Blennow M, Ingvar M et al. Acta Paediatr 1995; 84(11):1289–1295.
1190. Thorngren-Jerneck K, Ohlsson T et al. Pediatr Res 2001; 49(4):495–501.
1191. Suhonen-Polvi H, Kero P et al. Eur J Nucl Med 1993; 20(9):759–765.
1192. Volpe JJ, Herscovitch P et al. Ann Neurol 1985; 17(3):287–296.
1193. Volpe JJ, Herscovitch P et al. Pediatrics 1983; 72(5):589–601.
1194. Altman DI, Volpe JJ. Advances in Pediatrics 1987; 34:111–138.
1195. Altman DI, Powers WJ et al. Ann Neurol 1988; 24(2):218–226.
1196. Altman DI, Perlman JM et al. Pediatrics 1993; 92(1):99–104.
1197. Meyer MA, Hubner KF et al. J Neuroimaging 1994; 4(2):104–105.
1198. Tanaka M, Uesugi M et al. Ann Nucl Med 1995; 9(1):43–45.
1199. Kassubek J, Juengling FD et al. J Neuroimaging 2001; 11(1):55–59.
1200. Salmon E, Sadzot B et al. J Neuroimaging 2002; 12(3):282–283.
1201. Tsuyuguchi N, Sunada I et al. Ann Nucl Med 2003; 17(1):47–51.
1202. Mineura K, Ogawa T et al. Journal of Neuroimaging 1997; 7(1):8–15.
1203. Dethy S, Manto M et al. Clinical Neurology & Neurosurgery 1995; 97(4):349–353.
1204. Yoshikawa H, Fueki N et al. J Child Neurol 1990; 5(4):311–315.
1205. Huber M, Pawlik G et al. J Neurol 1992; 239(3):157–161.
1206. Huber M, Herholz K et al. Arch Neurol 1989; 46(1):97–100.
1207. Tien RD, Ashdown BC et al. AJR Am J Roentgenol 1992; 158(6):1329–1332.
1208. Chiapparini L, Granata T et al. Neuroradiology 2003; 45(3):171–183.
1209. Fiorella DJ, Provenzale JM et al. AJNR Am J Neuroradiol 2001; 22(7):1291–1299.
1210. Caplan R, Curtiss S et al. Brain Cogn 1996; 32(1):45–66.
1211. Cagnin A, Myers R et al. Brain 2001; 124(Pt 10):2014–2027.
1212. Blinkenberg M, Jensen CV et al. Neurology 1999; 53(1):149–153.
1213. Pozzilli C, Fieschi C et al. J Neurol Sci 1992; 112(1–2):51–57.
1214. Sun X, Tanaka M et al. Ann Nucl Med 1998; 12(2):89–94.
1215. Brooks DJ, Leenders KL et al. J Neurol Neurosurg Psychiatry 1984; 47(11):1182–1191.
1216. Roelcke U, Kappos L et al. Neurology 1997; 48(6):1566–1571.
1217. Pozzilli C, Bernardi S et al. J Neurol Neurosurg Psychiatry 1988; 51(8):1058–1062.
1218. Jansen HM, Willemsen AT et al. J Neurol Sci 1995; 132(2):139–145.
1219. Vowinckel E, Reutens D et al. J Neurosci Res 1997; 50(2):345–353.
1220. Debruyne JC, Versijpt J et al. Eur J Neurol 2003; 10(3):257–264.
1221. Schiepers C, Van Hecke P et al. Mult Scler 1997; 3(1):8–17.
1222. Newberg A, Hassan A et al. Nucl Med Commun 2002; 23(8):773–777.
1223. Volkow ND, Warner N et al. Am J Physiol Imaging 1988; 3(2):91–98.
1224. Lang AE, Garnett ES. J Neurol Neurosurg Psychiatry 1988; 51(7):1010–1011.
1225. Stoppe G, Wildhagen K et al. Neurology 1990; 40(2):304–308.
1226. Kao CH, Ho YJ et al. Arthritis Rheum 1999; 42(1):61–68.
1227. Weiner SM, Otte A et al. Ann Rheum Dis 2000; 59(5):377–385.
1228. Chen AC. Pain 1993; 54(2):133–144.
1229. Jones AK, Brown WD et al. Proc R Soc Lond B Biol Sci 1991; 244(1309):39–44.
1230. Talbot JD, Marrett S et al. Science 1991; 251(4999):1355–1358.
1231. Peyron R, Laurent B et al. Neurophysiol Clin 2000; 30(5):263–288.
1232. Coghill RC, Talbot JD et al. J Neurosci 1994; 14(7):4095–4108.
1233. Casey KL, Minoshima S et al. J Neurophysiol 1994; 71(2):802–807.

1234. Rainville P, Duncan GH et al. Science 1997; 277(5328):968–971.
1235. Casey KL, Minoshima S et al. J Neurophysiol 1996; 76(1):571–581.
1236. Xu X, Fukuyama H et al. Neuroreport 1997; 8(2):555–559.
1237. Svensson P, Minoshima S et al. J Neurophysiol 1997; 78(1):450–460.
1238. May A, Kaube H et al. Pain 1998; 74(1):61–66.
1239. Derbyshire SW, Jones AK. Pain 1998; 76(1–2):127–135.
1240. Tölle TR, Kaufmann T et al. Ann Neurol 1999; 45(1):40–47.
1241. Peyron R, Larrea L et al. Brain 1999; 122(Pt 9):1765–1780.
1242. Coghill RC, Sang CN et al. J Neurophysiol 1999; 82(4):1934–1943.
1243. Petrovic P, Petersson KM et al. Pain 2000; 85(1–2):19–30.
1244. Hofbauer RK, Rainville P et al. J Neurophysiol 2001; 86(1):402–411.
1245. Lorenz J, Minoshima S et al. Brain 2003; 126(5):1079.
1246. Coghill RC, Sang CN et al. J Cereb Blood Flow Metab 1998; 18(2):141–147.
1247. Paulson PE, Minoshima S et al. Pain 1998; 76(1–2):223–229.
1248. Vogt BA, Derbyshire S et al. Eur J Neurosci 1996; 8(7):1461–1473.
1249. Svensson P, Johannsen P et al. Eur J Pain 1998; 2(2):95–107.
1250. Craig AD, Reiman EM et al. Nature 1996; 384(6606):258–260.
1251. Petrovic P, Ingvar M et al. Pain 1999; 83(3):459–470.
1252. Korotkov A, Ljubisavljevic M et al. Neurosci Lett 2002; 335(2):119–123.
1253. Lorenz J, Cross D et al. Neuron 2002; 35(2):383–393.
1254. Willoch F, Rosen G et al. Ann Neurol 2000; 48(6):842–849.
1255. Derbyshire SW, Jones AK et al. Neuroimage 2002; 16(1):158–168.
1256. Di PV, Jones AK et al. Pain 1991; 46(1):9–12.
1257. Duncan GH, Kupers RC et al. J Neurophysiol 1998; 80(6):3326–3330.
1258. Kupers RC, Gybels JM et al. Pain 2000; 87(3):295–302.
1259. Derbyshire SW, Jones AK et al. J Neurol Neurosurg Psychiatry 1994; 57(10):1166–1172.
1260. Derbyshire SW, Jones AK et al. Eur J Pain 1999; 3(2):103–113.
1261. Jones AK, Derbyshire SW. Ann Rheum Dis 1997; 56(10):601–607.
1262. Hsieh JC, Belfrage M et al. Pain 1995; 63(2):225–236.
1263. Garcia-Larrea L, Peyron R et al. Stereotact Funct Neurosurg 1997; 68(1–4 Pt 1):141–148.
1264. Garcia-Larrea L, Peyron R et al. Pain 1999; 83(2):259–273.
1265. Davis KD, Taub E et al. J Neurosurg 2000; 92(1):64–69.
1266. Willoch F, Gamringer U et al. Pain 2003; 103(1–2):119–130.
1267. Petrovic P, Kalso E et al. Science 2002; 295(5560):1737–1740.
1268. Adler LJ, Gyulai FE et al. Anesth Analg 1997; 84(1):120–126.
1269. Hsieh JC, Meyerson BA et al. Eur J Pain 1999; 3(1):51–65.
1270. Zubieta JK, Smith YR et al. Science 2001; 293(5528):311–315.
1271. Bencherif B, Fuchs PN et al. Pain 2002; 99(3):589–598.
1272. Zubieta JK, Smith YR et al. J Neurosci 2002; 22(12):5100–5107.
1273. Hagelberg N, Kajander JK et al. Synapse 2002; 45(1):25–30.
1274. Jones AK, Kitchen ND et al. J Cereb Blood Flow Metab 1999; 19(7):803–808.
1275. Willoch F, Tolle TR et al. AJNR Am J Neuroradiol 1999; 20(4):686–690.
1276. Jaaskelainen SK, Rinne JO et al. Pain 2001; 90(3):257–260.
1277. Hagelberg N, Forssell H et al. Pain 2003; 101(1–2):149–154.
1278. Tack CJ, van Gurp PJ et al. Diabetes 2002; 51(12):3545–3553.
1279. Ferrari MD. Lancet 1998; 351(9108):1043–1051.
1280. Kullmann DM. Brain 2002; 125(Pt 6):1177–1195.
1281. Diener HC. Headache 1997; 37(10):622–625.
1282. Sadzot B, Maquet P et al. Cephalalgia 1995; 15(4):316–322.
1283. Chugani DC, Niimura K et al. Neurology 1999; 53(7):1473–1479.
1284. Pfund Z, Chugani DC et al. J Child Neurol 2002; 17(4):253–260.
1285. Chabriat H, Tehindrazanarivelo A et al. Cephalalgia 1995; 15(2):104–108.
1286. Andersson JL, Muhr C et al. Cephalalgia 1997; 17(5):570–579.

1287. Sachs H, Wolf A et al. Arch Neurol 1986; 43(11):1117–1123.
1288. Gutschalk A, Kollmar R et al. Neurosci Lett 2002; 332(2):115.
1289. Weiller C, May A et al. Nat Med 1995; 1(7):658–660.
1290. Goadsby PJ. Microsc Res Tech 2001; 53(3):179–187.
1291. Bahra A, Matharu MS et al. Lancet 2001; 357(9261):1016–1017.
1292. May A, Bahra A et al. Lancet 1998; 352(9124):275–278.
1293. Scammell TE. Ann Neurol 2003; 53(2):154–166.
1294. MacFarlane JG, List SJ et al. Biol Psychiatry 1997; 41(3):305–310.
1295. Rinne JO, Hublin C et al. Neurology 1995; 45(9):1735–1738.
1296. Khan N, Antonini A et al. Neurology 1994; 44(11):2102–2104.
1297. Rinne JO, Hublin C et al. J Sleep Res 1996; 5(4):262–264.
1298. Sudo Y, Suhara T et al. Neurology 1998; 51(5):1297–1302.
1299. Hempel A, Henze M et al. Psychiatry Res 2001; 108(2):133–140.
1300. Staffen W, Karbe H et al. Fortschritte der Neurologie-Psychiatrie 1994; 62(4):119–124.
1301. Rudolf J, Grond M et al. J Child Neurol 1999; 14(8):543–547.
1302. Lee JS, Pfund Z et al. Muscle Nerve 2002; 26(4):506–512.
1303. Fiorelli M, Duboc D et al. Neurology 1992; 42(1):91–94.
1304. Mielke R, Herholz K et al. Psychiatry Research 1993; 50(2):93–99.
1305. Annane D, Fiorelli M et al. Neuromuscul Disord 1998; 8(1):39–45.
1306. Yoshikawa H, Fueki N et al. Brain Dev 1991; 13(3):190–192.
1307. Murphy DG, Mentis MJ et al. Biol Psychiatry 1997; 41(3):285–298.
1308. Duncan DB, Herholz K et al. Annals of Neurology 1995; 37(3):351–358.
1309. Al Essa M, Bakheet S et al. Brain Dev 1999; 21(1):24–29.
1310. Al Essa M, Bakheet S et al. Brain Dev 1999; 21(5):312–317.
1311. Joseph R. Psychiatry 1999; 62(2):138–172.
1312. Bench CJ, Friston KJ et al. Psychological Medicine 1993; 23(3):579–590.
1313. Bench CJ, Friston KJ et al. Psychological Medicine 1992; 22(3):607–615.
1314. Mayberg HS. In: Brain mapping: The disorders (Mazziotta JC et al., Eds.). Academic Press, 2000; 485–507.
1315. Harrison PJ. Brain 2002; 125(Pt 7):1428–1449.
1316. Kennedy SH, Javanmard M et al. Can J Psychiatry 1997; 42(5):467–475.
1317. Soares JC, Mann JJ. Biological Psychiatry 1997; 41(1):86–106.
1318. Drevets WC. Biol Psychiatry 2000; 48(8):813–829.
1319. Bench CJ, Frackowiak RS et al. Psychological Medicine 1995; 25(2):247–261.
1320. Mayberg HS, Brannan SK et al. Neuroreport 1997; 8(4):1057–1061.
1321. Buchsbaum MS, Wu J et al. Biol Psychiatry 1997; 41(1):15–22.
1322. Brody AL, Saxena S et al. Psychiatry Res 1999; 91(3):127–139.
1323. Smith GS, Reynolds CF, III et al. Am J Psychiatry 1999; 156(5):683–689.
1324. Yatham LN, Clark CC et al. Journal of ECT 16(2):171–6, 2000.
1325. Mayberg HS, Brannan SK et al. Biological Psychiatry 48(8):830–43, 2000.
1326. Baxter LR, Jr., Schwartz JM et al. Arch Gen Psychiatry 1989; 46(3):243–250.
1327. Saxena S, Brody AL et al. Biol Psychiatry 2001; 50(3):159–170.
1328. Dunn RT, Kimbrell TA et al. Biological Psychiatry 51(5):387–99, 2002.
1329. Drevets WC, Videen TO et al. Journal of Neuroscience 1992; 12(9):3628–3641.
1330. Smith KA, Morris JS et al. Br J Psychiatry 1999; 174:525–529.
1331. Drevets WC, Price JL et al. Pharmacol Biochem Behav 2002; 71(3):431–447.
1332. Schwartz JM, Baxter LR, Jr. et al. JAMA 1987; 258:1368–1374.
1333. Bremner JD, Vythilingam M et al. JAMA 2003; 289(23):3125–3134.
1334. Mayberg HS. Journal of Neuropsychiatry & Clinical Neurosciences 6(4):428–42, 1994.
1335. Brody AL, Saxena S et al. Biol Psychiatry 2001; 50(3):171–178.
1336. Saxena S, Brody AL et al. Arch Gen Psychiatry 2002; 59(3):250–261.
1337. Dolan RJ, Bench CJ et al. Journal of Neurology, Neurosurgery & Psychiatry 1993; 56(12):1290–1294.

1338. Mayberg HS. J Neuropsychiatry Clin Neurosci 1997; 9(3):471–481.
1339. Pardo JV, Pardo PJ et al. Am J Psychiatry 1993; 150(5):713–719.
1340. Mayberg HS, Liotti M et al. Am J Psychiatry 1999; 156(5):675–682.
1341. Baker SC, Frith CD et al. Psychological Medicine 1997; 27:565–578.
1342. Mayberg HS. Br Med Bull 2003; 65:193–207.
1343. Drevets WC. Prog Brain Res 2000; 126:413–431.
1344. Liotti M, Mayberg HS. J Clin Exp Neuropsychol 2001; 23(1):121–136.
1345. Bench CJ, Friston KJ et al. Psychol Med 1993; 23(3):579–590.
1346. Drevets WC, Raichle ME. Psychopharmacology Bulletin 28(3):261–74, 1992.
1347. Ketter TA, Kimbrell TA et al. Biol Psychiatry 1999; 46(10):1364–1374.
1348. Wu J, Buchsbaum MS et al. American Journal of Psychiatry 1999; 156:1149–1158.
1349. Smith GS, Reynolds III CF et al. Am J Geriatr Psychiatry 2002; 10(5):561–567.
1350. Ketter TA, Wang PW. Journal of Clinical Psychiatry 2002; 63 Suppl. 3:21–25.
1351. Wu JC, Gillin JC et al. American Journal of Psychiatry 1992; 149(4):538–543.
1352. Saxena S, Brody AL et al. Am J Psychiatry 2003; 160(3):522–532.
1353. Kegeles LS, Malone KM et al. Am J Psychiatry 2003; 160(1):76–82.
1354. Beuthien-Baumann B, Zündorf G et al. J Cereb Blood Flow Metab 2003; 23 (Suppl. 1).
1355. Visser PJ, Verhey FR et al. J Am Geriatr Soc 2000; 48(5):479–484.
1356. Dolan RJ, Bench CJ et al. Journal of Neurology, Neurosurgery & Psychiatry 1992; 55(9):768–773.
1357. Dolan RJ, Bench CJ et al. Psychological Medicine 1994; 24(4):849–857.
1358. Jorm AF. Gerontology 2000; 46(4):219–227.
1359. Palsson S, Aevarsson O et al. Br J Psychiatry 1999; 174:249–253.
1360. Ritchie K, Gilham C et al. Age Ageing 1999; 28(4):385–391.
1361. Drevets WC, Frank E et al. Biol Psychiatry 1999; 46(10):1375–1387.
1362. Drevets WC, Frank E et al. Nucl Med Biol 2000; 27(5):499–507.
1363. Sargent PA, Kjaer KH et al. Arch Gen Psychiatry 2000; 57(2):174–180.
1364. Andree B, Thorberg SO et al. Psychopharmacology (Berl) 1999; 144(3):303–305.
1365. Rabiner EA, Gunn RN et al. Neuropsychopharmacology 2000; 23(3):285–293.
1366. Rabiner EA, Gunn RN et al. Nuclear Medicine & Biology 2000; 27:509–513.
1367. Martinez D, Mawlawi O et al. Nucl Med Biol 2000; 27(5):523–527.
1368. Hirani E, Opacka-Juffry J et al. Synapse 2000; 36(4):330–341.
1369. Martinez D, Hwang D et al. Neuropsychopharmacology 2001; 24(3):209–229.
1370. Rabiner EA, Wilkins MR et al. J Pharmacol Exp Ther 2002; 301(3):1144–1150.
1371. Massou JM, Trichard C et al. Psychopharmacology (Berl) 1997; 133(1):99–101.
1372. Meyer JH, Kapur S et al. American Journal of Psychiatry 2001; 158:78–85.
1373. Caspi A, Sugden K et al. Science 2003; 301(5631):386.
1374. Ichimiya T, Suhara T et al. Biol Psychiatry 2002; 51(9):715–722.
1375. Suhara T, Takano A et al. Arch Gen Psychiatry 2003; 60(4):386.
1376. Kent JM, Coplan JD et al. Psychopharmacology (Berl) 2002; 164(3):341–348.
1377. Martinot M, Bragulat V et al. Am J Psychiatry 2001; 158(2):314–316.
1378. Meyer JH, Goulding VS et al. Psychopharmacology (Berl) 2002; 163(1):102–105.
1379. Learned-Coughlin SM, Bergstrom M et al. Biological Psychiatry 2003; 54(8):800–805.
1380. Siessmeier T, Nix WA et al. J Neurol Neurosurg Psychiatry 2003; 74(7):922–928.
1381. MacHale SM, Lawrie SM et al. Br J Psychiatry 2000; 176:550–556.
1382. Tashiro M, Juengling FD et al. Psychooncology 2000; 9(2):157–163.
1383. Phillips PC, Dhawan V et al. Annals of Neurology 1987; 21:59–63.
1384. Phillips PC, Moeller JR et al. Annals of Neurology 1991; 29(3):263–271.
1385. Wik G, Lekander M et al. Brain Behav Immun 1998; 12(3):242–246.
1386. Tashiro M, Itoh M et al. Psychooncology 2001; 10(6):541–546.
1387. Juengling FD, Ebert D et al. Psychopharmacology 2000; 152:383–389.
1388. Davidson LL, Heinrichs RW. Psychiatry Res 2003; 122(2):69–87.
1389. Frankle WG, Laruelle M. Ann Nucl Med 2002; 16(7):437–446.

1390. Wolkin A, Angrist B et al. Am J Psychiatry 1988; 145(2):251–253.
1391. Weinberger DR, Berman KF. Schizophr Bull 1988; 14(2):157–168.
1392. Guich SM, Buchsbaum MS et al. Schizophr Res 1989; 2(6):439–448.
1393. Andreasen NC, Oleary DS et al. Lancet 1997; 349(9067):1730–1734.
1394. Schultz SK, O'Leary DS et al. J Neuropsychiatry Clin Neurosci 2002; 14(1):19–24.
1395. Buchsbaum MS, Nenadic I et al. Schizophr Res 2002; 54(1–2):141–150.
1396. Volkow ND, Brodie JD et al. J Cereb Blood Flow Metab 1986; 6(4):441–446.
1397. Cleghorn JM, Garnett ES et al. Psychiatry Res 1989; 28(2):119–133.
1398. Biver F, Delvenne V et al. Acta Psychiatr Belg 1992; 92(5):261–278.
1399. Volkow ND, Brodie JD et al. J Neurol Neurosurg Psychiatry 1986; 49(10):1199–1202.
1400. Gur RE, Resnick SM et al. Arch Gen Psychiatry 1987; 44(2):119–125.
1401. Widen L, Blomqvist G et al. AJNR Am J Neuroradiol 1983; 4(3):550–552.
1402. Wolkin A, Sanfilipo M et al. Arch Gen Psychiatry 1992; 49(12):959–965.
1403. Wik G, Wiesel FA. Psychiatry Res 1991; 40(2):101–114.
1404. Miller DD, Andreasen NC et al. Neuropsychopharmacology 1997; 17(4):230–240.
1405. Potkin SG, Buchsbaum MS et al. J Clin Psychiatry 1994; 55 Suppl B:63–66.
1406. Wolkin A, Angrist B et al. Psychopharmacology (Berl) 1987; 92(2):241–246.
1407. Wolkin A, Sanfilipo M et al. Biol Psychiatry 1994; 36(5):317–325.
1408. Wolkin A, Sanfilipo M et al. American Journal of Psychiatry 1996; 153(3):346–354.
1409. Hazlett EA, Buchsbaum MS et al. Am J Psychiatry 1999; 156(8):1190–1199.
1410. Kasai K, Iwanami A et al. Neurosci Res 2002; 43(2):93–110.
1411. Silbersweig DA, Stern E et al. Nature 1995; 378(6553):176–179.
1412. Cleghorn JM, Franco S et al. Am J Psychiatry 1992; 149(8):1062–1069.
1413. Hazlett EA, Buchsbaum MS. Front Biosci 2001; 6:D1069-D1072.
1414. Spence SA, Hirsch SR et al. Br J Psychiatry 1998; 172:316–323.
1415. Ganguli R, Carter C et al. Biological Psychiatry 1997; 41(1):33–42.
1416. Carter CS, Perlstein W et al. Am J Psychiatry 1998; 155(9):1285–1287.
1417. Ragland JD, Gur RC et al. Am J Psychiatry 2001; 158(7):1114–1125.
1418. Wiser AK, Andreasen NC et al. Neuroreport 1998; 9(8):1895–1899.
1419. Ragland JD, Gur RC et al. Neuropsychology 1998; 12(3):399–413.
1420. Kim JJ, Kwon JS et al. Am J Psychiatry 2003; 160(5):919–923.
1421. Frith CD, Friston KJ et al. Br J Psychiatry 1995; 167(3):343–349.
1422. Breier A, Malhotra AK et al. American Journal of Psychiatry 1997; 154(6):805–811.
1423. Buchsbaum MS, Hazlett EA et al. Acta Psychiatr Scand Suppl 1999; 395:129–137.
1424. Wong DF, Wagner HN, Jr. et al. Science 1986; 234(4783):1558–1563.
1425. Tune LE, Wong DF et al. Psychiatry Res 1993; 49(3):219–237.
1426. Farde L, Wiesel FA et al. Arch Gen Psychiatry 1990; 47(3):213–219.
1427. Martinot JL, Peron-Magnan P et al. Am J Psychiatry 1990; 147(1):44–50.
1428. Nordstrom AL, Farde L et al. Psychiatry Res 1995; 61(2):67–83.
1429. Hietala J, Syvalahti E et al. Arch Gen Psychiatry 1994; 51(2):116–123.
1430. Okubo Y, Suhara T et al. Nature 1997; 385(6617):634–636.
1431. Pearlson GD, Wong DF et al. Arch Gen Psychiatry 1995; 52(6):471–477.
1432. Abi-Dargham A, Mawlawi O et al. J Neurosci 2002; 22(9):3708–3719.
1433. Breier A, Su TP et al. Proc Natl Acad Sci U S A 1997; 94(6):2569–2574.
1434. Abi-Dargham A, Gil R et al. Am J Psychiatry 1998; 155(6):761–767.
1435. Laruelle M, Abi-Dargham A et al. Biol Psychiatry 1999; 46(1):56–72.
1436. Abi-Dargham A, Rodenhiser J et al. Proc Natl Acad Sci U S A 2000; 97(14):8104–8109.
1437. Reith J, Benkelfat C et al. Proc Natl Acad Sci USA 1994; 91(24):11651–11654.
1438. Hietala J, Syvalahti E et al. Lancet 1995; 346(8983):1130–1131.
1439. Hietala J, Syvalahti E et al. Schizophr Res 1999; 35(1):41–50.
1440. Lindstrom LH, Gefvert O et al. Biol Psychiatry 1999; 46(5):681–688.
1441. Dao-Castellana MH, Paillere-Martinot ML et al. Schizophr Res 1997; 23(2):167–174.
1442. Laakso A, Vilkman H et al. Am J Psychiatry 2000; 157(2):269–271.

1443. Vollenweider FX, Vontobel P et al. Neuropsychopharmacology 1999; 20(5):424–433.
1444. Lewis R, Kapur S et al. Am J Psychiatry 1999; 156(1):72–78.
1445. Trichard C, Paillere-Martinot ML et al. Schizophr Res 1998; 31(1):13–17.
1446. Okubo Y, Suhara T et al. Life Sci 2000; 66(25):2455–2464.
1447. Ngan ET, Yatham LN et al. Am J Psychiatry 2000; 157(6):1016–1018.
1448. Tauscher J, Kapur S et al. Arch Gen Psychiatry 2002; 59(6):514–520.
1449. Farde L, Wiesel FA et al. Arch Gen Psychiatry 1988; 45(1):71–76.
1450. Pickar D, Su TP et al. American Journal of Psychiatry 1996; 153(12):1571–1578.
1451. Talvik M, Nordstrom AL et al. Am J Psychiatry 2001; 158(6):926–930.
1452. Farde L, Nordstrom AL et al. Arch Gen Psychiatry 1992; 49(7):538–544.
1453. Kapur S, Zipursky R et al. Arch Gen Psychiatry 2000; 57(6):553–559.
1454. Kapur S, Zipursky RB et al. Am J Psychiatry 1998; 155(7):921–928.
1455. Nordstrom AL, Nyberg S et al. Arch Gen Psychiatry 1998; 55(3):283–284.
1456. Remington G, Kapur S et al. J Clin Psychopharmacol 1998; 18(1):82–83.
1457. Nyberg S, Farde L et al. Psychopharmacology 1993; 110(3):265–272.
1458. Remington G, Kapur S. J Clin Psychiatry 1999; 60 Suppl 10:15–19.
1459. Farde L, Wiesel FA et al. Psychopharmacology (Berl) 1989; 99 Suppl:S28-S31.
1460. Chou YH, Halldin C et al. Psychopharmacology (Berl) 2003; 166(3):234–240.
1461. Nordstrom AL, Farde L et al. Am J Psychiatry 1995; 152(10):1444–1449.
1462. Trichard C, Paillere-Martinot ML et al. Am J Psychiatry 1998; 155(4):505–508.
1463. Kapur S, Zipursky RB et al. Am J Psychiatry 1999; 156(2):286–293.
1464. Nyberg S, Eriksson B et al. Am J Psychiatry 1999; 156(6):869–875.
1465. Gefvert O, Lundberg T et al. Eur Neuropsychopharmacol 2001; 11(2):105–110.
1466. Gottschalk LA, Fronczek J et al. Psychother Psychosom 2001; 70(1):17–24.
1467. Simpson JR, Jr., Drevets WC et al. Proc Natl Acad Sci U S A 2001; 98(2):688–693.
1468. Chua P, Krams M et al. Neuroimage 1999; 9(6):563–571.
1469. Kimbrell TA, George MS et al. Biol Psychiatry 1999; 46(4):454–465.
1470. Servan-Schreiber D, Perlstein WM et al. J Neuropsychiatry Clin Neurosci 1998; 10(2):148–159.
1471. Reiman EM. J Clin Psychiatry 1997; 58 Suppl 16:4–12.
1472. Benkelfat C, Bradwejn J et al. Am J Psychiatry 1995; 152(8):1180–1184.
1473. Wik G, Fredrikson M et al. Psychiatry Res 1993; 50(1):15–24.
1474. Laakso A, Wallius E et al. Am J Psychiatry 2003; 160(5):904–910.
1475. Tauscher J, Bagby RM et al. Am J Psychiatry 2001; 158(8):1326–1328.
1476. Abadie P, Boulenger JP et al. Eur J Neurosci 1999; 11(4):1470–1478.
1477. Wu JC, Buchsbaum MS et al. Biol Psychiatry 1991; 29(12):1181–1199.
1478. Reiman EM, Raichle ME et al. Am J Psychiatry 1986; 143(4):469–477.
1479. Nordahl TE, Semple WE et al. Neuropsychopharmacology 1990; 3(4):261–272.
1480. Nordahl TE, Stein MB et al. Biol Psychiatry 1998; 44(10):998–1006.
1481. Meyer JH, Swinson R et al. Psychiatry Res 2000; 98(3):133–143.
1482. Boshuisen ML, Ter Horst GJ et al. Biol Psychiatry 2002; 52(2):126–135.
1483. Reiman EM, Raichle ME et al. Arch Gen Psychiatry 1989; 46(6):493–500.
1484. Reiman EM, Fusselman MJ et al. Science 1989; 243(4894 Pt 1):1071–1074.
1485. Drevets WC, Videen TQ et al. Science 1992; 256(5064):1696.
1486. Bisaga A, Katz JL et al. Am J Psychiatry 1998; 155(9):1178–1183.
1487. Fischer H, Andersson JL et al. Neurosci Lett 1998; 251(2):137–140.
1488. Malizia AL, Cunningham VJ et al. Arch Gen Psychiatry 1998; 55(8):715–720.
1489. Fredrikson M, Wik G et al. Psychophysiology 1995; 32(1):43–48.
1490. Wik G, Fredrikson M et al. Int J Neurosci 1997; 91(3–4):253–263.
1491. Tillfors M, Furmark T et al. Biol Psychiatry 2002; 52(11):1113–1119.
1492. Furmark T, Tillfors M et al. Arch Gen Psychiatry 2002; 59(5):425–433.
1493. Baxter LR. J Clin Psychiatry 1990; 51 Suppl:22–25.
1494. Laplane D, Levasseur M et al. Brain 1989; 112:699–725.

1495. Nordahl TE, Benkelfat C et al. Neuropsychopharmacology 1989; 2(1):23–28.
1496. Baxter LR, Jr., Phelps ME et al. Arch Gen Psychiatry 1987; 44(3):211–218.
1497. Baxter LR, Jr., Schwartz JM et al. Am J Psychiatry 1988; 145(12):1560–1563.
1498. Kwon JS, Kim JJ et al. Psychiatry Res 2003; 122(1):37–47.
1499. Gamazo-Garran P, Soutullo CA et al. J Child Adolesc Psychopharmacol 2002; 12(3):259–263.
1500. Rauch SL, Savage CR et al. J Neuropsychiatry Clin Neurosci 1997; 9(4):568–573.
1501. Martinot JL, Allilaire JF et al. Acta Psychiatr Scand 1990; 82(3):233–242.
1502. McGuire PK, Bench CJ et al. Br J Psychiatry 1994; 164(4):459–468.
1503. Cottraux J, Gerard D et al. Psychiatry Res 1996; 60(2–3):101–112.
1504. Volkow ND, Fowler JS. Cereb Cortex 2000; 10(3):318–325.
1505. Rauch SL, Jenike MA et al. Arch Gen Psychiatry 1994; 51(1):62–70.
1506. Baxter LR, Jr., Thompson JM et al. Psychopathology 1987; 20 Suppl 1:114–122.
1507. Benkelfat C, Nordahl TE et al. Arch Gen Psychiatry 1990; 47(9):840–848.
1508. Kang DH, Kwon JS et al. Acta Psychiatr Scand 2003; 107(4):291–297.
1509. Perani D, Colombo C et al. British Journal of Psychiatry 1995; 166(2):244–250.
1510. Baxter LR, Jr., Schwartz JM et al. Arch Gen Psychiatry 1992; 49(9):681–689.
1511. Schwartz JM, Stoessel PW et al. Arch Gen Psychiatry 1996; 53(2):109–113.
1512. Rauch SL, Shin LM et al. Neuropsychopharmacology 2002; 27(5):782–791.
1513. Brody AL, Saxena S et al. Psychiatry Res 1998; 84(1):1–6.
1514. Azari NP, Pietrini P et al. Biol Psychiatry 1993; 34(11):798–809.
1515. Rauch SL, Dougherty DD et al. Biol Psychiatry 2001; 50(9):659–667.
1516. Sachdev P, Trollor J et al. Aust N Z J Psychiatry 2001; 35(5):684–690.
1517. Nuttin BJ, Gabriels LA et al. Neurosurgery 2003; 52(6):1263–1272.
1518. Semple WE, Goyer PF et al. Psychiatry Res 1996; 67(1):17–28.
1519. Markowitsch HJ, Kessler J et al. Neuropsychologia 1998; 36(1):77–82.
1520. Bremner JD, Vythilingam M et al. Biol Psychiatry 2003; 53(10):879–889.
1521. Shaw ME, Strother SC et al. Neuroimage 2002; 15(3):661–674.
1522. Bremner JD, Vythilingam M et al. Am J Psychiatry 2003; 160(5):924–932.
1523. Shin LM, Kosslyn SM et al. Arch Gen Psychiatry 1997; 54(3):233–241.
1524. Huber M, Siol T et al. Traumatology 2001; 7(4).
1525. Fischer H, Wik G et al. Neuroreport 1996; 7(13):2081–2086.
1526. Shin LM, McNally RJ et al. Am J Psychiatry 1999; 156(4):575–584.
1527. Pissiota A, Frans O et al. Eur Arch Psychiatry Clin Neurosci 2002; 252(2):68–75.
1528. Bremner JD, Staib LH et al. Biol Psychiatry 1999; 45(7):806–816.
1529. Osuch EA, Benson B et al. Biol Psychiatry 2001; 50(4):246–253.
1530. Bremner JD, Narayan M et al. Am J Psychiatry 1999; 156(11):1787–1795.
1531. Volkow ND, Fowler JS et al. J Clin Invest 2003; 111(10):1444–1451.
1532. Pohjalainen T, Rinne JO et al. Mol Psychiatry 1998; 3(3):256–260.
1533. Noble EP, Gottschalk LA et al. Am J Med Genet 1997; 74(2):162–166.
1534. Goldstein RZ, Volkow ND et al. Neuroreport 2002; 13(17):2253–2257.
1535. Volkow ND, Mullani N et al. Psychiatry Res 1988; 24(2):201–209.
1536. Volkow ND, Hitzemann R et al. Psychiatry Res 1990; 35(1):39–48.
1537. Wang GJ, Volkow ND et al. Alcohol Clin Exp Res 2000; 24(6):822–829.
1538. Wang GJ, Volkow ND et al. Alcohol Clin Exp Res 2003; 27(6):909–917.
1539. Ingvar M, Ghatan PH et al. J Stud Alcohol 1998; 59(3):258–269.
1540. Sachs H, Russell JA et al. Arch Neurol 1987; 44(12):1242–1251.
1541. Wik G, Borg S et al. Acta Psychiatr Scand 1988; 78(2):234–241.
1542. Adams KM, Gilman S et al. Alcohol Clin Exp Res 1998; 22(1):105–110.
1543. Volkow ND, Hitzemann R et al. Am J Psychiatry 1992; 149(8):1016–1022.
1544. Samson Y, Baron JC et al. J Neurol Neurosurg Psychiatry 1986; 49(10):1165–1170.
1545. Adams KM, Gilman S et al. Alcohol Clin Exp Res 1993; 17(2):205–210.
1546. Martin PR, Rio D et al. J Neuropsychiatry Clin Neurosci 1992; 4(2):159–167.
1547. Johnson-Greene D, Adams KM et al. J Clin Exp Neuropsychol 1997; 19(3):378–385.

1548. Wang GJ, Volkow ND et al. Alcohol Clin Exp Res 1998; 22(8):1850–1854.
1549. Gilman S, Adams K et al. Ann Neurol 1990; 28(6):775–785.
1550. Volkow ND, Wang GJ et al. Am J Psychiatry 1993; 150(3):417–422.
1551. Volkow ND, Wang GJ et al. Alcohol Clin Exp Res 1997; 21(7):1278–1284.
1552. Volkow ND, Wang GJ et al. Alcohol Clin Exp Res 1995; 19(2):510–516.
1553. Gilman S, Koeppe RA et al. Ann Neurol 1996; 40(2):163–171.
1554. Litton JE, Neiman J et al. Psychiatry Res 1993; 50(1):1–13.
1555. Gilman S, Adams KM et al. Alcohol Clin Exp Res 1996; 20(8):1456–1461.
1556. Hietala J, West C et al. Psychopharmacology (Berl) 1994; 116(3):285–290.
1557. Volkow ND, Wang GJ et al. Alcohol Clin Exp Res 1996; 20(9):1594–1598.
1558. Volkow ND, Wang GJ et al. Psychiatry Res 2002; 116(3):163–172.
1559. Tiihonen J, Vilkman H et al. Mol Psychiatry 1998; 3(2):156–161.
1560. Gilman S, Koeppe RA et al. Ann Neurol 1998; 44(3):326–333.
1561. Hommer D, Andreasen P et al. J Neurosci 1997; 17(8):2796–2806.
1562. Schlaepfer TE, Pearlson GD et al. Am J Psychiatry 1997; 154(9):1209–1213.
1563. Volkow ND, Wang GJ et al. Proc Natl Acad Sci U S A 1996; 93(19):10388–10392.
1564. Volkow ND, Ding YS et al. Arch Gen Psychiatry 1995; 52(6):456–463.
1565. Gatley SJ, Volkow ND et al. Biochem Pharmacol 1997; 53(1):43–52.
1566. Volkow ND, Fowler JS et al. Synapse 1999; 31(1):59–66.
1567. Volkow ND, Ding YS et al. J Addict Dis 1996; 15(4):55–71.
1568. Volkow ND, Wang GJ et al. Life Sci 2000; 67(12):1507–1515.
1569. Volkow ND, Wang GJ et al. Nature 1997; 386(6627):827–830.
1570. Volkow ND, Wang GJ et al. Am J Psychiatry 1998; 155(10):1325–1331.
1571. Bartzokis G, Beckson M et al. Neuropsychopharmacology 1999; 20(6):582–590.
1572. Tsukada H, Harada N et al. Brain Res 2000; 860(1–2):141–148.
1573. Tsukada H, Harada N et al. Synapse 2000; 37(2):95–103.
1574. Tsukada H, Harada N et al. J Neurosci 2000; 20(18):7067–7073.
1575. Tsukada H, Nishiyama S et al. Synapse 2001; 42(4):273–280.
1576. Wilcox KM, Lindsey KP et al. Synapse 2002; 43(1):78–85.
1577. Villemagne VL, Rothman RB et al. Synapse 1999; 32(1):44–50.
1578. Villemagne VL, Wong DF et al. Synapse 1999; 33(4):268–273.
1579. Morgan D, Grant KA et al. Nat Neurosci 2002; 5(2):169–174.
1580. Howell LL, Hoffman JM et al. J Neurosci Methods 2001; 106(2):161–169.
1581. Howell LL, Hoffman JM et al. Psychopharmacology (Berl) 2002; 159(2):154–160.
1582. Volkow ND, Mullani N et al. Br J Psychiatry 1988; 152:641–648.
1583. Melon PG, Boyd CJ et al. J Nucl Med 1997; 38(3):451–456.
1584. Volkow ND, Fowler JS et al. Synapse 1993; 14(2):169–177.
1585. Bolla KI, Eldreth DA et al. Neuroimage 2003; 19(3):1085–1094.
1586. Drexler K, Schweitzer JB et al. Am J Addict 2000; 9(4):331–339.
1587. Wu JC, Bell K et al. Neuropsychopharmacology 1997; 17(6):402–409.
1588. Volkow ND, Wang GJ et al. Am J Psychiatry 1998; 155(2):200–206.
1589. Volkow ND, Wang GJ et al. Life Sci 2000; 66(12):L161-L167.
1590. Volkow ND, Fowler JS et al. Am J Psychiatry 1991; 148(5):621–626.
1591. Volkow ND, Wang GJ et al. Am J Psychiatry 1999; 156(1):19–26.
1592. Bonson KR, Grant SJ et al. Neuropsychopharmacology 2002; 26(3):376–386.
1593. Wang GJ, Volkow ND et al. Life Sci 1999; 64(9):775–784.
1594. Zubieta JK, Gorelick DA et al. Nat Med 1996; 2(11):1225–1229.
1595. Childress AR, Mozley PD et al. Am J Psychiatry 1999; 156(1):11–18.
1596. Kilts CD, Schweitzer JB et al. Arch Gen Psychiatry 2001; 58(4):334–341.
1597. Volkow ND, Wang GJ et al. Neuropsychopharmacology 1996; 14(3):159–168.
1598. Tsukada H, Kreuter J et al. J Neurosci 1996; 16(23):7670–7677.
1599. Maggos CE, Tsukada H et al. Neuropsychopharmacology 1998; 19(2):146–153.
1600. Volkow ND, Wang GJ et al. Nature 1997; 386(6627):830–833.

1601. Volkow ND, Fowler JS et al. Am J Psychiatry 1990; 147(6):719–724.
1602. Wang GJ, Volkow ND et al. Life Sci 1995; 56(16):L299-L303.
1603. Melega WP, Raleigh MJ et al. Brain Res 1997; 766(1–2):113–120.
1604. Hartvig P, Torstenson R et al. J Neural Transm 1997; 104(4–5):329–339.
1605. Villemagne V, Yuan J et al. J Neurosci 1998; 18(1):419–427.
1606. Melega WP, Lacan G et al. Synapse 2000; 35(4):243–249.
1607. Melega WP, Lacan G et al. Neurosci Lett 1998; 258(1):17–20.
1608. Melega WP, Raleigh MJ et al. Behav Brain Res 1997; 84(1–2):259–268.
1609. McCann UD, Wong DF et al. J Neurosci 1998; 18(20):8417–8422.
1610. Sekine Y, Iyo M et al. Am J Psychiatry 2001; 158(8):1206–1214.
1611. Volkow ND, Chang L et al. Am J Psychiatry 2001; 158(3):377–382.
1612. Volkow ND, Chang L et al. J Neurosci 2001; 21(23):9414–9418.
1613. Nakamura H, Hishinuma T et al. Annals of the New York Academy of Sciences 1996; 801:401–408.
1614. Nakamura H, Hishinuma T et al. Nucl Med Biol 1997; 24(2):165–169.
1615. Ginovart N, Farde L et al. Synapse 1999; 31(2):154–162.
1616. Volkow ND, Chang L et al. Am J Psychiatry 2001; 158(12):2015–2021.
1617. Volkow ND, Chang L et al. Am J Psychiatry 2001; 158(3):383–389.
1618. Kopelman MD, Reed LJ et al. Neurocase 2001; 7(5):423–432.
1619. Scheffel U, Szabo Z et al. Synapse 1998; 29(2):183–192.
1620. Ricaurte GA, McCann UD et al. Toxicol Lett 2000; 112–113(8):143–146.
1621. Schreckenberger M, Gouzoulis-Mayfrank E et al. Eur J Nucl Med 1999; 26(12):1572–1579.
1622. Gamma A, Buck A et al. J Clin Psychopharmacol 2001; 21(1):66–71.
1623. Mithoefer M, Jerome L et al. Science 2003; 300(5625):1504–1505.
1624. Ricaurte GA, Yuan J et al. Science 2003; 301(5639):1479.
1625. Mattay VS, Berman KF et al. J Neurosci 1996; 16(15):4816–4822.
1626. Servan-Schreiber D, Carter CS et al. Biol Psychiatry 1998; 43(10):723–729.
1627. Ernst M, Zametkin AJ et al. Neuropsychopharmacology 1997; 17(6):391–401.
1628. Vollenweider FX, Maguire RP et al. Psychiatry Res 1998; 83(3):149–162.
1629. Sell LA, Morris JS et al. Drug Alcohol Depend 2000; 60(2):207–216.
1630. Martin-Soelch C, Chevalley AF et al. Eur J Neurosci 2001; 14(8):1360–1368.
1631. Kling MA, Carson RE et al. J Pharmacol Exp Ther 2000; 295(3):1070–1076.
1632. Stapleton JM, Gilson SF et al. Neuropsychopharmacology 2003; 28(4):765–772.
1633. Domino EF, Minoshima S et al. Neuroscience 2000; 101(2):277–282.
1634. Domino EF, Minoshima S et al. Synapse 2000; 38(3):313–321.
1635. Zubieta J, Lombardi U et al. Biol Psychiatry 2001; 49(11):906–913.
1636. Rose JE, Behm FM et al. Am J Psychiatry 2003; 160(2):323–333.
1637. Brody AL, Mandelkern MA et al. Arch Gen Psychiatry 2002; 59(12):1162–1172.
1638. Ding YS, Volkow ND et al. Synapse 2000; 35(3):234–237.
1639. Tsukada H, Miyasato K et al. Synapse 2002; 45(4):207–212.
1640. Fowler JS, Volkow ND et al. J Addict Dis 1998; 17(1):23–34.
1641. Fowler JS, Volkow ND et al. Nature 1996; 379:733–736.
1642. Fowler JS, Volkow ND et al. Proc Natl Acad Sci USA 1996; 93(24):14065–14069.
1643. Fowler JS, Wang GJ et al. Nicotine Tob Res 1999; 1(4):325–329.
1644. Fowler JS, Volkow ND et al. Life Sci 1998; 63(2):L19-L23.
1645. Dagher A, Bleicher C et al. Synapse 2001; 42(1):48–53.
1646. Martin-Solch C, Magyar S et al. Exp Brain Res 2001; 139(3):278–286.
1647. Ernst M, Matochik JA et al. Proc Natl Acad Sci U S A 2001; 98(8):4728–4733.
1648. Gouzoulis-Mayfrank E, Schreckenberger M et al. Neuropsychopharmacology 1999; 20(6):565–581.
1649. Miyazawa H, Osmont A et al. J Neurosci Methods 1993; 50(3):263–272.
1650. Breier A, Malhotra AK et al. Am J Psychiatry 1997; 154(6):805–811.
1651. Schiffer WK, Logan J et al. Neuropsychopharmacology 2003; epub.

1652. Moresco FM, Dieci M et al. Neuroimage 2002; 17(3):1470–1478.
1653. Juengling FD, Schmahl C et al. J Psychiatr Res 2003; 37(2):109–115.
1654. de la Fuente JM, Goldman S et al. J Psychiatr Res 1997; 31(5):531–541.
1655. Goyer PF, Andreason PJ et al. Neuropsychopharmacology 1994; 10(1):21–28.
1656. Soloff PH, Meltzer CC et al. Psychiatry Research: Neuroimaging 2003; 123(3):153–163.
1657. Schmahl CG, Elzinga BM et al. Biol Psychiatry 2003; 54(2):142–151.
1658. Leyton M, Okazawa H et al. Am J Psychiatry 2001; 158(5):775–782.
1659. Soloff PH, Meltzer CC et al. Biol Psychiatry 2000; 47(6):540–547.
1660. Wong MT, Fenwick PB et al. Psychiatry Res 1997; 68(2–3):111–123.
1661. Volkow ND, Tancredi L. Br J Psychiatry 1987; 151:668–673.
1662. Volkow ND, Tancredi LR et al. Psychiatry Res 1995; 61(4):243–253.
1663. Seidenwurm D, Pounds TR et al. AJNR Am J Neuroradiol 1997; 18(4):625–631.
1664. Raine A, Buchsbaum M et al. Biol Psychiatry 1997; 42(6):495–508.
1665. Raine A, Buchsbaum MS et al. Biol Psychiatry 1994; 36(6):365–373.
1666. Mann JJ. Nat Rev Neurosci 2003; 4:819–828.
1667. Oquendo MA, Placidi GP et al. Arch Gen Psychiatry 2003; 60(1):14–22.
1668. Krieg JC, Lauer C et al. Psychiatry Research 1989; 27(1):39–48.
1669. Herholz K. Psychiatry Res 1996; 62(1):105–110.
1670. Herholz K, Krieg JC et al. Biol Psychiatry 1987; 22(1):43–51.
1671. Delvenne V, Lotstra F et al. Biological Psychiatry 1995; 37(3):161–169.
1672. Delvenne V, Goldman S et al. Biol Psychiatry 1996; 40(8):761–768.
1673. Krieg JC, Holthoff V et al. Eur Arch Psychiatry Clin Neurosci 1991; 240(6):331–333.
1674. Delvenne V, Goldman S et al. Int J Eat Disord 1997; 21(4):313–320.
1675. Wu JC, Hagman J et al. Am J Psychiatry 1990; 147(3):309–312.
1676. Hagman JO, Buchsbaum MS et al. J Affect Disord 1990; 19(3):153–162.
1677. Frank GK, Kaye WH et al. Psychiatry Res 2000; 100(1):31–39.
1678. Delvenne V, Goldman S et al. Psychiatry Res 1997; 74(2):83–92.
1679. Blomqvist G, Alvarsson M et al. Am J Physiol Endocrinol Metab 2002; 283(1):E20-E28.
1680. Blomqvist G, Thorell JO et al. Am J Physiol 1995; 269(5 Pt 1):E948-E959.
1681. Hasselbalch SG, Knudsen GM et al. J Cereb Blood Flow Metab 1994; 14(1):125–131.
1682. Ellison Z, Foong J et al. Lancet 1998; 352(9135):1192.
1683. Gordon CM, Dougherty DD et al. J Pediatr 2001; 139(1):51–57.
1684. Kaye WH, Frank GK et al. Am J Psychiatry 2001; 158(7):1152–1155.
1685. Frank GK, Kaye WH et al. Biol Psychiatry 2002; 52(9):896–906.
1686. Dishino DD, Welch MJ et al. J Nucl Med 1983; 24(11):1030–1038.
1687. Yamamoto YL, Thompson CJ et al. J Comput Assist Tomogr 1977; 1(1):43–56.
1688. Schlageter NL, Carson RE et al. J Cereb Blood Flow Metab 1987; 7(1):1–8.
1689. Dhawan V, Jarden JO et al. Physics in Medicine & Biology 1988; 33(1):61–74.
1690. Jansen HM, van der NJ et al. J Neurol Neurosurg Psychiatry 1996; 60(2):221–224.
1691. De Reuck J, Vonck K et al. J Neurol Sci 2000; 181(1–2):13–18.
1692. De Reuck J, Santens P et al. J Neurol Sci 2001; 193(1):1–6.
1693. Stevens H, Jansen HM et al. J Neurol Sci 1999; 171(1):11–18.
1694. Bruehlmeier M, Roelcke U et al. J Nucl Med 2003; 44(8):1210–1218.
1695. Herscovitch P, Raichle ME et al. J Cereb Blood Flow Metab 1987; 7(5):527–542.
1696. Herholz K, Pietrzyk U et al. Stroke 1989; 20(9):1174–1181.
1697. Yvert JP, Maziere B et al. Eur J Nucl Med 1979; 4(2):95–99.
1698. Brooks DJ, Frackowiak RS et al. Acta Neurologica Scandinavica 1986; 73(4):415–422.
1699. Lassen NA. J Cereb Blood Flow Metab 1985; 5(3):347–349.
1700. Madsen PL, Sperling BK et al. J Appl Physiol 1993; 74(1):245–250.
1701. Hurn PD, Traystman RJ.In: Cerebral blood flow and metabolism 2002; 2(24):384–394.
1702. Lammertsma AA, Brooks DJ et al. J Cereb Blood Flow Metab 1984; 4(3):317–322.
1703. Yamauchi H, Fukuyama H et al. J Cereb Blood Flow Metab 1999; 19(1):109–114.
1704. Martin WR, Powers WJ et al. J Cereb Blood Flow Metab 1987; 7(4):421–426.

1705. Pantano P, Baron JC et al. Eur J Nucl Med 1985; 10(9–10):387–391.

1706. Leenders KL, Perani D et al. Brain 1990; 113 (Pt 1):27–47.

1707. Pantano P, Baron JC et al. Stroke 1984; 15(4):635–641.

1708. Martin AJ, Friston KJ et al. J Cereb Blood Flow Metab 1991; 11(4):684–689.

1709. Marchal G, Rioux P et al. Arch Neurol 1992; 49(10):1013–1020.

1710. Takada H, Nagata K et al. Neurol Res 1992; 14(2 Suppl):128–131.

1711. Meltzer CC, Cantwell MN et al. J Nucl Med 2000; 41(11):1842–1848.

1712. Jones T, Chesler DA et al. Br J Radiol 1976; 49(580):339–343.

1713. Frackowiak RS, Lenzi GL et al. Journal of Computer Assisted Tomography 1980; 4(6):727–736.

1714. Mintun MA, Raichle ME et al. J Nucl Med 1984; 25(2):177–187.

1715. Lammertsma AA, Wise RJ et al. J Cereb Blood Flow Metab 1983; 3(4):425–431.

1716. Herscovitch P, Mintun MA et al. J Nucl Med 1985; 26(4):416–417.

1717. Meyer E, Tyler JL et al. J Cereb Blood Flow Metab 1987; 7(4):403–414.

1718. Huang SC, Feng DG et al. J Cereb Blood Flow Metab 1986; 6(1):105–119.

1719. Ohta S, Meyer E et al. J Cereb Blood Flow Metab 1992; 12(2):179–192.

1720. Seki C, Kershaw J et al. J Cereb Blood Flow Metab 2003; 23(7):838–844.

1721. Hayashi T, Watabe H et al. J Cereb Blood Flow Metab 2003; 23(11):1314–1323.

1722. Burns A, Tyrrell P. Age Ageing 1992; 21(5):316–320.

1723. Takahashi T, Shirane R et al. AJNR Am J Neuroradiol 1999; 20(5):917–922.

1724. Fox PT, Raichle ME. Proc Natl Acad Sci USA 1986; 83(4):1140–1144.

1725. Marrett S, Gjedde A. Adv Exp Med Biol 1997; 413:205–8:205–208.

1726. Fujita H, Kuwabara H et al. J Cereb Blood Flow Metab 1999; 19(3):266–271.

1727. Shulman RG, Hyder F et al. Proc Natl Acad Sci U S A 2001; 98(11):6417–6422.

1728. Schmalbruch IK, Linde R et al. Stroke 2002; 33(1):251–255.

1729. Madsen PL, Cruz NF et al. J Cereb Blood Flow Metab 1999; 19(4):393–400.

1730. Gjedde A, Marrett S. J Cereb Blood Flow Metab 2001; 21(12):1384–1392.

1731. Simpson IA, Appel NM et al. J Neurochem 1999; 72(1):238–247.

1732. Hasselbalch SG, Knudsen GM et al. J Clin Endocrinol Metab 2001; 86(5):1986–1990.

1733. Pascual JM, Van Heertum RL et al. Ann Neurol 2002; 52(4):458–464.

1734. Vannucci SJ, Clark RR et al. Dev Neurosci 1998; 20(4–5):369–379.

1735. Gatley SJ. J Nucl Med 2003; 44(7):1082–1086.

1736. Chih CP, Roberts Jr EL. J Cereb Blood Flow Metab 2003; 23(11):1263–1281.

1737. Magistretti PJ, Pellerin L. Mol Psychiatry 1996; 1(6):445–452.

1738. Redies C, Hoffer LJ et al. American Journal of Physiology 1989; 256(6 Pt 1):E805–10.

1739. Hasselbalch SG, Madsen PL et al. Am J Physiol 1996; 270(5 Pt 1):E746-E751.

1740. Reivich M, Kuhl D et al. Circulation Research 1979; 44(1):127–137.

1741. Reivich M, Alavi A et al. J Cereb Blood Flow Metab 1985; 5(2):179–192.

1742. Ehrin E, Stone-Elander S et al. J Nucl Med 1983; 24(4):326–331.

1743. Jones SC, Ackerman RH et al. International Journal of Nuclear Medicine & Biology 1983; 10(4):173–180.

1744. Powers WJ, Dagogo-Jack S et al. Annals of Neurology 1995; 38(4):599–609.

1745. Graham MM, Peterson LM et al. J Nucl Med 1998; 39(10):1805–1810.

1746. Blomqvist G, Stone-Elander S et al. J Cereb Blood Flow Metab 1990; 10(4):467–483.

1747. Hasselbalch SG, Madsen PL et al. J Cereb Blood Flow Metab 1998; 18(2):154–160.

1748. Loessner A, Alavi A et al. J Nucl Med 1995; 36(7):1141–1149.

1749. Duara R, Grady C et al. Annals of Neurology 1984; 16(6):703–713.

1750. Volkow ND, Logan J et al. Am J Psychiatry 2000; 157(1):75–80.

1751. Bentourkia M, Michel C et al. Brain Dev 1998; 20(7):524–529.

1752. Chugani HT, Phelps ME et al. Annals of Neurology 1987; 22(4):487–497.

1753. Moore AH, Hovda DA et al. Brain Res Dev Brain Res 2000; 120(2):141–150.

1754. Jacobs B, Chugani HT et al. Cerebral Cortex 1995; 5(3):222–233.

1755. Gur RC, Mozley LH et al. Science 1995; 267(5197):528–531.

1756. Volkow ND, Wang GJ et al. American Journal of Psychiatry 1997; 154(1):119–121.
1757. Murphy DGM, DeCarli C et al. Arch Gen Psychiatry 1996; 53(7):585–594.
1758. Kawachi T, Ishii K et al. J Neurol Sci 2002; 199(1–2):79–83.
1759. Andreason PJ, Zametkin AJ et al. Psychiatry Research 1994; 51:175–183.
1760. Baxter LR, Jr., Mazziotta JC et al. Psychiatry Res 1987; 21(3):237–245.
1761. Rasgon NL, Small GW et al. Psychiatry Res 2001; 107(1):11–18.
1762. Tsuchida T, Sadato N et al. Eur J Nucl Med Mol Imaging 2002; 29(2):248–250.
1763. Rothman DL, Behar KL et al. Annu Rev Physiol 2003; eprint.
1764. Sibson NR, Dhankhar A et al. Proc Natl Acad Sci USA 1998; 95(1):316–321.
1765. Sokoloff L. Developmental Neuroscience 1993; 15(3–5):194–206.
1766. Mata M, Fink DJ et al. Journal of Neurochemistry 1980; 34(1):213–215.
1767. Dienel GA, Wang RY et al. J Cereb Blood Flow Metab 2002; 22(12):1490–1502.
1768. Heiss WD, Pawlik G et al. Brain Research 1985; 327(1–2):362–366.
1769. Maquet P, Degueldre C et al. Journal of Neuroscience 1997; 17(8):2807–2812.
1770. Andersson JLR, Onoe H et al. J Cereb Blood Flow Metab 1998; 18(7):701–715.
1771. Kjaer TW, Law I et al. J Sleep Res 2002; 11(3):201–207.
1772. Maquet P, Peters JM et al. Nature 1996; 382:163–166.
1773. Braun AR, Balkin TJ et al. Science 1998; 279(5347):91–95.
1774. Nofzinger EA, Mintun MA et al. Brain Res 1997; 770(1–2):192–201.
1775. Gottschalk LA, Buchsbaum MS et al. Brain Res 1991; 538(1):107–110.
1776. Hong CC, Gillin JC et al. Sleep 1995; 18(7):570–580.
1777. Braun AR, Balkin TJ et al. Brain 1997; 120(Pt 7):1173–1197.
1778. Born AP, Law I et al. Neuroimage 2002; 17(3):1325–1335.
1779. Grond M, Pawlik G et al. Psychiatry Res 1995; 61(3):173–179.
1780. Alkire MT, Pomfrett CJ et al. Anesthesiology 1999; 90(3):701–709.
1781. Fiset P, Paus T et al. J Neurosci 1999; 19(13):5506–5513.
1782. Alkire MT, Haier RJ et al. Anesthesiology 1997; 86(3):549–557.
1783. Alkire MT, Haier RJ et al. Anesthesiology 1995; 82(2):393–403.
1784. Alkire MT, Haier RJ. Br J Anaesth 2001; 86(5):618–626.
1785. Kaisti KK, Metsahonkala L et al. Anesthesiology 2002; 96(6):1358–1370.
1786. Juengling FD, Kassubek J et al. Neuroscience Letters 2002; 335(2):79–82.
1787. Ogawa K, Uema T et al. Anesthesiology 2003; 98(5):1101–1111.
1788. Theodore WH. Epilepsia 1988; 29(Supplement 2):S48–55.
1789. Buchsbaum MS, Wu J et al. Life Sci 1987; 40(25):2393–2400.
1790. Gillin JC, Buchsbaum MS et al. Neuropsychopharmacology 1996; 15(3):302–313.
1791. Wang GJ, Volkow ND et al. Psychiatry Res 1998; 82(1):37–46.
1792. Moresco RM, Tettamanti M et al. Nucl Med Commun 2001; 22(4):399–404.
1793. Bagary M, Fluck E et al. Psychopharmacology (Berl) 2000; 150(3):292–299.
1794. London ED, Broussolle EP et al. Arch Gen Psychiatry 1990; 47(1):73–81.
1795. Walsh SL, Gilson SF et al. Neuropsychopharmacology 1994; 10(3):157–170.
1796. Firestone LL, Gyulai F et al. Anesth Analg 1996; 82(6):1247–1251.
1797. Bonhomme V, Fiset P et al. J Neurophysiol 2001; 85(3):1299–1308.
1798. Treyer V, Jobin M et al. Eur J Nucl Med Mol Imaging 2003; 30:572–580.
1799. Heiss WD, Podreka I. European Neurology 1978; 17 Suppl 1:135–143.
1800. Sollevi A, Ericson K et al. J Cereb Blood Flow Metab 1987; 7(6):673–678.
1801. Stange K, Greitz D et al. Acta Physiol Scand 1997; 160(2):117–122.
1802. Blin J, Piercey MF et al. Journal of the Neurological Sciences 1994; 123(1–2):44–51.
1803. Molchan SE, Matochik JA et al. Neuropsychopharmacology 1994; 10(3):191–198.
1804. Cohen RM, Gross M et al. Exp Brain Res 1994; 100(1):133–143.
1805. Tsukada H, Kakiuchi T et al. Brain Res 1997; 749(1):10–17.
1806. Bahro M, Molchan SE et al. Neuropsychobiology 1999; 39(4):187–195.
1807. Friston KJ, Grasby PM et al. Neuroscience Letters 1992; 141(1):106–110.
1808. Gottschalk LA, Buchsbaum MS et al. Compr Psychiatry 1992; 33(1):52–59.

1809. Mazziotta JC, Phelps ME et al. Annals of Neurology 1982; 12(5):435–444.
1810. Schmidt ME, Ernst M et al. J Nucl Med 1996; 37(7):1142–1149.
1811. Siegel BV, Jr., Nuechterlein KH et al. Schizophr Res 1995; 17(1):85–94.
1812. Duara R, Gross-Glenn K et al. J Cereb Blood Flow Metab 1987; 7(3):266–271.
1813. Takamatsu H, Noda A et al. J Nucl Med 2003; 44(9):1516–1521.
1814. Mountz JM, Modell JG et al. Arch Gen Psychiatry 1989; 46(6):501–504.
1815. Stapleton JM, Morgan MJ et al. J Cereb Blood Flow Metab 1997; 17(6):704–712.
1816. Bartlett EJ, Brodie JD et al. J Cereb Blood Flow Metab 1988; 8(4):502–512.
1817. Giordani B, Boivin MJ et al. Psychiatry Res 1990; 35(1):49–60.
1818. Gur RC, Gur RE et al. J Cereb Blood Flow Metab 1987; 7(2):173–177.
1819. Mazziotta JC, Phelps ME et al. Neurology 1982; 32(9):921–937.
1820. Phelps ME, Kuhl DE et al. Science 1981; 211(4489):1445–1448.
1821. Fox PT, Mintun MA et al. J Cereb Blood Flow Metab 1988; 8(5):642–653.
1822. Friston KJ, Frith CD et al. J Cereb Blood Flow Metab 1991; 11(4):690–699.
1823. Sidtis JJ, Strother SC et al. Neuroimage 2003; 20(2):615–624.
1824. Powers WJ, Fox PT et al. Neurology 1988; 38(9):1475–1478.
1825. Frackowiak RSJ, Friston KJ et al. Human brain function.Academic Press, San Diego, 1997.
1826. Mazziotta JC et al., Eds. Brain mapping: The disorders. Academic Press, 2000.
1827. Cabeza R, Kingstone A. Handbook of Functional Neuroimaging of Cognition. MIT Press, 2001.
1828. Buxton RB, Alpert NM et al. J Cereb Blood Flow Metab 1987; 7(6):709–719.
1829. Buxton RB, Wechsler LR et al. J Cereb Blood Flow Metab 1984; 4(1):8–16.
1830. Alpert NM, Buxton RB et al. J Cereb Blood Flow Metab 1988; 8(3):403–410.
1831. Kearfott KJ, Junck L et al. J Nucl Med 1983; 24(9):805–811.
1832. Boado RJ, Li JY et al. Proc Natl Acad Sci U S A 1999; 96(21):12079–12084.
1833. Matsuo H, Tsukada S et al. Neuroreport 2000; 11(16):3507–3511.
1834. Killian DM, Chikhale PJ. Neurosci Lett 2001; 306(1–2):1–4.
1835. Smith QR, Momma S et al. J Neurochem 1987; 49(5):1651–1658.
1836. Kanai Y, Segawa H et al. J Biol Chem 1998; 273(37):23629–23632.
1837. Pardridge WM. Neurochem Res 1998; 23(5):635–644.
1838. Romanowski CA, Leslie DF et al. Neuroradiology 1997; 39(6):389–393.
1839. Sun Y, Deibler GE et al. Journal of Neurochemistry 1992; 59(3):863–873.
1840. Vaalburg W, Coenen HH et al. International Journal of Radiation Applications & Instrumentation – Part B, Nuclear Medicine & Biology 1992; 19(2):227–237.
1841. Phelps ME, Barrio JR et al. Annals of Neurology 1984; 15 Suppl:S192–202.
1842. Keen RE, Barrio JR et al. J Cereb Blood Flow Metab 1989; 9(4):429–445.
1843. Smith CB, Deibler GE et al. Proc Natl Acad Sci USA 1988; 85(23):9341–9345.
1844. Ishiwata K, Kubota K et al. J Nucl Med 1993; 34(11):1936–1943.
1845. Wienhard K, Herholz K et al. J Nucl Med 1991; 32(7):1338–1346.
1846. Miyagawa T, Oku T et al. J Cereb Blood Flow Metab 1998; 18(5):500–509.
1847. Heiss WD, Wienhard K et al. J Nucl Med 1996; 37:1180–1182.
1848. Beuthien-Baumann B, Bredow J et al. Eur J Nucl Med Mol Imaging 2003.
1849. Coenen HH.In: PET studies on amino acid metabolism and protein synthesis 1993; 109–129.
1850. Wester HJ, Herz M et al. J Nucl Med 1999; 40(1):205–212.
1851. Pauleit D, Floeth F et al. Eur J Nucl Med Mol Imaging 2003.
1852. Weber WA, Wester HJ et al. Eur J Nucl Med 2000; 27(5):542–549.
1853. Ito H, Hatazawa J et al. J Nucl Med 1995; 36(7):1232–1237.
1854. McConathy J, Martarello L et al. J Med Chem 2002; 45(11):2240–2249.
1855. Sutinen E, Jyrkkio S et al. Eur J Nucl Med 2001; 28(7):847–854.
1856. Blasberg RG, Fenstermacher JD et al. J Cereb Blood Flow Metab 1983; 3(1):8–32.
1857. Koeppe RA, Mangner T et al. J Cereb Blood Flow Metab 1990; 10(5):727–739.
1858. O'Tuama LA, Guilarte TR et al. J Cereb Blood Flow Metab 1988; 8(3):341–345.

1859. Planas AM, Prenant C et al. In: PET studies on amino acid metabolism and protein synthesis (Mazoyer BM et al., Eds.). Kluwer Academic Publishers, 1993; 53–68.
1860. Sato K, Kameyama M et al. Eur J Nucl Med 1992; 19(6):426–430.
1861. O'Tuama LA, Phillips PC et al. J Nucl Med 1991; 32(1):16–22.
1862. Schober O, Duden C et al. Eur J Nucl Med 1987; 13(2):103–105.
1863. Bergstrom M, Ericson K et al. Journal of Computer Assisted Tomography 1987; 11(2):208–213.
1864. Hawkins RA, Huang SC et al. J Cereb Blood Flow Metab 1989; 9(4):446–460.
1865. Wiesel FA, Blomqvist G et al. J Nucl Med 1991; 32(11):2043–2049.
1866. Pruim J, Willemsen ATM et al. Radiology 1995; 197(1):221–226.
1867. Willemsen AT, van Waarde A et al. J Nucl Med 1995; 36(3):411–419.
1868. Kole AC, Pruim J et al. J Nucl Med 1997; 38(2):191–195.
1869. Paans AM, Pruim J et al. European Journal of Pediatrics 1996; 155 Suppl 1:S78–81.
1870. Ishiwata K, Kubota K et al. Nuclear Medicine & Biology 1993; 20(8):895–899.
1871. Li JY, Boado RJ et al. J Cereb Blood Flow Metab 2001; 21(8):929–936.
1872. Pardridge WM.In: Cerebral blood flow and metabolism 2002; 2(7):119–139.
1873. Betz AL, Goldstein GW et al.In: Basic neurochemistry 2003;(32):681–699.
1874. Sherley JL, Kelly TJ. J Biol Chem 1988; 263(17):8350–8358.
1875. Eriksson S, Arner E et al. Adv Enzyme Regul 1994; 34:13–25.
1876. Wells P, Aboagye E et al. J Natl Cancer Inst 2003; 95(9):675–682.
1877. Shields AF, Mankoff D et al. J Nucl Med 1996; 37(2):290–296.
1878. Mankoff DA, Shields AF et al. J Nucl Med 1998; 39(6):1043–1055.
1879. Eary JF, Mankoff DA et al. Cancer Res 1999; 59(3):615–621.
1880. De Reuck J, Santens P et al. Acta Neurol Belg 1999; 99(2):118–125.
1881. Grierson JR, Shields AF. Nucl Med Biol 2000; 27(2):143–156.
1882. Buck AK, Halter G et al. J Nucl Med 2003; 44(9):1426–1431.
1883. Wagner M, Seitz U et al. Cancer Res 2003; 63(10):2681.
1884. Vesselle H, Grierson J et al. J Nucl Med 2003; 44(9):1482–1488.
1885. Conti PS, Alauddin MM et al. Nucl Med Biol 1995; 22(6):783–789.
1886. Mangner TJ, Klecker RW et al. Nucl Med Biol 2003; 30(3):215–224.
1887. Lu L, Samuelsson L et al. J Nucl Med 2002; 43(12):1688–1698.
1888. Borbath I, Gregoire V et al. Eur J Nucl Med Mol Imaging 2002; 29(1):19–27.
1889. Tjuvajev JG, Stockhammer G et al. Cancer Res 1995; 55(24):6126–6132.
1890. Luker GD, Sharma V et al. Proc Natl Acad Sci U S A 2002; 99(10):6961–6966.
1891. Blasberg RG, Tjuvajev JG. J Clin Invest 2003; 111(11):1620–1629.
1892. Pardridge WM. J Cereb Blood Flow Metab 1997; 17(7):713–731.
1893. Shi N, Boado RJ et al. Proc Natl Acad Sci U S A 2000; 97(26):14709–14714.
1894. Lee HJ, Zhang Y et al. J Cereb Blood Flow Metab 2002; 22(2):223–231.
1895. Wu D, Yang J et al. J Clin Invest 1997; 100(7):1804–1812.
1896. Coloma MJ, Lee HJ et al. Pharm Res 2000; 17(3):266–274.
1897. Zhang Y, Jeong LH et al. J Gene Med 2002; 4(2):183–194.
1898. Wengenack TM, Curran GL et al. Nat Biotechnol 2000; 18(8):868–872.
1899. Doubrovin M, Ponomarev V et al. Mol Imaging 2003; 2(2):93–112.
1900. Kreuter J, Shamenkov D et al. J Drug Target 2002; 10(4):317–325.
1901. McKenzie R, Fried MW et al. N Engl J Med 1995; 333(17):1099–1105.
1902. Tjuvajev JG, Finn R et al. Cancer Res 1996; 56(18):4087–4095.
1903. Tjuvajev JG, Avril N et al. Cancer Res 1998; 58(19):4333–4341.
1904. Tjuvajev JG, Doubrovin M et al. J Nucl Med 2002; 43(8):1072–1083.
1905. Jacobs A, Braunlich I et al. J Nucl Med 2001; 42(3):467–475.
1906. Alauddin MM, Conti PS. Nucl Med Biol 1998; 25(3):175–180.
1907. Alauddin MM, Shahinian A et al. J Nucl Med 2001; 42(11):1682–1690.
1908. Yaghoubi S, Barrio JR et al. J Nucl Med 2001; 42(8):1225–1234.
1909. Min JJ, Iyer M et al. Eur J Nucl Med Mol Imaging 2003.

1910. Ponomarev V, Doubrovin M et al. Neoplasia 2001; 3(6):480–488.
1911. Mayer-Kuckuk P, Banerjee D et al. Proc Natl Acad Sci U S A 2002; 99(6):3400–3405.
1912. Ray P, Wu AM et al. Cancer Res 2003; 63(6):1160–1165.
1913. De A, Lewis XZ et al. Mol Ther 2003; 7(5):681–691.
1914. Tjuvajev JG, Joshi A et al. Neoplasia 1999; 1(4):315–320.
1915. Liang Q, Gotts J et al. Mol Ther 2002; 6(1):73–82.
1916. Zinn KR, Chaudhuri TR. Eur J Nucl Med Mol Imaging 2002; 29(3):388–399.
1917. Rogers BE, McLean SF et al. Clin Cancer Res 1999; 5(2):383–393.
1918. MacLaren DC, Gambhir SS et al. Gene Ther 1999; 6(5):785–791.
1919. Tavitian B, Terrazzino S et al. Nat Med 1998; 4(4):467–471.
1920. Tavitian B, Marzabal S et al. Pharm Res 2002; 19(4):367–376.
1921. Glaser M, Collingridge DR et al. Appl Radiat Isot 2003; 58(1):55–62.
1922. Doubrovin M, Ponomarev V et al. Proc Natl Acad Sci U S A 2001; 98(16):9300–9305.
1923. Adonai N, Nguyen KN et al. Proc Natl Acad Sci U S A 2002; 99(5):3030–3035.
1924. Koehne G, Doubrovin M et al. Nat Biotechnol 2003.
1925. Olasz EB, Lang L et al. J Immunol Methods 2002; 260(1–2):137–148.
1926. Collingridge DR, Carroll VA et al. Cancer Res 2002; 62(20):5912–5919.
1927. Brown WD, Oakes TR et al. J Nucl Med 1998; 39(11):1884–1891.
1928. Firnau G, Sood S et al. J Neurochem 1987; 48(4):1077–1082.
1929. Cumming P, Munk OL et al. Synapse 2001; 41(3):212–218.
1930. Yee RE, Huang SC et al. J Neurochem 2000; 74(3):1147–1157.
1931. Brown WD, Taylor MD et al. Neurology 1999; 53(6):1212–1218.
1932. Firnau G, Sood S et al. J Nucl Med 1988; 29(3):363–369.
1933. Wahl L, Chirakal R et al. J Cereb Blood Flow Metab 1994; 14(4):664–670.
1934. Doudet DJ, McLellan CA et al. J Cereb Blood Flow Metab 1991; 11(5):726–734.
1935. Wahl L, Nahmias C. J Nucl Med 1996; 37(3):432–437.
1936. Patlak CS, Blasberg RG et al. J Cereb Blood Flow Metab 1983; 3(1):1–7.
1937. Sossi V, Holden JE et al. J Cereb Blood Flow Metab 2003; 23(3):301–309.
1938. Nahmias C, Wahl L et al. Mov Disord 1995; 10(3):298–304.
1939. Jordan S, Bankiewicz KS et al. Neurochem Res 1998; 23(4):513–517.
1940. Brown WD, DeJesus OT et al. Synapse 1999; 34(2):111–123.
1941. DeJesus OT, Endres CJ et al. Synapse 2001; 39(1):58–63.
1942. Eberling JL, Jagust WJ et al. Brain Res 1998; 805(1–2):259–262.
1943. Jordan S, Eberling JL et al. Brain Res 1997; 750(1–2):264–276.
1944. Doudet DJ, Chan GL et al. J Cereb Blood Flow Metab 1999; 19(3):278–287.
1945. Barrio JR, et al. J Cereb Blood Flow Metab 1996; 16(4):667–678.
1946. Hayase N, Tomiyoshi K et al. Ann Nucl Med 1995; 9(3):119–123.
1947. DeJesus OT, Holden JE et al. Brain Research 1992; 597(1):151–154.
1948. Huang SC, Quintana J et al. Nucl Med Biol 1999; 26(4):365–370.
1949. Tedroff J, Aquilonius SM et al. Acta Neurol Scand 1992; 85(2):95–102.
1950. Torstenson R, Tedroff J et al. J Cereb Blood Flow Metab 1999; 19(10):1142–1149.
1951. Torres GE, Gainetdinov RR et al. Nat Rev Neurosci 2003; 4(1):13–25.
1952. Fowler JS, Volkow ND et al. Synapse 1989; 4(4):371–377.
1953. Yu DW, Gatley SJ et al. J Med Chem 1992; 35(12):2178–2183.
1954. Fowler JS, Volkow ND et al. Synapse 1992; 12(3):220–227.
1955. Gatley SJ, Yu DW et al. J Neurochem 1994; 62(3):1154–1162.
1956. Telang FW, Volkow ND et al. Synapse 1999; 31(4):290–296.
1957. Wang GJ, Volkow ND et al. Life Sci 1995; 57(14):L187–L191.
1958. Volkow ND, Wang GJ et al. Life Sci 1999; 65(1):L7–12.
1959. Volkow ND, Fowler JS et al. J Psychopharmacol 1999; 13(4):337–345.
1960. Volkow ND, Wang GJ et al. J Pharmacol Exp Ther 1999; 288(1):14–20.
1961. Ding YS, Fowler JS et al. Psychopharmacology (Berl) 1997; 131(1):71–78.
1962. Meltzer PC, Liang AY et al. J Med Chem 1993; 36(7):855–862.

1963. Neumeyer JL, Wang S et al. J Med Chem 1994; 37(11):1558–1561.
1964. Wilson AA, DaSilva JN et al. Nucl Med Biol 1996; 23(2):141–146.
1965. Hume SP, Luthra SK et al. Nucl Med Biol 1996; 23(3):377–384.
1966. Muller L, Halldin C et al. Journal of Radioanalytical & Nuclear Chemistry 1996; 206(1):133–144.
1967. Lundkvist C, Halldin C et al. Nuclear Medicine & Biology 1995; 22(7):905–913.
1968. Halldin C, Farde L et al. Synapse 1996; 22(4):386–390.
1969. Wong DF, Yung B et al. Synapse 1993; 15(2):130–142.
1970. Bergstrom KA, Halldin C et al. Eur J Nucl Med 1997; 24(6):596–601.
1971. Helfenbein J, Sandell J et al. Nucl Med Biol 1999; 26(5):491–499.
1972. Haaparanta M, Bergman J et al. Synapse 1996; 23(4):321–327.
1973. Stout D, Petric A et al. Nucl Med Biol 1999; 26(8):897–903.
1974. Petric A, Barrio JR et al. Nucl Med Biol 1999; 26(5):529–535.
1975. Gu XH, Zong R et al. Bioorg Med Chem Lett 2001; 11(23):3049–3053.
1976. Helfenbein J, Loc'h C et al. Life Sci 1999; 65(25):2715–2726.
1977. Hume SP, Brown DJ et al. J Neurosci Methods 1997; 76(1):45–51.
1978. Fowler JS, Volkow ND et al. Synapse 1998; 28(2):111–116.
1979. Lundkvist C, Halldin C et al. Nucl Med Biol 1997; 24(7):621–627.
1980. Schonbachler RD, Gucker PM et al. Nucl Med Biol 2002; 29(1):19–27.
1981. Halldin C, Erixon-Lindroth N et al. Eur J Nucl Med Mol Imaging 2003.
1982. Goodman MM, Kilts CD et al. Nuclear Medicine and Biology 2000; 27(1):1–12.
1983. Tsukada H, Nishiyama S et al. Brain Res 1999; 849(1–2):85–96.
1984. Votaw J, Byas-Smith M et al. Anesthesiology 2003; 98(2):404–411.
1985. Scheffel U, Steinert C et al. Synapse 1996; 23(2):61–69.
1986. Volkow ND, Fowler JS et al. Ann Neurol 1994; 36(2):237–239.
1987. Harada N, Nishiyama S et al. Synapse 2002; 45(1):38–45.
1988. DeJesus OT, Shelton SE et al. Synapse 2002; 44(4):246–251.
1989. Halldin C, Stone-Elander S et al. Int J Rad Appl Instrum [A] 1986; 37(10):1039–1043.
1990. Gifford AN, Gatley SJ et al. Synapse 1998; 28(2):167–175.
1991. Farde L, Halldin C et al. Psychopharmacology (Berl) 1987; 92(3):278–284.
1992. Farde L. Psychopharmacology (Berl) 1992; 107(1):23–29.
1993. Christian BT, Babich JW et al. Int J Mol Med 1998; 1(1):243–247.
1994. Rinne JO, Laihinen A et al. J Neurosci Res 1990; 27(4):494–499.
1995. Wang Y, Chan GL et al. Synapse 1998; 30(1):56–61.
1996. Abi-Dargham A, Martinez D et al. J Cereb Blood Flow Metab 2000; 20(2):225–243.
1997. Karlsson P, Farde L et al. Psychopharmacology 1995; 119(1):1–8.
1998. Laihinen AO, Rinne JO et al. J Nucl Med 1994; 35(12):1916–1920.
1999. Karlsson P, Farde L et al. Psychopharmacology 1993; 113(2):149–156.
2000. Sedvall G, Farde L et al. Psychopharmacology (Berl) 1991; 103(2):150–153.
2001. Karlsson P, Sedvall G et al. Psychopharmacology 1995; 121(3):300–308.
2002. Wong DF, Gjedde A et al. J Cereb Blood Flow Metab 1986; 6(2):137–146.
2003. Wagner HN, Jr., Burns HD et al. Annals of Neurology 1984; 15 Suppl:S79–84.
2004. Coenen HH, Laufer P et al. Life Sciences 1987; 40(1):81–88.
2005. Barrio JR, Satyamurthy N et al. J Cereb Blood Flow Metab 1989; 9(6):830–839.
2006. Jovkar S, Wienhard K et al. Eur J Nucl Med 1991; 18(3):158–163.
2007. Coenen HH, Wienhard K et al. Eur J Nucl Med 1988; 14(2):80–87.
2008. Swart JA, van der Werf JF et al. J Cereb Blood Flow Metab 1990; 10(3):297–306.
2009. Young LT, Wong DF et al. Synapse 1991; 9(3):188–194.
2010. Hagberg G, Gefvert O et al. Psychiatry Res 1998; 82(3):147–160.
2011. Huang SC, Bahn MM et al. J Cereb Blood Flow Metab 1989; 9(6):850–858.
2012. Bahn MM, Huang SC et al. J Cereb Blood Flow Metab 1989; 9(6):840–849.
2013. Maziere B, Loc'h C et al. Eur J Pharmacol 1985; 114(3):267–272.
2014. Farde L, Ehrin E et al. Proc Natl Acad Sci U S A 1985; 82(11):3863–3867.

2015. Farde L, Pauli S et al. Psychopharmacology (Berl) 1988; 94(4):471–478.
2016. Farde L, Hall H et al. Science 1986; 231(4735):258–261.
2017. Farde L, Eriksson L et al. J Cereb Blood Flow Metab 1989; 9(5):696–708.
2018. Ito H, Hietala J et al. J Cereb Blood Flow Metab 1998; 18(9):941–950.
2019. Logan J, Volkow ND et al. J Cereb Blood Flow Metab 1994; 14(6):995–1010.
2020. Bench CJ, Lammertsma AA et al. Psychopharmacology 1993; 112(2–3):308–314.
2021. Nordstrom AL, Farde L et al. Biological Psychiatry 1993; 33(4):227–235.
2022. Nordstrom AL, Farde L et al. Psychopharmacology 1992; 106(4):433–438.
2023. Farde L, Hall H et al. Science 1986; 231(4735):258–261.
2024. Lammertsma AA, Bench CJ et al. J Cereb Blood Flow Metab 1996; 16(1):42–52.
2025. Seeman P, Guan HC et al. Eur J Pharmacol 1992; 227(2):139–146.
2026. Wong DF, Gjedde A et al. J Cereb Blood Flow Metab 1986; 6(2):147–153.
2027. Rinne JO, Hietala J et al. J Cereb Blood Flow Metab 1993; 13(2):310–314.
2028. Antonini A, Leenders KL et al. Arch Neurol 1993; 50(5):474–480.
2029. Pohjalainen T, Rinne JO et al. Am J Psychiatry 1998; 155(6):768–773.
2030. Farde L, Hall H et al. Synapse 1995; 20(3):200–208.
2031. Nyberg S, Farde L et al. Psychiatry Research: Neuroimaging 1996; 67(3):163–171.
2032. Schlosser R, Brodie JD et al. Synapse 1998; 28(1):66–70.
2033. Wang GJ, Volkow ND et al. J Nucl Med 1999; 40(8):1285–1291.
2034. Jonsson EG, Nothen MM et al. Mol Psychiatry 1999; 4(3):290–296.
2035. Ishiwata K, Senda M. Nucl Med Biol 1999; 26(6):627–631.
2036. Mukherjee J, Yang ZY et al. Nucl Med Biol 1995; 22(3):283–296.
2037. Mukherjee J, Christian BT et al. Synapse 2002; 46(3):170–188.
2038. Mukherjee J, Yang ZY et al. Life Sci 1996; 59(8):669–678.
2039. Grunder G, Siessmeier T et al. J Nucl Med 2003; 44(1):109–116.
2040. Mukherjee J, Christian BT et al. Neuropsychopharmacology 2001; 25(4):476–488.
2041. Carson RE, Breier A et al. J Cereb Blood Flow Metab 1997; 17(4):437–447.
2042. Alpert NM, Badgaiyan RD et al. Neuroimage 2003; 19(3):1049–1060.
2043. Chugani DC, Ackermann RF et al. J Cereb Blood Flow Metab 1988; 8(3):291–303.
2044. Laruelle M. J Cereb Blood Flow Metab 2000; 20(3):423–451.
2045. Passchier J, Gee A et al. Methods 2002; 27(3):278–286.
2046. Dewey SL, Smith GS et al. Synapse 1993; 13(4):350–356.
2047. Endres CJ, Kolachana BS et al. J Cereb Blood Flow Metab 1997; 17(9):932–942.
2048. Volkow ND, Wang GJ et al. Synapse 1994; 16(4):255–262.
2049. Drevets WC, Gautier C et al. Biol Psychiatry 2001; 49(2):81–96.
2050. Leyton M, Boileau I et al. Neuropsychopharmacology 2002; 27(6):1027–1035.
2051. Price JC, Drevets WC et al. J Nucl Med 2002; 43(8):1090–1100.
2052. Dewey SL, Logan J et al. Synapse 1991; 7(4):324–327.
2053. Mukherjee J, Yang ZY et al. Synapse 1997; 27(1):1–13.
2054. Smith JG, Raper SE et al. Human Gene Therapy 1997; 8(8):943–954.
2055. Aalto S, Hirvonen J et al. Psychopharmacology (Berl) 2002; 164(4):401–406.
2056. Breier A, Adler CM et al. Synapse 1998; 29(2):142–147.
2057. Smith GS, Schloesser R et al. Neuropsychopharmacology 1998; 18(1):18–25.
2058. Dewey SL, Smith GS et al. J Neurosci 1992; 12(10):3773–3780.
2059. Smith GS, Dewey SL et al. Am J Psychiatry 1997; 154(4):490–496.
2060. Koepp MJ, Gunn RN et al. Nature 1998; 393(6682):266–268.
2061. Fuente-Fernandez R, Phillips AG et al. Behav Brain Res 2002; 136(2):359–363.
2062. Strafella AP, Paus T et al. J Neurosci 2001; 21: RC157(15):1–4.
2063. Logan J, Fowler JS et al. J Neural Transm 2001; 108(3):279–286.
2064. Ginovart N, Hassoun W et al. Neuropsychopharmacology 2002; 27(1):72–84.
2065. Cumming P, Wong DF et al. Ann N Y Acad Sci 2002; 965:440–450.
2066. Zijlstra S, van der WH et al. Nucl Med Biol 1993; 20(1):7–12.
2067. Hwang DR, Kegeles LS et al. Nucl Med Biol 2000; 27(6):533–539.

2068. Cumming P, Gillings NM et al. Nucl Med Biol 2003; 30(5):547–553.
2069. Thibaut F, Vaugeois JM et al. Neuropharmacology 1996; 35(3):267–272.
2070. Chou YH, Karlsson P et al. Psychopharmacology (Berl) 1999; 146(2):220–227.
2071. Fowler JS, MacGregor RR et al. Science 1987; 235(4787):481–485.
2072. Logan J, Fowler JS et al. J Cereb Blood Flow Metab 2002; 22(11):1367–1376.
2073. Fowler JS, Wang GJ et al. J Nucl Med 1995; 36(7):1255–1262.
2074. Logan J, Fowler JS et al. Nucl Med Biol 2000; 27(1):43–49.
2075. Fowler JS, Volkow ND et al. J Clin Pharmacol 1999; Suppl:13S–16S.
2076. Fowler JS, Volkow ND et al. Neurobiol Aging 1997; 18(4):431–435.
2077. Murakami M, Kondoh Y et al. International Journal of Radiation Applications
      & Instrumentation – Part B, Nuclear Medicine & Biology 1992; 19(6):619–626.
2078. MacGregor RR, Halldin C et al. Biochem Pharmacol 1985; 34(17):3207–3210.
2079. Mukherjee J, Yang ZY. Nucl Med Biol 1999; 26(6):619–625.
2080. Fowler JS, Logan J et al. J Neurochem 2001; 79(5):1039–1046.
2081. Bottlaender M, Dolle F et al. J Pharmacol Exp Ther 2003; 305(2):467–473.
2082. Gilissen C, De Groot TJ et al. J Nucl Med 2003; 44(2):269–275.
2083. Bravo D, Parsons SM. Neurochemistry International 2002; 41(5):285–289.
2084. Widen L, Eriksson L et al. Neurosci Lett 1992; 136(1):1–4.
2085. Yokoi F, Komiyama T et al. Eur J Nucl Med 1993; 20(1):46–52.
2086. Nordberg A, Hartvig P et al. J Neural Transm Park Dis Dement Sect 1989; 1(3):195–205.
2087. Saji H, Magata Y et al. Chem Pharm Bull (Tokyo) 1992; 40(3):734–736.
2088. Nyback H, Halldin C et al. Psychopharmacology (Berl) 1994; 115(1–2):31–36.
2089. Nordberg A, Lundqvist H et al. Alzheimer Dis Assoc Disord 1995; 9(1):21–27.
2090. Sihver W, Fasth KJ et al. Nucl Med Biol 1999; 26(6):633–640.
2091. Valette H, Bottlaender M et al. Life Sci 1999; 64(5):L93–L97.
2092. Chefer SI, London ED et al. Synapse 2003; 48(1):25–34.
2093. Bottlaender M, Valette H et al. J Nucl Med 2003; 44(4):596–601.
2094. Sihver W, Fasth KJ et al. J Neurochem 1999; 73(3):1264–1272.
2095. Frey KA, Hichwa RD et al. Proc Natl Acad Sci U S A 1985; 82(19):6711–6715.
2096. Frey KA, Ciliax B et al. Neurochem Res 1991; 16(9):1017–1023.
2097. Dewey SL, MacGregor RR et al. Synapse 1990; 5(3):213–223.
2098. Dewey SL, Volkow ND et al. J Neurosci Res 1990; 27(4):569–575.
2099. Varastet M, Brouillet E et al. European Journal of Pharmacology 1992; 213(2):275–284.
2100. Mulholland GK, Otto CA et al. J Nucl Med 1992; 33(3):423–430.
2101. Frey KA, Koeppe RA et al. J Cereb Blood Flow Metab 1992; 12(1):147–154.
2102. Suhara T, Inoue O et al. Psychopharmacology (Berl) 1994; 113(3–4):311–317.
2103. Shinotoh H, Asahina M et al. J Neural Transm Park Dis Dement Sect 1994; 7(1):35–46.
2104. Mulholland GK, Kilbourn MR et al. Nuclear Medicine & Biology 1995; 22(1):13–17.
2105. Koeppe RA, Frey KA et al. J Cereb Blood Flow Metab 1994; 14(1):85–99.
2106. Lee KS, Frey KA et al. J Cereb Blood Flow Metab 1996; 16(2):303–310.
2107. Hartvig P, Torstenson R et al. Dementia & Geriatric Cognitive Disorders 1997;
      8(5):259–266.
2108. Zubieta JK, Koeppe RA et al. J Cereb Blood Flow Metab 1998; 18(6):619–631.
2109. Sihver S, Sihver W et al. J Pharmacol Exp Ther 1999; 290(2):917–922.
2110. Jagoda EM, Kiesewetter DO et al. Neuropharmacology 2003; 44(5):653–661.
2111. Carson RE, Kiesewetter DO et al. J Cereb Blood Flow Metab 1998; 18(10):1130–1142.
2112. Podruchny TA, Connolly C et al. Synapse 2003; 48(1):39–44.
2113. Mesulam MM, Geula C. Brain Res 1992; 577(1):112–120.
2114. Tavitian B, Pappata S et al. Neuroreport 1993; 4(5):535–538.
2115. Snyder SE, Tluczek L et al. Nucl Med Biol 1998; 25(8):751–754.
2116. Snyder SE, Gunupudi N et al. J Cereb Blood Flow Metab 2001; 21(2):132–143.
2117. Namba H, Irie T et al. Brain Research 1994; 667(2):278–282.
2118. Iyo M, Namba H et al. Lancet 1997; 349(9068):1805–1809.

2119. Namba H, Iyo M et al. Eur J Nucl Med 1999; 26(2):135–143.
2120. Koeppe RA, Frey KA et al. J Cereb Blood Flow Metab 1999; 19(10):1150–1163.
2121. Herholz K, Lercher M et al. Eur J Nucl Med 2001; 28:472–477.
2122. Zündorf G, Herholz K et al.In: Brain imaging using PET 2002;(7):41–46.
2123. Nagatsuka S, Fukushi K et al. J Cereb Blood Flow Metab 2001; 21(11):1354–1366.
2124. Musachio JL, Flesher JE et al. Nucl Med Biol 2002; 29(5):547–552.
2125. Funaki Y, Kato M et al. J Pharmacol Sci 2003; 91(2):105–112.
2126. De Vos F, Santens P et al. Nucl Med Biol 2000; 27(8):745–747.
2127. Mann JJ, Malone KM et al. J Cereb Blood Flow Metab 1996; 16(3):418–426.
2128. Diksic M, Grdisa M. Neurochemical Research 1995; 20(11):1353–1360.
2129. Chugani DC, Muzik O. J Cereb Blood Flow Metab 2000; 20(1):2–9.
2130. Chugani DC, Muzik O et al. Synapse 1998; 28(1):33–43.
2131. Chugani DC, Muzik O et al. Ann Neurol 1999; 45(3):287–295.
2132. Shoaf SE, Carson R et al. Neuropsychopharmacology 1998; 19(5):345–353.
2133. Chugani DC, Chugani HT et al. Annals of Neurology 1998; 44(6):858–866.
2134. Hartvig P, Lindner KJ et al. Journal of Neural Transmission – General Section 1992; 88(1):1–10.
2135. Hagberg GE, Torstenson R et al. J Cereb Blood Flow Metab 2002; 22(11):1352–1366.
2136. Suehiro M, Scheffel U et al. Life Sci 1993; 53(11):883–892.
2137. Szabo Z, Scheffel U et al. J Cereb Blood Flow Metab 1999; 19(9):967–981.
2138. Szabo Z, Kao PF et al. Synapse 1995; 20(1):37–43.
2139. Szabo Z, Scheffel U et al. J Cereb Blood Flow Metab 1995; 15(5):798–805.
2140. Parsey RV, Kegeles LS et al. J Nucl Med 2000; 41(9):1465–1477.
2141. Marjamaki P, Zessin J et al. Synapse 2003; 47(1):45–53.
2142. Brust P, Zessin J et al. Synapse 2003; 47(2):143–151.
2143. Houle S, Ginovart N et al. Eur J Nucl Med 2000; 27(11):1719–1722.
2144. Huang Y, Hwang DR et al. J Cereb Blood Flow Metab 2002; 22(11):1377–1398.
2145. Wilson AA, Ginovart N et al. Nucl Med Biol 2002; 29(5):509–515.
2146. Ginovart N, Wilson AA et al. Synapse 2003; 47(2):123–133.
2147. Meyer JH, Wilson AA et al. American Journal of Psychiatry 2001; 158:1843–1849.
2148. Ichise M, Liow JS et al. J Cereb Blood Flow Metab 2003; 23(9):1096–1112.
2149. Passchier J, van Waarde A. Eur J Nucl Med 2001; 28(1):113–129.
2150. Pike VW, McCarron JA et al. Eur J Pharmacol 1996; 301(1–3):R5–R7.
2151. Pike VW, Halldin C et al. Nucl Med Biol 2000; 27(5):449–455.
2152. Parsey RV, Slifstein M et al. J Cereb Blood Flow Metab 2000; 20(7):1111–1133.
2153. Rabiner EA, Messa C et al. Neuroimage 2002; 15(3):620–632.
2154. Parsey RV, Oquendo MA et al. Brain Res 2002; 954(2):173–182.
2155. Tauscher J, Verhoeff NP et al. Neuropsychopharmacology 2001; 24(5):522–530.
2156. Meltzer CC, Drevets WC et al. Brain Res 2001; 895(1–2):9–17.
2157. Tsukada H, Kakiuchi T et al. Synapse 2001; 42(4):242–251.
2158. Farde L, Andree B et al. Neuropsychopharmacology 2000; 22(4):422–429.
2159. Yasuno F, Suhara T et al. Am J Psychiatry 2003; 160(2):334–340.
2160. Sandell J, Halldin C et al. Nucl Med Biol 2002; 29(1):39–45.
2161. Sandell J, Halldin C et al. Nucl Med Biol 2001; 28(2):177–185.
2162. Shiue CY, Shiue GG et al. Synapse 1997; 25(2):147–154.
2163. Le Bars D, Lemaire C et al. Nucl Med Biol 1998; 25(4):343–350.
2164. Sanabria-Bohorquez SM, Biver F et al. European Journal of Nuclear Medicine & Molecular Imaging 2002; 29:76–81.
2165. Zimmer L, Mauger G et al. J Neurochem 2002; 80(2):278–286.
2166. Costes N, Merlet I et al. J Cereb Blood Flow Metab 2002; 22(6):753–765.
2167. Hume S, Hirani E et al. Synapse 2001; 41(2):150–159.
2168. Carson RE, Lang L et al. Nucl Med Biol 2000; 27(5):493–497.
2169. Carson RE, Wu Y et al. J Cereb Blood Flow Metab 2003; 23(2):249–260.

2170. Hall H, Farde L et al. Synapse 2000; 38(4):421–431.
2171. Kugaya A, Epperson CN et al. Am J Psychiatry 2003; 160(8):1522–1524.
2172. Smith GS, Price JC et al. Synapse 1998; 30(4):380–392.
2173. Price JC, Lopresti BJ et al. Synapse 2001; 41(1):11–21.
2174. Tan PZ, Baldwin RM et al. Nucl Med Biol 1999; 26(6):601–608.
2175. Pinborg LH, Adams KH et al. J Cereb Blood Flow Metab 2003; 23(8):985–996.
2176. Rosier A, Dupont P et al. Psychiatry Res 1996; 68(1):11–22.
2177. Meltzer CC, Smith G et al. Brain Res 1998; 813(1):167–171.
2178. Sheline YI, Mintun MA et al. American Journal of Psychiatry 2002; 159:430–435.
2179. Grunder G, Yokoi F et al. Neuropsychopharmacology 1997; 17(3):175–185.
2180. Kanerva H, Vilkman H et al. Psychopharmacology (Berl) 1999; 145(1):76–81.
2181. Blin J, Pappata S et al. European Journal of Pharmacology 1988; 147(1):73–82.
2182. Petit-Taboue MC, Landeau B et al. J Nucl Med 1996; 37(1):95–104.
2183. Kapur S, Jones C et al. Nucl Med Commun 1997; 18(5):395–399.
2184. Petit-Taboue MC, Landeau B et al. J Nucl Med 1999; 40(1):25–32.
2185. Meyer JH, Cho R et al. Psychopharmacology (Berl) 1999; 144(3):279–281.
2186. Lundkvist C, Halldin C et al. Life Sciences 1996; 58(10):PL 187–92.
2187. Andree B, Nyberg S et al. J Clin Psychopharmacol 1998; 18(4):317–323.
2188. Talvik-Lotfi M, Nyberg S et al. Psychopharmacology (Berl) 2000; 148(4):400–403.
2189. Kakiuchi T, Nishiyama S et al. Brain Res 2000; 883(1):135–142.
2190. Persson A, Ehrin E et al. J Psychiatr Res 1985; 19(4):609–622.
2191. Lingford-Hughes A, Hume SP et al. J Cereb Blood Flow Metab 2002; 22(7):878–889.
2192. Pappata S, Samson Y et al. J Cereb Blood Flow Metab 1988; 8(3):304–313.
2193. Samson Y, Hantraye P et al. European Journal of Pharmacology 1985; 110(2):247–251.
2194. Koeppe RA, Holthoff VA et al. J Cereb Blood Flow Metab 1991; 11(5):735–744.
2195. Holthoff VA, Koeppe RA et al. J Cereb Blood Flow Metab 1991; 11(5):745–752.
2196. Debruyne D, Abadie P et al. European Journal of Drug Metabolism & Pharmacokinetics 1991; 16(2):141–152.
2197. Halldin C, Stone-Elander S et al. Int J Rad Appl Instrum [A] 1988; 39(9):993–997.
2198. Niimura K, Muzik O et al. J Nucl Med 1999; 40(12):1985–1991.
2199. Price JC, Mayberg HS et al. J Cereb Blood Flow Metab 1993; 13(4):656–667.
2200. Delforge J, Syrota A et al. J Cereb Blood Flow Metab 1993; 13(3):454–468.
2201. Lassen NA, Bartenstein PA et al. J Cereb Blood Flow Metab 1995; 15(1):152–165.
2202. Delforge J, Pappata S et al. J Cereb Blood Flow Metab 1995; 15(2):284–300.
2203. Delforge J, Spelle L et al. J Nucl Med 1996; 37(1):5–11.
2204. Chugani DC, Muzik O et al. Ann Neurol 2001; 49(5):618–626.
2205. Pauli S, Liljequist S et al. Psychopharmacology 1992; 107(2–3):180–185.
2206. Halldin C, Farde L et al. Psychopharmacology (Berl) 1992; 108(1–2):16–22.
2207. Maeda J, Suhara T et al. Synapse 2003; 47(3):200–208.
2208. Grunder G, Siessmeier T et al. Eur J Nucl Med 2001; 28(10):1463–1470.
2209. Leveque P, Sanabria-Bohorquez S et al. Eur J Nucl Med Mol Imaging 2003; epub.
2210. Black KL, Ikezaki K et al. Cancer 1990; 65(1):93–97.
2211. Petit-Taboue MC, Baron JC et al. European Journal of Pharmacology 1991; 200(2–3):347–351.
2212. Banati RB, Newcombe J et al. Brain 2000; 123 (Pt 11):2321–2337.
2213. Lockhart A, Davis B et al. Nucl Med Biol 2003; 30(2):199–206.
2214. Bressan RA, Pilowsky LS. Eur J Nucl Med 2000; 27(11):1723–1731.
2215. Ametamey SM, Bruehlmeier M et al. Nucl Med Biol 2002; 29(2):227–231.
2216. Blin J, Denis A et al. Neurosci Lett 1991; 121(1–2):183–186.
2217. Shiue CY, Vallabhahosula S et al. Nucl Med Biol 1997; 24(2):145–150.
2218. Haradahira T, Okauchi T et al. Synapse 2002; 43(2):131–133.
2219. Holschbach MH, Olsson RA. Curr Pharm Des 2002; 8(26):2345–2352.
2220. Ishiwata K, Nariai T et al. Annals of Nuclear Medicine 2002; 16(6):377–382.

2221. Bauer A, Holschbach MH et al. Neuroimage 2003; 19(4):1760–1769.
2222. Ishiwata K, Ogi N et al. Annals of Nuclear Medicine 2002; 16(7):467–475.
2223. Tagawa M, Kano M et al. Br J Clin Pharmacol 2001; 52(5):501–509.
2224. Higuchi M, Yanai K et al. Neuroscience 2000; 99(4):721–729.
2225. Charalambous A, Marciniak G et al. Pharmacology, Biochemistry & Behavior 1991; 40(3):503–507.
2226. Gifford AN, Makriyannis A et al. Chem Phys Lipids 2002; 121(1–2):65–72.
2227. Katoch-Rouse R, Pavlova OA et al. J Med Chem 2003; 46(4):642–645.
2228. Hartvig P, Bergstrom K et al. J Pharmacol Exp Ther 1984; 230(1):250–255.
2229. Dannals RF, Ravert HT et al. Int J Appl Radiat Isot 1985; 36(4):303–306.
2230. Saji H, Tsutsumi D et al. Ann Nucl Med 1992; 6(1):63–67.
2231. Frost JJ, Wagner HN, Jr. et al. J Comput Assist Tomogr 1985; 9(2):231–236.
2232. Frost JJ, Douglass KH et al. J Cereb Blood Flow Metab 1989; 9(3):398–409.
2233. Zubieta JK, Dannals RF et al. Am J Psychiatry 1999; 156(6):842–848.
2234. Sadzot B, Price JC et al. J Cereb Blood Flow Metab 1991; 11(2):204–219.
2235. Cunningham VJ, Hume SP et al. J Cereb Blood Flow Metab 1991; 11(1):1–9.
2236. Jones AK, Cunningham VJ et al. J Neurosci Methods 1994; 51(2):123–134.
2237. Frost JJ, Mayberg HS et al. J Cereb Blood Flow Metab 1990; 10(4):484–492.
2238. Vogt BA, Watanabe H et al. Human Brain Mapping 1995; 3(1):1–12.
2239. Jones AK, Qi LY et al. Neurosci Lett 1991; 126(1):25–28.
2240. Chesis PL, Hwang DR et al. J Med Chem 1990; 33(5):1482–1490.
2241. Shiue CY, Bai LQ et al. International Journal of Radiation Applications & Instrumentation – Part B, Nuclear Medicine & Biology 1991; 18(3):281–288.
2242. Galynker I, Schlyer DJ et al. Nucl Med Biol 1996; 23(3):325–331.
2243. Pert CB, Danks JA et al. FEBS Lett 1984; 177(2):281–286.
2244. Sawada Y, Kawai R et al. J Cereb Blood Flow Metab 1991; 11(2):183–203.
2245. Carson RE, Channing MA et al. J Cereb Blood Flow Metab 1993; 13(1):24–42.
2246. Cohen RM, Carson RE et al. Synapse 2000; 38(2):226–229.
2247. Lever JR, Scheffel U et al. Eur J Pharmacol 1992; 216(3):459–460.
2248. Madar I, Lever JR et al. Synapse 1996; 24(1):19–28.
2249. Smith JS, Zubieta JK et al. J Cereb Blood Flow Metab 1999; 19(9):956–966.
2250. Kawamura K, Ishiwata K et al. Ann Nucl Med 2000; 14(4):285–292.
2251. Ishiwata K, Kobayashi T et al. Nucl Med Biol 2001; 28(7):787–792.
2252. Hoyte RM, Zhang JX et al. J Med Chem 2002; 45(24):5397–5405.
2253. Visser GM, Krugers HJ et al. Nucl Med Biol 1995; 22(7):915–920.
2254. Pomper MG, Kochanny MJ et al. Int J Rad Appl Instrum B 1992; 19(4):461–480.
2255. Moresco RM, Casati R et al. J Cereb Blood Flow Metab 1995; 15(2):301–311.
2256. Moresco RM, Scheithauer BW et al. Nucl Med Commun 1997; 18(7):606–615.
2257. Hargreaves R. Journal of Clinical Psychiatry 2002; 63 Suppl 11:18–24.
2258. Chang MC, Arai T et al. Brain Res 1997; 755(1):74–83.
2259. Giovacchini G, Chang MC et al. J Cereb Blood Flow Metab 2002; 22(12):1453–1462.
2260. Imahori Y, Fujii R et al. J Nucl Med 1993; 34(9):1543–1551.
2261. Gatley SJ. J Nucl Med 1993; 34(12):2208–2215.
2262. Budinger TF, Derenzo SE et al. J Comput Assist Tomogr 1977; 1(1):131–145.
2263. Stöcklin G, Pike VW. Radiopharmaceuticals for positron emission tomography. Kluwer, Dordrecht, 1993.
2264. Stöcklin G. Eur J Nucl Med 1992; 19(7):527–551.
2265. Langstrom B, Kihlberg T et al. Acta Chem Scand 1999; 53(9):651–669.
2266. Valk PE, Bailey DL et al. Positron Emission Tomography. Basic Science and Clinical Practice. Springer Verlag, 2003.
2267. Wolf AP. Annals of Neurology 1984; 15 Suppl:S19–24.
2268. Halldin C, Gulyas B et al. Quarterly Journal of Nuclear Medicine 2001; 45(2):139–152.
2269. Elsinga PH. Methods 2002; 27(3):208–217.

2270. Hume SP, Gunn RN et al. Eur J Nucl Med 1998; 25(2):173–176.
2271. Alexoff DL, Vaska P et al. J Nucl Med 2003; 44(5):815–822.
2272. Wienhard K, Dahlbom M et al. Journal of Computer Assisted Tomography 1994; 18(1):110–118.
2273. Spinks TJ, Jones T et al. Physics in Medicine and Biology 2000; 45(9):2601–2618.
2274. Eriksson L, Wienhard K et al. IEEE Transactions on Nuclear Science 2002; 49(3):640–643.
2275. Brix G, Zaers J et al. J Nucl Med 1997; 38(10):1614–1623.
2276. DeGrado TR, Turkington TG et al. J Nucl Med 1994; 35(8):1398–1406.
2277. Karp JS, Freifelder R et al. J Nucl Med 1997; 38(4):636–643.
2278. Terpogossian MM, Ficke DC et al. Journal of Computer Assisted Tomography 1994; 18(4):661–669.
2279. Cleon G, Allemand R et al. IEEE Transactions on Nuclear Science 2002; 49(1):139–140.
2280. Wong WH, Mullani NA et al. IEEE Transactions on Nuclear Science 1984; 31(1):381–386.
2281. Karp JS, Surti S et al. J Nucl Med 2003; 44(8):1340–1349.
2282. Nahmias C, Nutt R et al. J Nucl Med 2002; 43(5):36.
2283. Wienhard K, Schmand M et al. IEEE Transactions on Nuclear Science 2002; 49(1):104–110.
2284. Nutt R. Eur J Nucl Med Mol Imaging 2002; 29(11):1523–1525.
2285. Knoess C, Boellaard R et al. IEEE Transactions on Nuclear Science 2003; MIC.
2286. Casey ME, Nutt R. IEEE Transactions on Nuclear Science 1986; 33(1):460–463.
2287. Wienhard K, Eriksson L et al. Journal of Computer Assisted Tomography 1992; 16(5):804–813.
2288. Budinger TF. Seminars in Nuclear Medicine 1998; 28(3):247–267.
2289. Ishii K, Orihara H et al. Review of Scientific Instruments 1990; 61(12):3755–3762.
2290. Townsend DW, Geissbuhler A et al. IEEE Transactions on Medical Imaging 1991; 10(4):505–512.
2291. Cherry SR, Dahlbom M et al. Journal of Computer Assisted Tomography 1991; 15(4):655–668.
2292. Spinks TJ, Jones T et al. Physics in Medicine & Biology 1992; 37(8):1637–1655.
2293. Karp JS, Daube-Witherspoon ME et al. Journal of Cerebral Blood Flow and Metabolism 1991; 11(2):A38–A44.
2294. Townsend D, Frey P et al. Nuclear Instruments & Methods in Physics Research Section A-Accelerators Spectrometers Detectors and Associated Equipment 1984; 221(1):105–112.
2295. Bendriem B, Townsend DW. The theory and practice of 3D PET. Kluwer Academic Publishers, Dordrecht, 1998.
2296. Karp JS, Kinahan PE et al. IEEE Transactions on Medical Imaging 1993; 12(2):299–306.
2297. Ter Pogossian MM, Mullani NA et al. Radiology 1978; 128(2):477–484.
2298. Boren EL, Jr., Delbeke D et al. Eur J Nucl Med 1999; 26(4):379–387.
2299. Pitman AG, Hicks RJ et al. Br J Radiol 2002; 75(890):114–121.
2300. Kuikka JT, Sohlberg A et al. Clin Physiol Funct Imaging 2002; 22(5):328–331.
2301. Townsend DW, Wensveen M et al. J Nucl Med 1993; 34(8):1367–1376.
2302. Bailey DL, Young H et al. Eur J Nucl Med 1997; 24(1):6–15.
2303. Townsend D, Frey P et al. J Nucl Med 1987; 28(10):1554–1562.
2304. Marsden PK, Ott RJ et al. Physics in Medicine and Biology 1989; 34(8):1043–1062.
2305. Cherry SR, Marsden PK et al. Eur J Nucl Med 1989; 15(11):694–700.
2306. Schafers KP. Nuklearmedizin-Nuclear Medicine 2003; 42(3):86–89.
2307. Chatziioannou AF. Eur J Nucl Med 2002; 29(1):98–114.
2308. Jeavons AP. J Nucl Med 2000; 41(8):1442–1443.
2309. Ziegler SI, Pichler BJ et al. Eur J Nucl Med 2001; 28(2):136–143.
2310. Del Guerra A, Belcari N. Q J Nucl Med 2002; 46(1):35–47.
2311. Weber S, Bauer A et al. IEEE Transactions on Nuclear Science 2000; 47(4):1665–1669.
2312. Tai C, Chatziioannou A et al. Phys Med Biol 2001; 46(7):1845–1862.
2313. Cherry SR, Shao Y et al. IEEE Transactions on Nuclear Science 1997; 44(3):1161–1166.
2314. Knoess C, Siegel S et al. Eur J Nucl Med Mol Imaging 2003; 30:737–747.

2315. Jeavons AP, Chandler RA et al. IEEE Transactions on Nuclear Science 1999; 46(3):468–473.
2316. Valk PE, Jagust WJ et al. Radiology 1990; 176(3):783–790.
2317. Aston JA, Cunningham VJ et al. J Cereb Blood Flow Metab 2002; 22(8):1019–1034.
2318. Strul D, Bendriem B. J Cereb Blood Flow Metab 1999; 19(5):547–559.
2319. Meltzer CC, Kinahan PE et al. J Nucl Med 1999; 40(12):2053–2065.
2320. Jagust WJ, Eberling JL et al. Brain Research 1993; 629(2):189–198.
2321. Eriksson L, Wienhard K et al. IEEE Transactions on Nuclear Science 2002; 49(5):2085–2088.
2322. Schoder H, Erdi YE et al. Eur J Nucl Med Mol Imaging 2003; epub.
2323. Raylman RR, Hammer BE et al. IEEE Transactions on Nuclear Science 1996; 43(4):2406–2412.
2324. Ranicar AS, Williams CW et al. Medical Progress through Technology 1991; 17(3–4):259–264.
2325. Lapointe D, Cadorette J et al. IEEE Transactions on Nuclear Science 1998; 45(4):2195–2199.
2326. Wollenweber SD, Hichwa RD et al. IEEE Transactions on Nuclear Science 1997; 44(4):1613–1617.
2327. Wollenweber SD, Hichwa RD et al. IEEE Transactions on Nuclear Science 1997; 44(3):1417–1419.
2328. Eriksson L, Ingvar M et al. IEEE Transactions on Nuclear Science 1995; 42(4):1007–1011.
2329. Nelson AD, Miraldi F et al. J Nucl Med 1993; 34(6):1000–1006.
2330. Defrise M, Kinahan P.In: The theory and practice of 3D PET 1998;(2):11–53.
2331. Reader AJ, Ally S et al. IEEE Transactions on Nuclear Science 2002; 49(3):693–699.
2332. Zaidi H, Hasegawa B. J Nucl Med 2003; 44(2):291–315.
2333. Townsend DW, Cherry SR. Eur Radiol 2001; 11(10):1968–1974.
2334. Bailey DL. In: The theory and practice of 3D PET 1998;(3):55–109.
2335. Hutton BF, Baccarne V. Eur J Nucl Med 1998; 25(12):1658–1665.
2336. Cherry SR, Meikle SR et al. J Nucl Med 1993; 34(4):671–678.
2337. Mckee BTA, Gurvey AT et al. IEEE Transactions on Medical Imaging 1992; 11(4):560–569.
2338. Werling A, Bublitz O et al. Phys Med Biol 2002; 47(16):2947–2960.
2339. Sossi V, Oakes TR et al. J Nucl Med 1998; 39(10):1714–1719.
2340. Sossi V, Barney JS et al. IEEE Transactions on Nuclear Science 1995; 42(4):1157–1161.
2341. Kinahan PE, Rogers JG. IEEE Transactions on Nuclear Science 1989; 36:964–968
2342. Daube-Witherspoon ME, Muehllehner G. J Nucl Med 1987; 28(11):1717–1724.
2343. Defrise M. Inverse Problems 1995; 11(5):983–994.
2344. Defrise M, Liu XA. Inverse Problems 1999; 15(4):1047–1065.
2345. Vandenberghe S, D'Asseler Y et al. Comput Med Imaging Graph 2001; 25(2):105–111.
2346. Son HK, Yun MJ et al. IEEE Transactions on Nuclear Science 2003; 50(1):37–41.
2347. Paul AK, Tatsumi M et al. Nuclear Medicine Communications 2002; 23(1):103–110.
2348. Reinders AA, Paans AM et al. Neuroimage 2002; 15(1):175–181.
2349. Liow JS, Strother SC et al. J Nucl Med 1997; 38(10):1623–1631.
2350. Hudson H, Larkin R. IEEE Trans Med Imaging 1994; 13:601–609.
2351. Liu X, Comtat C et al. IEEE Trans Med Imaging 2001; 20(8):804–814.
2352. Krzywinski M, Sossi V et al. IEEE Transactions on Nuclear Science 1999; 46(4):1114–1120.
2353. Vollmar S, Michel C et al. Physics in Medicine and Biology 2002; 47(15):2651–2658.
2354. Lipinski B, Herzog H et al. IEEE Trans Med Imaging 1997; 16(2):129–136.
2355. Brix G, Doll J et al. Eur J Nucl Med 1997; 24(7):779–786.
2356. Hutton BF, Kyme AZ et al. IEEE Transactions on Nuclear Science 2002; 49(1):188–194.
2357. Comtat C, Kinahan PE et al. IEEE Transactions on Nuclear Science 1998; 45(3):1083–1089.
2358. Buvat I. Physics in Medicine and Biology 2002; 47(10):1761–1775.
2359. Dahlbom M. IEEE Transactions on Nuclear Science 2002; 49(5):2062–2066.
2360. Chatziioannou A, Qi J et al. IEEE Trans Med Imaging 2000; 19(5):507–512.
2361. Lopresti BJ, Russo A et al. IEEE Transactions on Nuclear Science 1999; 46(6):2059–2067.
2362. Fulton RR, Meikle SR et al. IEEE Transactions on Nuclear Science 2002; 49(1):116–123.

2363. Bloomfield PM, Spinks TJ et al. Physics in Medicine and Biology 2003; 48(8):959–978.
2364. Thurfjell L, Pagani M et al. Journal of Neuroimaging 2000; 10(1):39–46.
2365. Andersson JLR, Thurfjell L. Journal of Computer Assisted Tomography 1997; 21(1):136–144.
2366. Nelson SJ, Day MR et al. J Comput Assist Tomogr 1997; 21(2):183–191.
2367. Alpert NM, Berdichevsky D et al. Neuroimage 1996; 3(1):10–18.
2368. Eberl S, Kanno I et al. J Nucl Med 1996; 37(1):137–145.
2369. Phillips RL, London ED et al. J Nucl Med 1990; 31(12):2052–2057.
2370. Pisani P, Guzzardi R et al. Medical Progress through Technology 1991; 17(3–4):205–209.
2371. Grzeszczuk R, Tan KK et al. Journal of Computer Assisted Tomography 1992; 16(5):764–773.
2372. Pietrzyk U.In: Medical image registration 2001;(9):199–216.
2373. Woods RP, Grafton ST et al. J Comput Assist Tomogr 1998; 22(1):139–152.
2374. Grosu AL, Lachner R et al. Int J Radiat Oncol Biol Phys 2003; 56(5):1450–1463.
2375. Viergever MA, Maintz JB et al. Comput Med Imaging Graph 2001; 25(2):147–151.
2376. von Stockhausen HM.Universität zu Köln 1998.
2377. von Stockhausen HM, Pietrzyk U et al. Neuroimage 1998; 7 (Suppl.)(4):S799.
2378. Wu TH, Wang JK et al. Eur J Nucl Med Mol Imaging 2003.
2379. Junck L, Moen JG et al. J Nucl Med 1990; 31(7):1220–1226.
2380. Hoh CK, Dahlbom M et al. J Nucl Med 1993; 34(11):2009–2018.
2381. Kiebel SJ, Ashburner J et al. Neuroimage 1997; 5(4 Pt 1):271–279.
2382. Levin DN, Pelizzari CA et al. Radiology 1988; 169(3):817–823.
2383. Andersson JLR, Thurfjell L. Eur J Nucl Med 1999; 26(7):718–733.
2384. Studholme C, Hill DL et al. Med Phys 1997; 24(1):25–35.
2385. Maes F, Collignon A et al. IEEE Trans Med Imaging 1997; 16(2):187–198.
2386. Collins DL, Zijdenbos AP et al. Information Processing in Medical Imaging, Proceedings 1999; 1613:210–223.
2387. Pietrzyk U, Herholz K et al. European Journal of Radiology 1996; 21:174–182.
2388. Pietrzyk U, Thiel A et al. Neuroimage 1998; 7 (Suppl.)(4):S789.
2389. Maes F, Vandermeulen D et al. Medical Image Analysis 1999; 3:373–368.
2390. Goertzen AL, Meadors AK et al. Phys Med Biol 2002; 47(24):4315–4328.
2391. Jacobs RE, Cherry SR. Curr Opin Neurobiol 2001; 11(5):621–629.
2392. Talairach J, Tournoux P. Co-planar stereotaxic atlas of the human brain. Georg Thieme Verlag, Stuttgart, 1988.
2393. Minoshima S, Koeppe RA et al. J Nucl Med 1993; 34(2):322–329.
2394. Friston KJ, Passingham RE et al. J Cereb Blood Flow Metab 1989; 9(5):690–695.
2395. Tzourio-Mazoyer N, Landeau B et al. Neuroimage 2002; 15(1):273–289.
2396. Meyer JH, Gunn RN et al. Neuroimage 1999; 9(5):545–553.
2397. Gispert JD, Pascau J et al. Neuroimage 2003; 19(3):601–612.
2398. Dann R, HOFORD J et al. Journal of Computer Assisted Tomography 1989; 13(4):603–611.
2399. Friston KJ, Frith CD et al. Journal of Computer Assisted Tomography 1991; 15(4):634–639.
2400. Minoshima S, Koeppe RA et al. J Nucl Med 1994; 35(9):1528–1537.
2401. Woods RP, Grafton ST et al. J Comput Assist Tomogr 1998; 22(1):153–165.
2402. Friston KJ, Ashburner J et al. Hum Brain Mapp 1995; 2:165–189.
2403. Grachev ID, Berdichevsky D et al. Neuroimage 1999; 9:250–268.
2404. Ashburner J, Neelin P et al. Neuroimage 1997; 6(4):344–352.
2405. Fischl B, Salat DH et al. Neuron 2002; 33(3):341–355.
2406. Ashburner J, Csernansky JG et al. The Lancet Neurology 2003; 2(2):79–88.
2407. Dinov ID, Mega MS et al. J Comput Assist Tomogr 2000; 24(1):128–138.
2408. Kriegeskorte N, Goebel R. Neuroimage 2001; 14(2):329–346.
2409. Shen D, Davatzikos C. Neuroimage 2003; 18(1):28–41.
2410. Ashburner J, Andersson JL et al. Hum Brain Mapp 2000; 9(4):212–225.
2411. Bajcsy R, Lieberson R et al. Journal of Computer Assisted Tomography 1983; 7(4):618–625.
2412. Greitz T, Bohm C et al. Journal of Computer Assisted Tomography 1991; 15(1):26–38.

2413. Minoshima S, Koeppe RA et al. J Nucl Med 1994; 35(6):949–954.
2414. Rizzo G, Gilardi MC et al. Eur J Nucl Med 1995; 22(11):1313–1318.
2415. Evans AC, Marrett S et al. J Cereb Blood Flow Metab 1991; 11(2):A69–78.
2416. Steinmetz H, Furst G et al. J Comput Assist Tomogr 1989; 13(1):10–19.
2417. Rademacher J, Caviness VS, Jr. et al. Cerebral Cortex 1993; 3(4):313–329.
2418. Evans AC, Collins DL. J Nucl Med 1993; 34(5):70–71.
2419. Mazziotta JC, Toga AW et al. Neuroimage 1995; 2(2 Part 1):89–101.
2420. Mazziotta J, Toga A et al. Philos Trans R Soc Lond B Biol Sci 2001; 356(1412):1293–1322.
2421. Thompson PM, Woods RP et al. Hum Brain Mapp 2000; 9(2):81–92.
2422. Fox PT, Perlmutter JS et al. Journal of Computer Assisted Tomography 1985; 9(1):141–153.
2423. Brett M, Johnsrude IS et al. Nat Rev Neurosci 2002; 3(3):243–249.
2424. Greer P, Villemagne V et al. Brain Res Bull 2002; 58(4):429.
2425. Jons PH, Ernst M et al. Hum Brain Mapp 1997; 5(2):119–123.
2426. Wong KP, Feng D et al. IEEE Trans Inf Technol Biomed 2001; 5(1):67–76.
2427. Herscovitch P, Raichle ME. J Cereb Blood Flow Metab 1985; 5(1):65–69.
2428. Kanno I, Lassen NA. J Comput Assist Tomogr 1979; 3(1):71–76.
2429. Eriksson L, Kanno I. Medical Progress through Technology 1991; 17(3–4):249–257.
2430. Graham MM, Lewellen BL. J Nucl Med 1993; 34(8):1357–1360.
2431. Iida H, Kanno I et al. J Cereb Blood Flow Metab 1986; 6(5):536–545.
2432. Meyer E. J Nucl Med 1989; 30(6):1069–1078.
2433. Wakita K, Imahori Y et al. J Nucl Med 2000; 41(9):1484–1490.
2434. Madsen PL, Holm S et al. J Cereb Blood Flow Metab 1993; 13(4):646–655.
2435. Lassen NA, Ingvar DH. Arch Neurol 1963; 9:615–622.
2436. Raichle ME, Martin WR et al. J Nucl Med 1983; 24(9):790–798.
2437. Quarles RP, Mintun MA et al. J Cereb Blood Flow Metab 1993; 13(5):733–747.
2438. Larson KB, Markham J et al. J Cereb Blood Flow Metab 1987; 7(4):443–463.
2439. Herholz K, Patlak CS. J Cereb Blood Flow Metab 1987; 7(2):214–229.
2440. Baron JC, Frackowiak RS et al. J Cereb Blood Flow Metab 1989; 9(6):723–742.
2441. Herscovitch P, Markham J et al. J Nucl Med 1983; 24(9):782–789.
2442. Kanno I, Iida H et al. J Cereb Blood Flow Metab 1987; 7(2):143–153.
2443. Koeppe RA, Hutchins GD et al. J Nucl Med 1987; 28(11):1695–1703.
2444. Fox PT, Mintun MA et al. J Cereb Blood Flow Metab 1984; 4(3):329–333.
2445. Cherry SR, Woods RP et al. J Cereb Blood Flow Metab 1993; 13(4):630–638.
2446. Volkow ND, Mullani N et al. J Nucl Med 1991; 32(1):58–61.
2447. Koeppe RA, Holden JE et al. J Cereb Blood Flow Metab 1985; 5(2):224–234.
2448. Kanno I, Iida H et al. J Nucl Med 1991; 32(10):1931–1934.
2449. Holden JE, Gatley SJ et al. J Nucl Med 1981; 22(12):1084–1088.
2450. Kanno I, Lammertsma AA et al. J Cereb Blood Flow Metab 1984; 4(2):224–234.
2451. Ingvar M, Eriksson L et al. J Cereb Blood Flow Metab 1994; 14(4):628–638.
2452. Herzog H, Seitz RJ et al. J Cereb Blood Flow Metab 1996; 16(4):645–649.
2453. Watabe H, Itoh M et al. Journal of Cerebral Blood Flow and Metabolism 1996; 16(2):311–319.
2454. Lammertsma AA, Jones T et al. Journal of Computer Assisted Tomography 1981; 5(4):544–550.
2455. Crone C. Acta Physiol Scand 1963; 58:292–305.
2456. Lammertsma AA, Brooks DJ et al. J Cereb Blood Flow Metab 1984; 4(4):523–534.
2457. Webb S, Ott RJ et al. Phys Med Biol 1989; 34(12):1767–1771.
2458. Herholz K, Wienhard K et al. J Cereb Blood Flow Metab 1989; 9(1):104–110.
2459. Brooks DJ, Beaney RP et al. J Cereb Blood Flow Metab 1986; 6(2):230–239.
2460. Huang SC, Phelps ME et al. American Journal of Physiology 1980; 238(1):E69–82.
2461. Danielsen EH, Smith DF et al. Synapse 1999; 33(4):247–258.
2462. Sokoloff L, Reivich M et al. Journal of Neurochemistry 1977; 28(5):897–916.
2463. Dhawan V, Moeller JR et al. J Nucl Med 1989; 30(9):1483–1488.

2464. Dienel GA, Cruz NF et al. Journal of Neurochemistry 1992; 59(4):1430–1436.
2465. Phelps ME, Huang SC et al. Annals of Neurology 1979; 6(5):371–388.
2466. Kuwabara H, Evans AC et al. J Cereb Blood Flow Metab 1990; 10(2):180–189.
2467. Wu HM, Bergsneider M et al. Molecular Imaging & Biology 2003; 5(1):32–41.
2468. Wienhard K, Pawlik G et al. J Cereb Blood Flow Metab 1991; 11(3):485–491.
2469. Suda S, Shinohara M et al. J Cereb Blood Flow Metab 1990; 10(4):499–509.
2470. Gjedde A, Wienhard K et al. J Cereb Blood Flow Metab 1985; 5(2):163–178.
2471. Lucignani G, Schmidt KC et al. J Nucl Med 1993; 34(3):360–369.
2472. Wienhard K, Pawlik G et al. J Cereb Blood Flow Metab 1985; 5(1):115–125.
2473. Hutchins GD, Holden JE et al. J Cereb Blood Flow Metab 1984; 4(1):35–40.
2474. Jons PH, Ernst M et al. Human Brain Mapping 1997; 5(2):119–123.
2475. Takikawa S, Dhawan V et al. Radiology 1993; 188(1):131–136.
2476. Chen K, Bandy D et al. J Cereb Blood Flow Metab 1998; 18(7):716–723.
2477. Cumming P, Leger GC et al. J Cereb Blood Flow Metab 1993; 13(4):668–675.
2478. Patlak CS, Blasberg RG. J Cereb Blood Flow Metab 1985; 5(4):584–590.
2479. Huang SC, Barrio JR et al. J Cereb Blood Flow Metab 1986; 6(5):515–521.
2480. Videbaek C, Ott P et al. J Cereb Blood Flow Metab 1999; 19(9):948–955.
2481. Mintun MA, Raichle ME et al. Annals of Neurology 1984; 15(3):217–227.
2482. Wienhard K, Coenen HH et al. Journal of Neural Transmission – General Section 1990; 81(3):195–213.
2483. Koeppe RA, Frey KA et al. J Cereb Blood Flow Metab 1997; 17(9):919–931.
2484. Kawai R, Carson RE et al. J Cereb Blood Flow Metab 1991; 11(4):529–544.
2485. Logan J, Fowler JS et al. J Cereb Blood Flow Metab 1990; 10(5):740–747.
2486. Logan J, Fowler JS et al. J Cereb Blood Flow Metab 1996; 16(5):834–840.
2487. Gunn RN, Lammertsma AA et al. Neuroimage 1997; 6(4):279–287.
2488. Blomqvist G, Pauli S et al. Eur J Nucl Med 1990; 16(4–6):257–265.
2489. Ichise M, Toyama H et al. J Cereb Blood Flow Metab 2002; 22(10):1271–1281.
2490. Huesman RH, Coxson PG. IEEE Trans Med Imaging 1997; 16(5):675–683.
2491. Thie JA, Smith GT et al. IEEE Trans Med Imaging 1997; 16(1):11–16.
2492. Blomqvist G. J Cereb Blood Flow Metab 1984; 4(4):629–632.
2493. Alpert NM, Eriksson L et al. J Cereb Blood Flow Metab 1984; 4(1):28–34.
2494. Huang SC, Carson RE et al. J Cereb Blood Flow Metab 1982; 2(1):99–108.
2495. Carson RE, Huang SC et al. J Cereb Blood Flow Metab 1986; 6(2):245–258.
2496. Yokoi T, Kanno I et al. J Cereb Blood Flow Metab 1991; 11(3):492–501.
2497. Herholz K, Dickhoven S et al.In: Quantification of brain function using PET 1996;(34):175–180.
2498. Suhara T, Nakayama K et al. Psychopharmacology 1992; 106(1):14–18.
2499. Herholz K, Pawlik G et al. Journal of Computer Assisted Tomography 1985; 9(1):154–161.
2500. Hammers A, Koepp MJ et al. Hum Brain Mapp 2002; 15(3):165–174.
2501. Yasuno F, Hasnine AH et al. Neuroimage 2002; 16(1):577–586.
2502. Klein GJ, Teng X et al. IEEE Trans Med Imaging 1997; 16(4):405–415.
2503. Muller-Gartner HW, Links JM et al. J Cereb Blood Flow Metab 1992; 12(4):571–583.
2504. Steel RDG, Torrie JH. Principles and procedures of statistics. A biometrical approach. McGraw-Hill, New York, 1980.
2505. Litton JE, Hall H et al. J Cereb Blood Flow Metab 1994; 14(2):358–361.
2506. Andersson JL, Ashburner J et al. Neuroimage 2001; 13(6):1193–1206.
2507. Pawlik G. J Cereb Blood Flow Metab 1991; 11(2):A136–9.
2508. Herholz K. Eur J Nucl Med 1988; 14(9–10):477–484.
2509. Zhou Y, Huang SC et al. Neuroimage 2002; 15(3):697–707.
2510. Herholz K, Wagner R et al. Nuklearmedizin 1990; 29(2):84.
2511. Messa C, Choi Y et al. Journal of Computer Assisted Tomography 1992; 16(5):684–689.
2512. Wong WH, Hicks K. J Nucl Med 1994; 35(7):1206–1212.
2513. Gunn RN, Lammertsma AA et al. Neuroimage 1997; 6(4):279–287.

2514. Delforge J, Spelle L et al. J Cereb Blood Flow Metab 1997; 17(3):343–355.
2515. Frey KA, Holthoff VA et al. Annals of Neurology 1991; 30(5):663–672.
2516. Cunningham VJ, Jones T. J Cereb Blood Flow Metab 1993; 13(1):15–23.
2517. Meikle SR, Matthews JC et al. Phys Med Biol 1998; 43(3):651–666.
2518. Herholz K, Zündorf G et al. J Cereb Blood Flow Metab 2001; 21 (Suppl. 1):S528.
2519. Friston KJ, Holmes AP et al. Human Brain Mapping 1995; 2(4):189–210.
2520. Friston KJ, Frith CD et al. J Cereb Blood Flow Metab 1990; 10(4):458–466.
2521. Wang GJ, Volkow ND et al. J Nucl Med 1999; 40(5):715–720.
2522. Petersson KM, Nichols TE et al. Philos Trans R Soc Lond B Biol Sci 1999;
       354(1387):1261–1281.
2523. Petersson KM, Nichols TE et al. Philos Trans R Soc Lond B Biol Sci 1999;
       354(1387):1239–1260.
2524. Friston KJ, Worsley KJ et al. Human Brain Mapping 1994; 1:210–220.
2525. Friston KJ, Penny W. Neuroimage 2003; 19(3):1240–1249.
2526. Clark C, Carson R. J Cereb Blood Flow Metab 1993; 13(6):1038–1040.
2527. Worsley KJ, Marrett S et al. Human Brain Mapping 1996; 4(1):58–73.
2528. Poline JB, Worsley KJ et al. Neuroimage 1997; 5(2):83–96.
2529. Kiebel SJ, Poline JB et al. Neuroimage 1999; 10(6):756–766.
2530. Poline JB, Worsley KJ et al. Journal of Computer Assisted Tomography 1995; 19(5):788–796.
2531. Van Horn JD, Ellmore TM et al. Neuroimage 1998; 7(2):97–107.
2532. Turkheimer FE, Brett M et al. J Cereb Blood Flow Metab 2000; 20(11):1610–1618.
2533. Friston KJ, Malizia AL et al. J Cereb Blood Flow Metab 1997; 17(1):80–93.
2534. Piccini P, Weeks RA et al. Ann Neurol 1997; 42(5):720–726.
2535. Richardson MP, Koepp MJ et al. Neurology 1998; 51(2):485–492.
2536. Horwitz B. Neuroimage 2003; 19(2):466–470.
2537. Friston KJ, Frith CD et al. J Cereb Blood Flow Metab 1993; 13(1):5–14.
2538. Friston K, Phillips J et al. Hum Brain Mapp 1999; 8(2–3):92–97.
2539. Strother SC, Anderson JR et al. J Cereb Blood Flow Metab 1995; 15(5):738–753.
2540. Horwitz B, McIntosh AR et al. Behavioural Brain Research 1995; 66(1–2):187–193.
2541. Horwitz B. J Cereb Blood Flow Metab 1991; 11(2):A114–20.
2542. Horwitz B, Tagamets MA. Hum Brain Mapp 1999; 8(2–3):137–142.
2543. Holmes AP, Blair RC et al. J Cereb Blood Flow Metab 1996; 16(1):7–22.
2544. Halber M, Herholz K et al. J Cereb Blood Flow Metab 1997; 17(10):1033–1039.
2545. Hansen LK, Larsen J et al. Neuroimage 1999; 9(5):534–544.
2546. al-Mousawi AH, Evans N et al. British Journal of Psychiatry 1996; 169(4):509–516.
2547. Moeller JR, Strother SC. J Cereb Blood Flow Metab 1991; 11(2):A121–A135.
2548. Zündorf G, Holthoff V et al. Human Brain Mapping 2002 Meeting 2002;741.
2549. Azari NP, Pettigrew KD et al. J Cereb Blood Flow Metab 1993; 13(3):438–447.
2550. Kippenhan JS, Barker WW et al. J Nucl Med 1992; 33(8):1459–1467.
2551. Boudraa AE, Champier J et al. Comput Med Imaging Graph 1996; 20(1):31–41.

Subject Index